Graduate Texts in Mathematics 86

Springer-Verlag Berlin Heidelberg GmbH

Graduate Texts in Mathematics

1 TAKEUTI/ZARING. Introduction to Axiomatic Set Theory. 2nd ed.
2 OXTOBY. Measure and Category. 2nd ed.
3 SCHAEFFER. Topological Vector Spaces.
4 HILTON/STAMMBACH. A Course in Homological Algebra. 2nd ed.
5 MAC LANE. Categories for the Working Mathematician. 2nd ed.
6 HUGHES/PIPER. Projective Planes.
7 SERRE. A Course in Arithmetic.
8 TAKEUTI/ZARING. Axiomatic Set Theory.
9 HUMPHREYS. Introduction to Lie Algebras and Representation Theory.
10 COHEN. A Course in Simple Homotopy Theory.
11 CONWAY. Functions of One Complex Variable I. 2nd ed.
12 BEALS. Advanced Mathematical Analysis.
13 ANDERSON/FULLER. Rings and Categories of Modules. 2nd ed.
14 GOLUBITSKY/GUILLEMIN. Stable Mappings and Their Singularities.
15 BERBERIAN. Lectures in Functional Analysis and Operator Theory.
16 WINTER. The Structure of Fields.
17 ROSENBLATT. Random Processes. 2nd ed.
18 HALMOS. Measure Theory.
19 HALMOS. A Hilbert Space Problem Book. 2nd ed., revised.
20 HUSEMÖLLER. Fibre Bundles. 2nd ed.
21 HUMPHREYS. Linear Algebraic Groups.
22 BARNES/MACK. An Algebraic Introduction to Mathematical Logic.
23 GREUB. Linear Algebra. 4th ed.
24 HOLMES. Geometric Functional Analysis and its Applications.
25 HEWITT/STROMBERG. Real and Abstract Analysis.
26 MANES. Algebraic Theories.
27 KELLEY. General Topology.
28 ZARISKI/SAMUEL. Commutative Algebra. Vol. I.
29 ZARISKI/SAMUEL. Commutative Algebra. Vol. II.
30 JACOBSON. Lectures in Abstract Algebra I. Basic Concepts.
31 JACOBSON. Lectures in Abstract Algebra II. Linear Algebra.
32 JACOBSON. Lectures in Abstract Algebra III. Theory of Fields and Galois Theory.
33 HIRSCH. Differential Topology.
34 SPITZER. Principles of Random Walk. 2nd ed.
35 WERMER. Banach Algebras and Several Complex Variables. 2nd ed.
36 KELLEY/NAMIOKA et al. Linear Topological Spaces.
37 MONK. Mathematical Logic.
38 GRAUERT/FRITZSCHE. Several Complex Variables.
39 ARVESON. An Invitation to C^*-Algebras.
40 KEMENY/SNELL/KNAPP. Denumerable Markov Chains. 2nd ed.
41 APOSTOL. Modular Functions and Dirichlet Series in Number Theory. 2nd ed.
42 SERRE. Linear Representations of Finite Groups.
43 GILLMAN/JERISON. Rings of Continuous Functions.
44 KENDIG. Elementary Algebraic Geometry.
45 LOÈVE. Probability Theory I. 4th ed.
46 LOÈVE. Probability Theory II. 4th ed.
47 MOISE. Geometric Topology in Dimensions 2 and 3.
48 SACHS/WU. General Relativity for Mathematicians.
49 GRUENBERG/WEIR. Linear Geometry. 2nd ed.
50 EDWARDS. Fermat's Last Theorem.
51 KLINGENBERG. A Course in Differential Geometry.
52 HARTSHORNE. Algebraic Geometry.
53 MANIN. A Course in Mathematical Logic.
54 GRAVER/WATKINS. Combinatorics with Emphasis on the Theory of Graphs.
55 BROWN/PEARCY. Introduction to Operator Theory I: Elements of Functional Analysis.
56 MASSEY. Algebraic Topology: An Introduction
57 CROWELL/FOX. Introduction to Knot Theory.
58 KOBLITZ. p-adic Numbers, p-adic Analysis, and Zeta-Functions. 2nd ed.
59 LANG. Cyclotomic Fields.
60 ARNOLD. Mathematical Methods in Classical Mechanics. 2nd ed.
61 WHITEHEAD. Elements of Homotopy Theory.
62 KARGAPOLOV/MERZIJAKOV. Fundamentals of the Theory of Groups.
63 BOLLOBÁS. Graph Theory.
64 EDWARDS. Fourier Series. Vol. I. 2nd ed.

continued after index

J. H. van Lint

Introduction to Coding Theory

Third Revised and Expanded Edition

Springer

J.H. van Lint
Eindhoven University of Technology
Department of Mathematics
Den Dolech 2, P.O. Box 513
5600 MB Eindhoven
The Netherlands

Library of Congress Cataloging-in-Publication Data

Lint, Jacobus Hendricus van, 1932-
Introduction to coding theory / J.H. van Lint. -- 3rd rev. and expanded ed.
p. cm. -- (Graduate texts in mathematics, 0072-5285 ; 86)
Includes bibliographical references and index.
ISBN 978-3-642-63653-0 ISBN 978-3-642-58575-3 (eBook)
DOI 10.1007/978-3-642-58575-3
1. Coding theory. I. Title. II. Series.
QA268 .L57 1998
003'.54--dc21
98-48080
CIP

Mathematics Subject Classification (1991): 94-01, 94B, 11T71

ISSN 0072-5285

ISBN 978-3-642-63653-0

Typesetting: Asco Trade Typesetting Ltd., Hong Kong
46/3111 – 5 4 3 2 – Printed on acid-free paper SPIN 11358084

Preface to the Third Edition

It is gratifying that this textbook is still sufficiently popular to warrant a third edition. I have used the opportunity to improve and enlarge the book.

When the second edition was prepared, only two pages on algebraic geometry codes were added. These have now been removed and replaced by a relatively long chapter on this subject. Although it is still only an introduction, the chapter requires more mathematical background of the reader than the remainder of this book.

One of the very interesting recent developments concerns binary codes defined by using codes over the alphabet $\mathbb{Z}_4$. There is so much interest in this area that a chapter on the essentials was added. Knowledge of this chapter will allow the reader to study recent literature on $\mathbb{Z}_4$-codes.

Furthermore, some material has been added that appeared in my Springer Lecture Notes **201**, but was not included in earlier editions of this book, e. g. Generalized Reed-Solomon Codes and Generalized Reed-Muller Codes. In Chapter 2, a section on "Coding Gain" (the engineer's justification for using error-correcting codes) was added.

For the author, preparing this third edition was a most welcome return to mathematics after seven years of administration. For valuable discussions on the new material, I thank C. P. J. M. Baggen, I. M. Duursma, H. D. L. Hollmann, H. C. A. van Tilborg, and R. M. Wilson. A special word of thanks to R. A. Pellikaan for his assistance with Chapter 10.

Eindhoven J.H. VAN LINT
November 1998

Preface to the Second Edition

The first edition of this book was conceived in 1981 as an alternative to outdated, oversized, or overly specialized textbooks in this area of discrete mathematics—a field that is still growing in importance as the need for mathematicians and computer scientists in industry continues to grow.

The body of the book consists of two parts: a rigorous, mathematically oriented first course in coding theory followed by introductions to special topics. The second edition has been largely expanded and revised. The main editions in the second edition are:

(1) a long section on the binary Golay code;
(2) a section on Kerdock codes;
(3) a treatment of the Van Lint-Wilson bound for the minimum distance of cyclic codes;
(4) a section on binary cyclic codes of even length;
(5) an introduction to algebraic geometry codes.

Eindhoven
November 1991

J.H. VAN LINT

Preface to the First Edition

Coding theory is still a young subject. One can safely say that it was born in 1948. It is not surprising that it has not yet become a fixed topic in the curriculum of most universities. On the other hand, it is obvious that discrete mathematics is rapidly growing in importance. The growing need for mathematicians and computer scientists in industry will lead to an increase in courses offered in the area of discrete mathematics. One of the most suitable and fascinating is, indeed, coding theory. So, it is not surprising that one more book on this subject now appears. However, a little more justification and a little more history of the book are necessary. At a meeting on coding theory in 1979 it was remarked that there was no book available that could be used for an introductory course on coding theory (mainly for mathematicians but also for students in engineering or computer science). The best known textbooks were either too old, too big, too technical, too much for specialists, etc. The final remark was that my Springer Lecture Notes (#201) were slightly obsolete and out of print. Without realizing what I was getting into I announced that the statement was not true and proved this by showing several participants the book *Inleiding in de Coderingstheorie*, a little book based on the syllabus of a course given at the Mathematical Centre in Amsterdam in 1975 (M.C. Syllabus 31). The course, which was a great success, was given by M.R. Best, A.E. Brouwer, P. van Emde Boas, T.M.V. Janssen, H.W. Lenstra Jr., A. Schrijver, H.C.A. van Tilborg and myself. Since then the book has been used for a number of years at the Technological Universities of Delft and Eindhoven.

The comments above explain why it seemed reasonable (to me) to translate the Dutch book into English. In the name of Springer-Verlag I thank the Mathematical Centre in Amsterdam for permission to do so. Of course it turned out to be more than a translation. Much was rewritten or expanded, problems were changed and solutions were added, and a new chapter and several new proofs were included. Nevertheless the M.C. Syllabus (and the Springer Lecture Notes 201) are the basis of this book.

The book consists of three parts. Chapter 1 contains the prerequisite mathematical knowledge. It is written in the style of a memory-refresher. The reader who discovers topics that he does not know will get some idea about them but it is recommended that he also looks at standard textbooks on those topics. Chapters 2 to 6 provide an introductory course in coding theory. Finally, Chapters 7 to 11 are introductions to special topics and can be used as supplementary reading or as a preparation for studying the literature.

Despite the youth of the subject, which is demonstrated by the fact that the papers mentioned in the references have 1974 as the average publication year, I have not considered it necessary to give credit to every author of the theorems, lemmas, etc. Some have simply become standard knowledge.

It seems appropriate to mention a number of textbooks that I use regularly and that I would like to recommend to the student who would like to learn more than this introduction can offer. First of all F.J. MacWilliams and N.J.A. Sloane, *The Theory of Error-Correcting Codes* (reference [46]), which contains a much more extensive treatment of most of what is in this book and has 1500 references! For the more technically oriented student with an interest in decoding, complexity questions, etc. E.R. Berlekamp's *Algebraic Coding Theory* (reference [2]) is a must. For a very well-written mixture of information theory and coding theory I recommend: R.J. McEliece, *The Theory of Information and Coding* (reference [51]). In the present book very little attention is paid to the relation between coding theory and combinatorial mathematics. For this the reader should consult P.J. Cameron and J.H. van Lint, *Designs, Graphs, Codes and their Links* (reference [11]).

I sincerely hope that the time spent writing this book (instead of doing research) will be considered well invested.

Eindhoven J.H. VAN LINT
July 1981

Second edition comments: Apparently the hope expressed in the final line of the preface of the first edition came true: a second edition has become necessary. Several misprints have been corrected and also some errors. In a few places some extra material has been added.

Contents

Preface to the Third Edition V

Preface to the Second Edition VII

Preface to the First Edition IX

CHAPTER 1
Mathematical Background 1

1.1. Algebra 1
1.2. Krawtchouk Polynomials 14
1.3. Combinatorial Theory 17
1.4. Probability Theory 19

CHAPTER 2
Shannon's Theorem 22

2.1. Introduction 22
2.2. Shannon's Theorem 27
2.3. On Coding Gain 29
2.4. Comments 31
2.5. Problems 32

CHAPTER 3
Linear Codes 33

3.1. Block Codes 33
3.2. Linear Codes 35
3.3. Hamming Codes 38

3.4. Majority Logic Decoding 39
3.5. Weight Enumerators 40
3.6. The Lee Metric 42
3.7. Comments 44
3.8. Problems 45

CHAPTER 4
Some Good Codes 47

4.1. Hadamard Codes and Generalizations 47
4.2. The Binary Golay Code 48
4.3. The Ternary Golay Code 51
4.4. Constructing Codes from Other Codes 51
4.5. Reed–Muller Codes 54
4.6. Kerdock Codes 60
4.7. Comments 61
4.8. Problems 62

CHAPTER 5
Bounds on Codes 64

5.1. Introduction: The Gilbert Bound 64
5.2. Upper Bounds 67
5.3. The Linear Programming Bound 74
5.4. Comments 78
5.5. Problems 79

CHAPTER 6
Cyclic Codes 81

6.1. Definitions 81
6.2. Generator Matrix and Check Polynomial 83
6.3. Zeros of a Cyclic Code 84
6.4. The Idempotent of a Cyclic Code 86
6.5. Other Representations of Cyclic Codes 89
6.6. BCH Codes 91
6.7. Decoding BCH Codes 98
6.8. Reed–Solomon Codes 99
6.9. Quadratic Residue Codes 103
6.10. Binary Cyclic Codes of Length $2n$ (n odd) 106
6.11. Generalized Reed–Muller Codes 108
6.12. Comments 110
6.13. Problems 111

CHAPTER 7
Perfect Codes and Uniformly Packed Codes 112

7.1. Lloyd's Theorem 112
7.2. The Characteristic Polynomial of a Code 115

7.3. Uniformly Packed Codes . . . 118
7.4. Examples of Uniformly Packed Codes . . . 120
7.5. Nonexistence Theorems . . . 123
7.6. Comments . . . 127
7.7. Problems . . . 127

CHAPTER 8
Codes over $\mathbb{Z}_4$ . . . 128

8.1. Quaternary Codes . . . 128
8.2. Binary Codes Derived from Codes over $\mathbb{Z}_4$ . . . 129
8.3. Galois Rings over $\mathbb{Z}_4$ . . . 132
8.4. Cyclic Codes over $\mathbb{Z}_4$ . . . 136
8.5. Problems . . . 138

CHAPTER 9
Goppa Codes . . . 139

9.1. Motivation . . . 139
9.2. Goppa Codes . . . 140
9.3. The Minimum Distance of Goppa Codes . . . 142
9.4. Asymptotic Behaviour of Goppa Codes . . . 143
9.5. Decoding Goppa Codes . . . 144
9.6. Generalized BCH Codes . . . 145
9.7. Comments . . . 146
9.8. Problems . . . 147

CHAPTER 10
Algebraic Geometry Codes . . . 148

10.1. Introduction . . . 148
10.2. Algebraic Curves . . . 149
10.3. Divisors . . . 155
10.4. Differentials on a Curve . . . 156
10.5. The Riemann–Roch Theorem . . . 158
10.6. Codes from Algebraic Curves . . . 160
10.7. Some Geometric Codes . . . 162
10.8. Improvement of the Gilbert–Varshamov Bound . . . 165
10.9. Comments . . . 165
10.10. Problems . . . 166

CHAPTER 11
Asymptotically Good Algebraic Codes . . . 167

11.1. A Simple Nonconstructive Example . . . 167
11.2. Justesen Codes . . . 168
11.3. Comments . . . 172
11.4. Problems . . . 172

CHAPTER 12
Arithmetic Codes 173

12.1. AN Codes 173
12.2. The Arithmetic and Modular Weight 175
12.3. Mandelbaum–Barrows Codes 179
12.4. Comments 180
12.5. Problems 180

CHAPTER 13
Convolutional Codes 181

13.1. Introduction 181
13.2. Decoding of Convolutional Codes 185
13.3. An Analog of the Gilbert Bound for Some Convolutional Codes 187
13.4. Construction of Convolutional Codes from Cyclic Block Codes 188
13.5. Automorphisms of Convolutional Codes 191
13.6. Comments 193
13.7. Problems 194

Hints and Solutions to Problems 195

References 218

Index 223

CHAPTER 1

Mathematical Background

In order to be able to read this book a fairly thorough mathematical background is necessary. In different chapters many different areas of mathematics play a rôle. The most important one is certainly algebra but the reader must also know some facts from elementary number theory, probability theory and a number of concepts from combinatorial theory such as designs and geometries. In the following sections we shall give a brief survey of the prerequisite knowledge. Usually proofs will be omitted. For these we refer to standard textbooks. In some of the chapters we need a large number of facts concerning a not too well-known class of orthogonal polynomials, called Krawtchouk polynomials. These properties are treated in Section 1.2. The notations that we use are fairly standard. We mention a few that may not be generally known. If C is a finite set we denote the number of elements of C by $|C|$. If the expression B is the definition of concept A then we write $A := B$. We use "iff" for "if and only if". An identity matrix is denoted by I and the matrix with all entries equal to one is J. Similarly we abbreviate the vector with all coordinates 0 (resp. 1) by $\mathbf{0}$ (resp. $\mathbf{1}$). Instead of using $[x]$ we write $\lfloor x \rfloor := \max\{n \in \mathbb{Z} \mid n \leq x\}$ and we use the symbol $\lceil x \rceil$ for rounding upwards.

§1.1. Algebra

We need only very little from elementary number theory. We assume known that in $\mathbb{N}$ every number can be written in exactly one way as a product of prime numbers (if we ignore the order of the factors). If a divides b, then we write $a|b$. If p is a prime number and $p^r|a$ but $p^{r+1} \nmid a$, then we write $p^r \| a$. If

$k \in \mathbb{N}$, $k > 1$, then a representation of n in the base k is a representation

$$n = \sum_{i=0}^{l} n_i k^i,$$

$0 \le n_i < k$ for $0 \le i \le l$. The largest integer n such that $n|a$ and $n|b$ is called the greatest common divisor of a and b and denoted by g.c.d.(a, b) or simply (a, b). If $m|(a-b)$ we write $a \equiv b \pmod{m}$.

(1.1.1) Theorem. *If*

$$\varphi(n) := |\{m \in \mathbb{N} \mid 1 \le m \le n, (m, n) = 1\}|,$$

then

(i) $\varphi(n) = n \prod_{p|n} (1 - 1/p)$,
(ii) $\sum_{d|n} \varphi(d) = n$.

The function φ is called the *Euler indicator*.

(1.1.2) Theorem. *If* $(a, m) = 1$ *then* $a^{\varphi(m)} \equiv 1 \pmod{m}$.

Theorem 1.1.2 is called the Euler–Fermat theorem.

(1.1.3) Definition. The *Möbius function* μ is defined by

$$\mu(n) := \begin{cases} 1, & \text{if } n = 1, \\ (-1)^k, & \text{if } n \text{ is the product of } k \text{ distinct prime factors,} \\ 0, & \text{otherwise.} \end{cases}$$

(1.1.4) Theorem. *If f and g are functions defined on $\mathbb{N}$ such that*

$$g(n) = \sum_{d|n} f(d),$$

then

$$f(n) = \sum_{d|n} \mu(d) g\left(\frac{n}{d}\right).$$

Theorem 1.1.4 is known as the *Möbius inversion formula*.

Algebraic Structures

We assume that the reader is familiar with the basic ideas and theorems of linear algebra although we do refresh his memory below. We shall first give a sequence of definitions of algebraic structures with which the reader must be familiar in order to appreciate algebraic coding theory.

(1.1.5) Definition. A *group* $(G, \ )$ is a set G on which a product operation has been defined satisfying

(i) $\forall_{a \in G} \forall_{b \in G}[ab \in G]$,
(ii) $\forall_{a \in G} \forall_{b \in G} \forall_{c \in G}[(ab)c = a(bc)]$,
(iii) $\exists_{e \in G} \forall_{a \in G}[ae = ea = a]$,
(the element e is unique),
(iv) $\forall_{a \in G} \exists_{b \in G}[ab = ba = e]$,
(b is called the inverse of a and also denoted by a^{-1}).

If furthermore

(v) $\forall_{a \in G} \forall_{b \in G}[ab = ba]$,

then the group is called *abelian* or *commutative*.

If $(G, \ )$ is a group and $H \subset G$ such that $(H, \ )$ is also a group, then $(H, \ )$ is called a subgroup of $(G, \ )$. Usually we write G instead of $(G, \ )$. The number of elements of a finite group is called the *order* of the group. If $(G, \ )$ is a group and $a \in G$, then the smallest positive integer n such that $a^n = e$ (if such an n exists) is called the *order* of a. In this case the elements $e, a, a^2, \ldots, a^{n-1}$ form a so-called *cyclic* subgroup with a as *generator*. If $(G, \ )$ is abelian and $(H, \ )$ is a subgroup then the sets $aH := \{ah \mid h \in H\}$ are called *cosets* of H. Since two cosets are obviously disjoint or identical, the cosets form a partition of G. An element chosen from a coset is called a *representative* of the coset. It is not difficult to show that the cosets again form a group if we define multiplication of cosets by $(aH)(bH) := abH$. This group is called the *factor group* and indicated by G/H. As a consequence note that if $a \in G$, then the order of a divides the order of G (also if G is not abelian).

A fundamental theorem of group theory states that a finite abelian group is a direct sum of cyclic groups.

(1.1.6) Definition. A set R with two operations, usually called addition and multiplication, denoted by $(R, +, \ )$, is called a *ring* if

(i) $(R, +)$ is an abelian group,
(ii) $\forall_{a \in R} \forall_{b \in R} \forall_{c \in R}[(ab)c = a(bc)]$,
(iii) $\forall_{a \in R} \forall_{b \in R} \forall_{c \in R}[a(b + c) = ab + ac \wedge (a + b)c = ac + bc]$.

The identity element of $(R, +)$ is usually denoted by 0.

If the additional property

(iv) $\forall_{a \in R} \forall_{b \in R}[ab = ba]$

holds, then the ring is called *commutative*.

The integers $\mathbb{Z}$ are the best known example of a ring.

If $(R, +, \ )$ is a commutative ring, a nonzero element $a \in R$ is called a *zero divisor* if there exists a nonzero element $b \in R$ such that $ab = 0$. If a nontrivial ring has no zero divisors, it is called an *integral domain*. In the same way that $\mathbb{Z}$ is extended to $\mathbb{Q}$, an integral domain can be embedded in its *field of fractions* or *quotient field*.

(1.1.7) Definition. If $(R, +, \ )$ is a ring and $\emptyset \neq S \subseteq R$, then S is called an *ideal* if

(i) $\forall_{a \in S} \forall_{b \in S}[a - b \in S]$,
(ii) $\forall_{a \in S} \forall_{b \in R}[ab \in S \wedge ba \in S]$.

It is clear that if S is an ideal in R, then $(S, +, \ )$ is a subring, but requirement (ii) says more than that.

(1.1.8) Definition. A *field* is a ring $(R, +, \ )$ for which $(R \setminus \{0\}, \ )$ is an abelian group.

(1.1.9) Theorem. *Every finite ring R with at least two elements such that*

$$\forall_{a \in R} \forall_{b \in R}[ab = 0 \Rightarrow (a = 0 \vee b = 0)]$$

is a field.

(1.1.10) Definition. Let $(V, +)$ be an abelian group, $\mathbb{F}$ a field and let a multiplication $\mathbb{F} \times V \to V$ be defined satisfying

(i) $\forall_{\mathbf{a} \in V}[1\mathbf{a} = \mathbf{a}]$,
$\quad \forall_{\alpha \in \mathbb{F}} \forall_{\beta \in \mathbb{F}} \forall_{\mathbf{a} \in V}[\alpha(\beta\mathbf{a}) = (\alpha\beta)\mathbf{a}]$,
(ii) $\forall_{\alpha \in \mathbb{F}} \forall_{\mathbf{a} \in V} \forall_{\mathbf{b} \in V}[\alpha(\mathbf{a} + \mathbf{b}) = \alpha\mathbf{a} + \alpha\mathbf{b}]$,
$\quad \forall_{\alpha \in \mathbb{F}} \forall_{\beta \in \mathbb{F}} \forall_{\mathbf{a} \in V}[(\alpha + \beta)\mathbf{a} = \alpha\mathbf{a} + \beta\mathbf{a}]$.

Then the triple $(V, +, \mathbb{F})$ is called a *vector space* over the field $\mathbb{F}$. The identity element of $(V, +)$ is denoted by $\mathbf{0}$.

We assume the reader to be familiar with the vector space $\mathbb{R}^n$ consisting of all n-tuples $(a_1, a_2, \ldots, a_n)$ with the obvious rules for addition and multiplication. We remind him of the fact that a *k-dimensional subspace* C of this vector space is a vector space with a *basis* consisting of vectors $\mathbf{a}_1 := (a_{11}, a_{12}, \ldots, a_{1n})$, $\mathbf{a}_2 := (a_{21}, a_{22}, \ldots, a_{2n})$, $\ldots$, $\mathbf{a}_k := (a_{k1}, a_{k2}, \ldots, a_{kn})$, where the word basis means that every $\mathbf{a} \in C$ can be written in a unique way as $\alpha_1\mathbf{a}_1 + \alpha_2\mathbf{a}_2 + \cdots + \alpha_k\mathbf{a}_k$. The reader should also be familiar with the process of going from one basis of C to another by taking combinations of basis vectors, etc. We shall usually write vectors as *row vectors* as we did above. The *inner product* $\langle \mathbf{a}, \mathbf{b} \rangle$ of two vectors $\mathbf{a}$ and $\mathbf{b}$ is defined by

$$\langle \mathbf{a}, \mathbf{b} \rangle := a_1 b_1 + a_2 b_2 + \cdots + a_n b_n.$$

The elements of a basis are called *linearly independent*. In other words this means that a linear combination of these vectors is $\mathbf{0}$ iff all the coefficients are 0. If $\mathbf{a}_1, \ldots, \mathbf{a}_k$ are k linearly independent vectors, i.e. a basis of a k-dimensional subspace C, then the system of equations $\langle \mathbf{a}_i, \mathbf{y}\rangle = 0$ $(i = 1, 2, \ldots, k)$ has as its solution all the vectors in a subspace of dimension $n - k$ which we denote by $C^\perp$. So,

$$C^\perp := \{\mathbf{y} \in \mathbb{R}^n | \forall_{\mathbf{x}\in C}[\langle \mathbf{x}, \mathbf{y}\rangle = 0]\}.$$

These ideas play a fundamental role later on, where $\mathbb{R}$ is replaced by a finite field $\mathbb{F}$. The theory reviewed above goes through in that case.

(1.1.11) Definition. Let $(V, +)$ be a vector space over $\mathbb{F}$ and let a multiplication $V \times V \to V$ be defined that satisfies

(i) $(V, +, \ )$ is a ring,
(ii) $\forall_{\alpha\in\mathbb{F}}\forall_{\mathbf{a}\in V}\forall_{\mathbf{b}\in V}[(\alpha\mathbf{a})\mathbf{b} = \mathbf{a}(\alpha\mathbf{b})]$.

Then we say that the system is an *algebra* over $\mathbb{F}$.

Suppose we have a finite group $(G, \cdot)$ and we consider the elements of G as basis vectors for a vector space $(V, +)$ over a field $\mathbb{F}$. Then the elements of V are represented by linear combinations $\alpha_1 g_1 + \alpha_2 g_2 + \cdots + \alpha_n g_n$, where

$$\alpha_i \in \mathbb{F}, \qquad g_i \in G, \qquad (1 \le i \le n = |G|).$$

We can define a multiplication $*$ for these vectors in the obvious way, namely

$$\left(\sum_i \alpha_i g_i\right) * \left(\sum_j \beta_j g_j\right) := \sum_i \sum_j (\alpha_i\beta_j)(g_i \cdot g_j),$$

which can be written as $\sum_k \gamma_k g_k$, where γ_k is the sum of the elements $\alpha_i\beta_j$ over all pairs (i, j) such that $g_i \cdot g_j = g_k$. This yields an algebra which is called the *group algebra* of G over $\mathbb{F}$ and denoted by $\mathbb{F}G$.

EXAMPLES. Let us consider a number of examples of the concepts defined above.

If $A := \{a_1, a_2, \ldots, a_n\}$ is a finite set, we can consider all one-to-one mappings of S onto S. These are called *permutations*. If σ_1 and σ_2 are permutations we define $\sigma_1\sigma_2$ by $(\sigma_1\sigma_2)(a) := \sigma_1(\sigma_2(a))$ for all $a \in A$. It is easy to see that the set S_n of all permutations of A with this multiplication is a group, known as the *symmetric group of degree n*. In this book we shall often be interested in special permutation groups. These are subgroups of S_n. We give one example. Let C be a k-dimensional subspace of $\mathbb{R}^n$. Consider all permutations σ of the integers $1, 2, \ldots, n$ such that for every vector $\mathbf{c} = (c_1, c_2, \ldots, c_n) \in C$ the vector $(c_{\sigma(1)}, c_{\sigma(2)}, \ldots, c_{\sigma(n)})$ is also in C. These clearly form a subgroup of S_n. Of course C will often be such that this subgroup of S consists of the identity only but there are more interesting examples! Another example of a permutation group which will turn up later is the *affine permutation group* defined as follows. Let $\mathbb{F}$ be a (finite) field. The mapping $f_{u,v}$, when $u \in \mathbb{F}$, $v \in \mathbb{F}$, $u \ne 0$, is defined on $\mathbb{F}$ by $f_{u,v}(x) := ux + v$ for all $x \in \mathbb{F}$. These mappings are permutations of $\mathbb{F}$ and clearly they form a group under composition of functions.

A *permutation matrix* P is a (0, 1)-matrix that has exactly one 1 in each row and column. We say that P corresponds to the permutation σ of $\{1, 2, \ldots, n\}$ if $p_{ij} = 1$ iff $i = \sigma(j)$ $(i = 1, 2, \ldots, n)$. With this convention the product of permutations corresponds to the product of their matrices. In this way one obtains the so-called matrix representation of a group of permutations.

A group G of permutations acting on a set Ω is called *k-transitive* on Ω if for every ordered k-tuple $(a_1, \ldots, a_k)$ of distinct elements of Ω and for every k-tuple $(b_1, \ldots, b_k)$ of distinct elements of Ω, there is an element $\sigma \in G$ such that $b_i = \sigma(a_i)$ for $1 \le i \le k$. If $k = 1$ we call the group transitive.

Let S be an ideal in the ring $(R, +, \ )$. Since $(S, +)$ is a subgroup of the abelian group $(R, +)$, we can form the factor group. The cosets are now called *residue classes mod S*. For these classes we introduce a multiplication in the obvious way: $(a + S)(b + S) := ab + S$. The reader who is not familiar with this concept should check that this definition makes sense (i.e. it does not depend on the choice of representatives a resp. b). In this way we have constructed a ring, called the *residue class ring* R mod S and denoted by R/S. The following example will surely be familiar. Let $R := \mathbb{Z}$ and let p be a prime. Let S be $p\mathbb{Z}$, the set of all multiples of p, which is sometimes also denoted by (p). Then R/S is the ring of integers mod p. The elements of R/S can be represented by $0, 1, \ldots, p - 1$ and then addition and multiplication are the usual operations in $\mathbb{Z}$ followed by a reduction mod p. For example, if we take $p = 7$, then $4 + 5 = 2$ because in $\mathbb{Z}$ we have $4 + 5 \equiv 2 \pmod 7$. In the same way $4 \cdot 5 = 6$ in $\mathbb{Z}/7\mathbb{Z} = \mathbb{Z}/(7)$. If S is an ideal in $\mathbb{Z}$ and $S \ne \{0\}$, then there is a smallest positive integer k in S. Let $s \in S$. We can write s as $ak + b$, where $0 \le b < k$. By the definition of ideal we have $ak \in S$ and hence $b = s - ak \in S$ and then the definition of k implies that $b = 0$. Therefore $S = (k)$. An ideal consisting of all multiples of a fixed element is called a *principal ideal.* If a ring R has no other ideals than principal ideals, it is called a *principal ideal ring.* Therefore $\mathbb{Z}$ is such a ring.

An ideal S is called a *prime ideal* if $ab \in S$ implies $a \in S$ or $b \in S$. An ideal S in a ring R is called *maximal* if for every ideal I with $S \subset I \subset R$, $I = S$ or $I = R$ $(S \ne R)$. If a ring has a unique maximal ideal, it is called a *local ring*.

(1.1.12) Theorem. *If p is a prime then* $\mathbb{Z}/p\mathbb{Z}$ *is a field.*

This is an immediate consequence of Theorem 1.1.9 but also obvious directly. A finite field with n elements is denoted by $\mathbb{F}_n$ or GF(n) (Galois field).

Rings and Finite Fields

More about finite fields will follow below. First some more about rings and ideals. Let $\mathbb{F}$ be a finite field. Consider the set $\mathbb{F}[x]$ consisting of all polynomials $a_0 + a_1x + \cdots + a_nx^n$, where n can be any integer in $\mathbb{N}$ and $a_i \in \mathbb{F}$ for $0 \le i \le n$. With the usual definition of addition and multiplication of polynomials this yields a ring ($\mathbb{F}[x]$, +,), which is usually just denoted by $\mathbb{F}[x]$. The set of all polynomials that are multiples of a fixed polynomial $g(x)$, i.e. all polynomials of the form $a(x)g(x)$ where $a(x) \in \mathbb{F}[x]$, is an ideal in $\mathbb{F}[x]$.

As before, we denote this ideal by $(g(x))$. The following theorem states that there are no other types.

(1.1.13) Theorem. *$\mathbb{F}[x]$ is a principal ideal ring.*

The residue class ring $\mathbb{F}[x]/(g(x))$ can be represented by the polynomials whose degree is less than the degree of $g(x)$. In the same way as our example $\mathbb{Z}/7\mathbb{Z}$ given above, we now multiply and add these representatives in the usual way and then reduce mod $g(x)$. For example, we take $\mathbb{F} = \mathbb{F}_2 = \{0, 1\}$ and $g(x) = x^3 + x + 1$. Then $(x + 1)(x^2 + 1) = x^3 + x^2 + x + 1 = x^2$. This example is a useful one to study carefully if one is not familiar with finite fields. First observe that $g(x)$ is *irreducible*, i.e., there do not exist polynomials $a(x)$ and $b(x) \in \mathbb{F}[x]$, both of degree less than 3, such that $g(x) = a(x)b(x)$. Next, realize that this means that in $\mathbb{F}_2[x]/(g(x))$ the product of two elements $a(x)$ and $b(x)$ is 0 iff $a(x) = 0$ or $b(x) = 0$. By Theorem 1.1.9 this means that $\mathbb{F}_2[x]/(g(x))$ is a field. Since the representatives of this residue class ring all have degrees less than 3, there are exactly eight of them. So we have found a field with eight elements, i.e. $\mathbb{F}_{2^3}$. This is an example of the way in which finite fields are constructed.

(1.1.14) Theorem. *Let p be a prime and let $g(x)$ be an irreducible polynomial of degree r in the ring $\mathbb{F}_p[x]$. Then the residue class ring $\mathbb{F}_p[x]/(g(x))$ is a field with p^r elements.*

Proof. The proof is the same as the one given for the example $p = 2$, $r = 3$, $g(x) = x^3 + x + 1$. □

(1.1.15) Theorem. *Let $\mathbb{F}$ be a field with n elements. Then n is a power of a prime.*

Proof. By definition there is an identity element for multiplication in $\mathbb{F}$. We denote this by 1. Of course $1 + 1 \in \mathbb{F}$ and we denote this element by 2. We continue in this way, i.e. $2 + 1 = 3$, etc. After a finite number of steps we encounter a field element that already has a name. Suppose, e.g. that the sum of k terms 1 is equal to the sum of l terms 1 ($k > l$). Then the sum of $(k - l)$ terms 1 is 0, i.e. the first time we encounter an element that already has a name, this element is 0. Say 0 is the sum of k terms 1. If k is composite, $k = ab$, then the product of the elements which we have called a resp. b is 0, a contradiction. So k is a prime and we have shown that $\mathbb{F}_p$ is a subfield of $\mathbb{F}$. We define linear independence of a set of elements of $\mathbb{F}$ with respect to (coefficients from) $\mathbb{F}_p$ in the obvious way. Among all linearly independent subsets of $\mathbb{F}$ let $\{x_1, x_2, \ldots, x_r\}$ be one with the maximal number of elements. If x is any element of $\mathbb{F}$ then the elements $x, x_1, x_2, \ldots, x_r$ are not linearly independent, i.e. there are coefficients $0 \neq \alpha, \alpha_1, \ldots, \alpha_r$ such that $\alpha x + \alpha_1 x_1 + \cdots + \alpha_r x_r = 0$ and hence x is a linear combination of x_1 to x_r. Since there are obviously p^r distinct linear combinations of x_1 to x_r the proof is complete. □

From the previous theorems we now know that a field with n elements exists iff n is a prime power, providing we can show that for every $r \geq 1$ there is an irreducible polynomial of degree r in $\mathbb{F}_p[x]$. We shall prove this by calculating the number of such polynomials. Fix p and let I_r denote the number of irreducible polynomials of degree r that are *monic*, i.e. the coefficient of x^r is 1. We claim that

(1.1.16) $$(1 - pz)^{-1} = \prod_{r=1}^{\infty} (1 - z^r)^{-I_r}.$$

In order to see this, first observe that the coefficient of z^n on the left-hand side is p^n, which is the number of monic polynomials of degree n with coefficients in $\mathbb{F}_p$. We know that each such polynomial can be factored uniquely into irreducible factors and we must therefore convince ourselves that these products are counted on the right-hand side of (1.1.16). To show this we just consider two irreducible polynomials $a_1(x)$ of degree r and $a_2(x)$ of degree s. There is a 1–1 correspondence between products $(a_1(x))^k(a_2(x))^l$ and terms $z_1^{kr} z_2^{ls}$ in the product of $(1 + z_1^r + z_1^{2r} + \cdots)$ and $(1 + z_2^s + z_2^{2s} + \cdots)$. If we identify z_1 and z_2 with z, then the exponent of z is the degree of $(a_1(x))^k(a_2(x))^l$. Instead of two polynomials $a_1(x)$ and $a_2(x)$, we now consider all irreducible polynomials and (1.1.16) follows.

In (1.1.16) we take logarithms on both sides, then differentiate, and finally multiply by z to obtain

(1.1.17) $$\frac{pz}{1 - pz} = \sum_{r=1}^{\infty} I_r \frac{rz^r}{1 - z^r}.$$

Comparing coefficients of z^n on both sides of (1.1.17) we find

(1.1.18) $$p^n = \sum_{r|n} rI_r.$$

Now apply Theorem 1.1.4 to (1.1.18). We find

(1.1.19) $$I_r = \frac{1}{r} \sum_{d|r} \mu(d) p^{r/d} > \frac{1}{r}\{p^r - p^{r/2} - p^{r/3} - \cdots\}$$
$$> \frac{1}{r}\left(p^r - \sum_{i=0}^{r/2} p^i\right) > \frac{1}{r} p^r (1 - p^{-r/2+1}) > 0.$$

Now that we know for which values of n a field with n elements exists, we wish to know more about these fields. The structure of $\mathbb{F}_{p^r}$ will play a very important role in many chapters of this book. As a preparation consider a finite field $\mathbb{F}$ and a polynomial $f(x) \in \mathbb{F}[x]$ such that $f(a) = 0$, where $a \in \mathbb{F}$. Then by dividing we find that there is a $g(x) \in \mathbb{F}[x]$ such that $f(x) = (x - a)g(x)$.

Continuing in this way we establish the trivial fact that a polynomial $f(x)$ of degree r in $\mathbb{F}[x]$ has at most r zeros in $\mathbb{F}$.

If α is an element of order e in the multiplicative group $(\mathbb{F}_{p^r}\backslash\{0\},\ )$, then α is a zero of the polynomial $x^e - 1$. In fact, we have

$$x^e - 1 = (x - 1)(x - \alpha)(x - \alpha^2)\cdots(x - \alpha^{e-1}).$$

It follows that the only elements of order e in the group are the powers α^i where $1 \le i < e$ and $(i, e) = 1$. There are $\varphi(e)$ such elements. Hence, for every e which divides $p^r - 1$ there are either 0 or $\varphi(e)$ elements of order e in the field. By (1.1.1) the possibility 0 never occurs. As a consequence there are elements of order $p^r - 1$, in fact exactly $\varphi(p^r - 1)$ such elements. We have proved the following theorem.

(1.1.20) Theorem. *In $\mathbb{F}_q$ the multiplicative group $(\mathbb{F}_q\backslash\{0\},\ )$ is a cyclic group.*

This group is often denoted by $\mathbb{F}_q^*$.

(1.1.21) Definition. A generator of the multiplicative group of $\mathbb{F}_q$ is called a *primitive element* of the field.

Note that Theorem 1.1.20 states that the elements of $\mathbb{F}_q$ are exactly the q distinct zeros of the polynomial $x^q - x$. An element β such that $\beta^k = 1$ but $\beta^l \neq 1$ for $0 < l < k$ is called a *primitive* kth root of unity. Clearly a primitive element α of $\mathbb{F}_q$ is a primitive $(q - 1)$th root of unity. If e divides $q - 1$ then α^e is a primitive $((q - 1)/e)$th root of unity. Furthermore a consequence of Theorem 1.1.20 is that $\mathbb{F}_{p^r}$ is a subfield of $\mathbb{F}_{p^s}$ iff r divides s. Actually this statement could be slightly confusing to the reader. We have been suggesting by our notation that for a given q the field $\mathbb{F}_q$ is unique. This is indeed true. In fact this follows from (1.1.18). We have shown that for $q = p^n$ every element of $\mathbb{F}_q$ is a zero of some irreducible factor of $x^q - x$ and from the remark above and Theorem 1.1.14 we see that this factor must have a degree r such that $r|n$. By (1.1.18) this means we have used all irreducible polynomials of degree r where $r|n$. In other words, the product of these polynomials is $x^q - x$. This establishes the fact that two fields $\mathbb{F}$ and $\mathbb{F}'$ of order q are isomorphic, i.e. there is a mapping φ: $\mathbb{F} \to \mathbb{F}'$ which is one-to-one and such that φ preserves addition and multiplication.

The following theorem is used very often in this book.

(1.1.22) Theorem. *Let $q = p^r$ and $0 \neq f(x) \in \mathbb{F}_q[x]$.*

(i) *If $\alpha \in \mathbb{F}_{q^k}$ and $f(\alpha) = 0$, then $f(\alpha^q) = 0$.*
(ii) *Conversely: Let $g(x)$ be a polynomial with coefficients in an extension field of $\mathbb{F}_q$. If $g(\alpha^q) = 0$ for every α for which $g(\alpha) = 0$, then $g(x) \in \mathbb{F}_q[x]$.*

PROOF.

(i) By the binomial theorem we have $(a+b)^p = a^p + b^p$ because p divides $\binom{p}{k}$ for $1 \le k \le p-1$. It follows that $(a+b)^q = a^q + b^q$. If $f(x) = \sum a_i x^i$ then $(f(x))^q = \sum a_i^q (x^q)^i$.

Because $a_i \in \mathbb{F}_q$ we have $a_i^q = a_i$. Substituting $x = \alpha$ we find $f(\alpha^q) = (f(\alpha))^q = 0$.

(ii) We already know that in a suitable extension field of $\mathbb{F}_q$ the polynomial $g(x)$ is a product of factors $x - \alpha_i$ (all of degree 1, that is) and if $x - \alpha_i$ is one of these factors, then $x - \alpha_i^q$ is also one of them. If $g(x) = \sum_{k=0}^{n} a_k x^k$ then a_k is a symmetric function of the zeros α_i and hence $a_k = a_k^q$, i.e. $a_k \in \mathbb{F}_q$.

If $\alpha \in \mathbb{F}_q$, where $q = p^r$, then the *minimal polynomial* of α over $\mathbb{F}_p$ is the irreducible polynomial $f(x) \in \mathbb{F}_p[x]$ such that $f(\alpha) = 0$. If α has order e then from Theorem 1.1.22 we know that this minimal polynomial is $\prod_{i=0}^{m-1} (x - \alpha^{p^i})$, where m is the smallest integer such that $p^m \equiv 1 \pmod{e}$. Sometimes we shall consider a field $\mathbb{F}_q$ with a fixed primitive element α. In that case we use $m_i(x)$ to denote the minimal polynomial of α^i. An irreducible polynomial which is the minimal polynomial of a primitive element in the corresponding field is called a *primitive polynomial*. Such polynomials are the most convenient ones to use in the construction of Theorem 1.1.14. We give an example in detail.

(1.1.23) EXAMPLE. The polynomial $x^4 + x + 1$ is primitive over $\mathbb{F}_2$. The field $\mathbb{F}_{2^4}$ is represented by polynomials of degree <4. The polynomial x is a primitive element. Since we prefer to use the symbol x for other purposes, we call this primitive element α. Note that $\alpha^4 + \alpha + 1 = 0$. Every element in $\mathbb{F}_{2^4}$ is a linear combination of the elements $1, \alpha, \alpha^2$, and α^3. We get the following table for $\mathbb{F}_{2^4}$. The reader should observe that this is the equivalent of a table of logarithms for the case of the field $\mathbb{R}$.

The representation on the right demonstrates again that $\mathbb{F}_{2^4}$ can be interpreted as the vector space $(\mathbb{F}_2)^4$, where $\{1, \alpha, \alpha^2, \alpha^3\}$ is the basis. The left-hand column is easiest for multiplication (add exponents, mod 15) and the right-hand column for addition (add vectors). It is now easy to check that

$$
\begin{aligned}
m_1(x) &= (x-\alpha)(x-\alpha^2)(x-\alpha^4)(x-\alpha^8) &&= x^4 + x + 1,\\
m_3(x) &= (x-\alpha^3)(x-\alpha^6)(x-\alpha^{12})(x-\alpha^9) &&= x^4 + x^3 + x^2 + x + 1,\\
m_5(x) &= (x-\alpha^5)(x-\alpha^{10}) &&= x^2 + x + 1,\\
m_7(x) &= (x-\alpha^7)(x-\alpha^{14})(x-\alpha^{13})(x-\alpha^{11}) &&= x^4 + x^3 + 1,
\end{aligned}
$$

and the decomposition of $x^{16} - x$ into irreducible factors is

$$x^{16} - x = x(x-1)(x^2+x+1)(x^4+x+1) \times (x^4+x^3+1)(x^4+x^3+x^2+x+1).$$

Note that $x^4 - x = x(x-1)(x^2+x+1)$ corresponding to the elements 0, 1, α^5, α^{10} which form the subfield $\mathbb{F}_4 = \mathbb{F}_2[x]/(x^2+x+1)$. The polynomial $m_3(x)$ is irreducible but not primitive.

Table of $\mathbb{F}_{2^4}$

0	$=$	$= (0\ 0\ 0\ 0)$
1	$= 1$	$= (1\ 0\ 0\ 0)$
α	$= \alpha$	$= (0\ 1\ 0\ 0)$
α^2	$= \alpha^2$	$= (0\ 0\ 1\ 0)$
α^3	$= \alpha^3$	$= (0\ 0\ 0\ 1)$
α^4	$= 1 + \alpha$	$= (1\ 1\ 0\ 0)$
α^5	$= \alpha + \alpha^2$	$= (0\ 1\ 1\ 0)$
α^6	$= \alpha^2 + \alpha^3$	$= (0\ 0\ 1\ 1)$
α^7	$= 1 + \alpha + \alpha^3$	$= (1\ 1\ 0\ 1)$
α^8	$= 1 + \alpha^2$	$= (1\ 0\ 1\ 0)$
α^9	$= \alpha + \alpha^3$	$= (0\ 1\ 0\ 1)$
α^{10}	$= 1 + \alpha + \alpha^2$	$= (1\ 1\ 1\ 0)$
α^{11}	$= \alpha + \alpha^2 + \alpha^3$	$= (0\ 1\ 1\ 1)$
α^{12}	$= 1 + \alpha + \alpha^2 + \alpha^3$	$= (1\ 1\ 1\ 1)$
α^{13}	$= 1 + \alpha^2 + \alpha^3$	$= (1\ 0\ 1\ 1)$
α^{14}	$= 1 + \alpha^3$	$= (1\ 0\ 0\ 1)$

The reader who is not familiar with finite fields should study (1.1.14) to (1.1.23) thoroughly and construct several examples such as $\mathbb{F}_9$, $\mathbb{F}_{27}$, $\mathbb{F}_{64}$ with the corresponding minimal polynomials, subfields, etc. For tables of finite fields see references [9] and [10].

Polynomials

We need a few more facts about polynomials. If $f(x) \in \mathbb{F}_q[x]$ we can define the *derivative* $f'(x)$ in a purely formal way by

$$\left(\sum_{k=0}^{n} a_k x^k\right)' := \sum_{k=1}^{n} k a_k x^{k-1}.$$

The usual rules for differentiation of sums and products go through and one finds for instance that the derivative of $(x-\alpha)^2 f(x)$ is $2(x-\alpha)f(x) + (x-\alpha)^2 f'(x)$. Therefore the following theorem is obvious.

(1.1.24) Theorem. *If $f(x) \in \mathbb{F}_q[x]$ and α is a multiple zero of $f(x)$ in some extension field of $\mathbb{F}_q$, then α is also a zero of the derivative $f'(x)$.*

Note however, that if $q = 2^r$, then the second derivative of any polynomial in $\mathbb{F}_q[x]$ is identically 0. This tells us nothing about the multiplicity of zeros of the polynomial. In order to get complete analogy with the theory of polynomials over $\mathbb{R}$, we introduce the so-called *Hasse derivative* of a polynomial $f(x) \in \mathbb{F}_q[x]$ by

$$f^{[k]}(x) := \frac{1}{k!} f^{(k)}(x);$$

$\left(\text{so the } k\text{-th Hasse derivative of } x^n \text{ is } \binom{n}{k} x^{n-k}\right).$

The reader should have no difficulty proving that α is a zero of $f(x)$ with multiplicity k iff it is a zero of $f^{[i]}(x)$ for $0 \le i < k$ and not a zero of $f^{[k]}(x)$.

Another result to be used later is the fact that if $f(x) = \prod_{i=1}^{n} (x - \alpha_i)$ then $f'(x) = \sum_{i=1}^{n} f(x)/(x - \alpha_i)$.

The following theorem is well known.

(1.1.25) Theorem. *If the polynomials* $a(x)$ *and* $b(x)$ *in* $\mathbb{F}[x]$ *have greatest common divisor* 1, *then there are polynomials* $p(x)$ *and* $q(x)$ *in* $\mathbb{F}[x]$ *such that*

$$a(x)p(x) + b(x)q(x) = 1.$$

PROOF. This is an immediate consequence of Theorem 1.1.13. □

Although we know from (1.1.19) that irreducible polynomials of any degree r exist, it sometimes takes a lot of work to find one. The proof of (1.1.19) shows one way to do it. One starts with all possible polynomials of degree 1 and forms all reducible polynomials of degree 2. Any polynomial of degree 2 not in the list is irreducible. Then one proceeds in the obvious way to produce irreducible polynomials of degree 3, etc. In Section 9.2 we shall need irreducible polynomials over $\mathbb{F}_2$ of arbitrarily high degree. The procedure sketched above is not satisfactory for that purpose. Instead, we proceed as follows.

(1.1.26) Lemma.

$$3^{\beta+1} \| (2^{3^\beta} + 1).$$

PROOF.

(i) For $\beta = 0$ and $\beta = 1$ the assertion is true.
(ii) Suppose $3^t \| (2^{3^\beta} + 1)$. Then from

$$(2^{3^{\beta+1}} + 1) = (2^{3^\beta} + 1)\{(2^{3^\beta} + 1)(2^{3^\beta} - 2) + 3\},$$

it follows that if $t \ge 2$, then $3^{t+1} \| (2^{3^{\beta+1}} + 1)$. □

(1.1.27) Lemma. *If* m *is the order of* 2 (mod 3^l), *then*

$$m = \varphi(3^l) = 2 \cdot 3^{l-1}.$$

PROOF. If $2^\alpha \equiv 1 \pmod 3$ then α is even. Therefore $m = 2s$. Hence $2^s + 1 \equiv 0 \pmod{3^l}$. The result follows from Theorem 1.1.2 and Lemma 1.1.26. □

(1.1.28) Theorem. *Let* $m = 2 \cdot 3^{l-1}$. *Then*

$$x^m + x^{m/2} + 1$$

is irreducible over $\mathbb{F}_2$.

PROOF. Consider $\mathbb{F}_{2^m}$. In this field let ξ be a primitive (3^l)th root of unity. The minimal polynomial of ξ then is, by Lemma 1.1.27

$$f(x) = (x - \xi)(x - \xi^2)(x - \xi^4)\cdots(x - \xi^{2^{m-1}}),$$

a polynomial of degree m. Note that

$$x^{3^l} + 1 = (1 + x)(1 + x + x^2)(1 + x^3 + x^6)\cdots(1 + x^{3^{l-1}} + x^{2\cdot 3^{l-1}}),$$

a factorization which contains only one polynomial of degree m, so the last factor must be $f(x)$, i.e. it is irreducible. □

Quadratic Residues

A consequence of the existence of a primitive element in any field $\mathbb{F}_q$ is that it is easy to determine the squares in the field. If q is even then every element is a square. If q is odd then $\mathbb{F}_q$ consists of 0, $\frac{1}{2}(q-1)$ nonzero squares and $\frac{1}{2}(q-1)$ nonsquares. The integers k with $1 \le k \le p-1$ which are squares in $\mathbb{F}_p$ are usually called *quadratic residues* (mod p). By considering $k \in \mathbb{F}_p$ as a power of a primitive element of this field, we see that k is a quadratic residue (mod p) iff $k^{(p-1)/2} \equiv 1 \pmod p$. For the element $p - 1 = -1$ we find: -1 is a square in $\mathbb{F}_p$ iff $p \equiv 1 \pmod 4$. In Section 6.9 we need to know whether 2 is a square in $\mathbb{F}_p$. To decide this question we consider the elements $1, 2, \ldots, (p-1)/2$ and let a be their product. Multiply each of the elements by 2 to obtain $2, 4, \ldots, p-1$. This sequence contains $\lfloor (p-1)/4 \rfloor$ factors which are factors of a and for any other factor k of a we see that $-k$ is one of the even integers $> (p-1)/2$. It follows that in $\mathbb{F}_p$ we have $2^{(p-1)/2}\ a = (-1)^{(p-1)/2 - \lfloor (p-1)/4 \rfloor}\, a$ and since $a \neq 0$ we see that 2 is a square iff

$$\frac{p-1}{2} - \left\lfloor \frac{p-1}{4} \right\rfloor$$

is even, i.e. $p \equiv \pm 1 \pmod 8$.

The Trace

Let $q = p^r$. We define a mapping $\mathrm{Tr}\colon \mathbb{F}_q \to \mathbb{F}_p$, which is called the *trace*, as follows.

(1.1.29) Definition. If $\xi \in \mathbb{F}_q$ then

$$\operatorname{Tr}(\xi) := \xi + \xi^p + \xi^{p^2} + \cdots + \xi^{p^{r-1}}.$$

(1.1.30) Theorem. *The trace function has the following properties:*

(i) *For every* $\xi \in \mathbb{F}_q$ *the trace* $\operatorname{Tr}(\xi)$ *is in* $\mathbb{F}_p$;
(ii) *There are elements* $\xi \in \mathbb{F}_q$ *such that* $\operatorname{Tr}(\xi) \neq 0$;
(iii) Tr *is a linear mapping.*

PROOF.

(i) By definition $(\operatorname{Tr}(\xi))^p = \operatorname{Tr}(\xi)$.
(ii) The equation $x + x^p + \cdots + x^{p^{r-1}} = 0$ cannot have q roots in $\mathbb{F}_q$,
(iii) Since $(\xi + \eta)^p = \xi^p + \eta^p$ and for every $a \in \mathbb{F}_p$ we have $a^p = a$, this is obvious. □

Of course the theorem implies that the trace takes every value $p^{-1}q$ times and we see that the polynomial $x + x^p + \cdots + x^{p^{r-1}}$ is a product of minimal polynomials (check this for Example 1.1.23).

Characters

Let $(G, +)$ be a group and let $(T, \ )$ be the group of complex numbers with absolute value 1 with multiplication as operation. A *character* is a homomorphism $\chi: G \to T$, i.e.

(1.1.31) $$\forall_{g_1 \in G} \forall_{g_2 \in G}[\chi(g_1 + g_2) = \chi(g_1)\chi(g_2)].$$

From the definition it follows that $\chi(0) = 1$ for every character χ. If $\chi(g) = 1$ for all $g \in G$ then χ is called the *principal character.*

(1.1.32) Lemma. *If* χ *is a character for* $(G, +)$ *then*

$$\sum_{g \in G} \chi(g) = \begin{cases} |G|, & \text{if } \chi \text{ is the principal character,} \\ 0, & \text{otherwise.} \end{cases}$$

PROOF. Let $h \in G$. Then

$$\chi(h) \sum_{g \in G} \chi(g) = \sum_{g \in G} \chi(h + g) = \sum_{k \in G} \chi(k).$$

If χ is not the principal character we can choose h such that $\chi(h) \neq 1$. □

§1.2. Krawtchouk Polynomials

In this section we introduce a sequence of polynomials which play an important role in several parts of coding theory, the so-called *Krawtchouk polyno-*

mials. These polynomials are an example of orthogonal polynomials and most of the theorems that we mention are special cases of general theorems that are valid for any sequence of orthogonal polynomials. The reader who does not know this very elegant part of analysis is recommended to consult one of the many textbooks about orthogonal polynomials (e.g. G. Szegö [67], D. Jackson [36], F. G. Tricomi [70]). In fact, for some of the proofs of theorems that we mention below, we refer the reader to the literature. Because of the great importance of these polynomials in the sequel, we treat them more extensively than most other subjects in this introduction.

Usually the Krawtchouk polynomials will appear in situations where two parameters n and q have already been fixed. These are usually omitted in the notation for the polynomials.

(1.2.1) Definition. For $k = 0, 1, 2, \ldots$, we define the *Krawtchouk polynomial* $K_k(x)$ by

$$K_k(x; n, q) := K_k(x) := \sum_{j=0}^{k} (-1)^j \binom{x}{j}\binom{n-x}{k-j}(q-1)^{k-j},$$

where

$$\binom{x}{j} := \frac{x(x-1)\cdots(x-j+1)}{j!}, \qquad (x \in \mathbb{R}).$$

Observe that for the special case $q = 2$ we have

$$K_k(x) = \sum_{j=0}^{k} (-1)^j \binom{x}{j}\binom{n-x}{k-j} = (-1)^k K_k(n-x). \tag{1.2.2}$$

By multiplying the Taylor series for $(1 + (q-1)z)^{n-x}$ and $(1-z)^x$ we find

$$\sum_{k=0}^{\infty} K_k(x) z^k = (1 + (q-1)z)^{n-x}(1-z)^x. \tag{1.2.3}$$

It is clear from (1.2.1) that $K_k(x)$ is a polynomial of degree k in x with leading coefficient $(-q)^k/k!$ The name orthogonal polynomial is connected with the following "orthogonality relation":

$$\sum_{i=0}^{n} \binom{n}{i}(q-1)^i K_k(i) K_l(i) = \delta_{kl}\binom{n}{k}(q-1)^k q^n. \tag{1.2.4}$$

The reader can easily prove this relation by multiplying both sides by $x^k y^l$ and summing over k and l (0 to ∞), using (1.2.3). Since the two sums are equal, the assertion is true. From (1.2.1) we find

$$(q-1)^i\binom{n}{i} K_k(i) = (q-1)^k \binom{n}{k} K_i(k), \tag{1.2.5}$$

which we substitute in (1.2.4) to find a second kind of orthogonality relation:

$$(1.2.6) \qquad \sum_{i=0}^{n} K_l(i)K_i(k) = \delta_{lk}q^n.$$

We list a few of the Krawtchouk polynomials ($k \le 2$)

$$(1.2.7) \qquad K_0(n, x) = 1,$$

$$K_1(n, x) = n(q-1) - qx, \qquad (= n - 2x \text{ if } q = 2),$$

$$K_2(n, x) = \tfrac{1}{2}\{q^2x^2 - q(2qn - q - 2n + 2)x + (q-1)^2 n(n-1)\},$$

$$\left(= 2x^2 - 2nx + \binom{n}{2} \text{ if } q = 2\right).$$

In Chapter 7 we shall need the coefficients of x^k, x^{k-1}, x^{k-2}, and x^0 in the expression of $K_k(x)$. If $K_k(x) = \sum_{i=0}^{k} c_i x^i$, then for $q = 2$ we have:

$$(1.2.8) \qquad c_k = (-2)^k/k!,$$

$$c_{k-1} = (-2)^{k-1} n/(k-1)!,$$

$$c_{k-2} = \tfrac{1}{6}(-2)^{k-2}\{3n^2 - 3n + 2k - 4\}/(k-2)!.$$

For several purposes we need certain recurrence relations for the Krawtchouk polynomials. The most important one is

$$(1.2.9) \qquad \begin{aligned}&(k+1)K_{k+1}(x)\\ &\quad = \{k + (q-1)(n-k) - qx\}K_k(x) - (q-1)(n-k+1)K_{k-1}(x).\end{aligned}$$

This is easily proved by differentiating both sides of (1.2.3) with respect to z and multiplying the result by $(1 + (q-1)z)(1-z)$. Comparison of coefficients yields the result. An even easier exercise is replacing x by $x - 1$ in (1.2.3) to obtain

$$(1.2.10) \qquad K_k(i) = K_k(i-1) - (q-1)K_{k-1}(i) - K_{k-1}(i-1),$$

which is an easy way to calculate the numbers $K_k(i)$ recursively.

If $P(x)$ is any polynomial of degree l then there is a unique expansion

$$(1.2.11) \qquad P(x) = \sum_{k=0}^{l} \alpha_k K_k(x),$$

which is called the *Krawtchouk expansion* of $P(x)$.

We mention without proof a few properties that we need later. They are special cases of general theorems on orthogonal polynomials. The first is the Christoffel–Darboux formula

$$(1.2.12) \qquad \frac{K_{k+1}(x)K_k(y) - K_k(x)K_{k+1}(y)}{y - x} = \frac{2}{k+1}\binom{n}{k}\sum_{i=0}^{k}\frac{K_i(x)K_i(y)}{\binom{n}{i}}.$$

The recurrence relation (1.2.9) and an induction argument show the very important interlacing property of the zeros of $K_k(x)$:

(1.2.13) $K_k(x)$ has k distinct real zeros on $(0, n)$; if these are $v_1 < v_2 < \cdots < v_k$ and if $u_1 < u_2 < \cdots < u_{k-1}$ are the zeros of K_{k-1}, then

$$0 < v_1 < u_1 < v_2 < \cdots < v_{k-1} < u_{k-1} < v_k < n.$$

The following property once again follows from (1.2.3) (where we now take $q = 2$) by multiplying two power series: If $x = 0, 1, 2, \ldots, n$, then

$$K_i(x)K_j(x) = \sum_{k=0}^{n} \alpha_k K_k(x), \tag{1.2.14}$$

where

$$\alpha_k := \binom{n-k}{(i+j-k)/2}\binom{k}{(i-j+k)/2}.$$

In Chapter 7 we shall need the relation

$$\sum_{k=0}^{l} K_k(x) = K_l(x-1; n-1, q). \tag{1.2.15}$$

This is easily proved by substituting (1.2.1) on the left-hand side, changing the order of summation and then using $\binom{x}{j} = \binom{x-1}{j-1} + \binom{x-1}{j}$ $(j \geq 1)$. We shall denote $K_l(x-1; n-1, q)$ by $\Psi_l(x)$.

§1.3. Combinatorial Theory

In several chapters we shall make use of notions and results from combinatorial theory. In this section we shall only recall a number of definitions and one theorem. The reader who is not familiar with this area of mathematics is referred to the book [93].

(1.3.1) Definition. Let S be a set with v elements and let $\mathscr{B}$ be a collection of subsets of S (which we call *blocks*) such that:

(i) $|B| = k$ for every $B \in \mathscr{B}$,
(ii) for every $T \subset S$ with $|T| = t$ there are exactly λ blocks B such that $T \subset B$.

Then the pair $(S, \mathscr{B})$ is called a *t-design* (notation $t - (v, k, \lambda)$). The elements of S are called the *points* of the design. If $\lambda = 1$ the design is called a *Steiner system.*

A t-design is often represented by its *incidence matrix* A which has $|\mathscr{B}|$ rows and $|S|$ columns and which has the characteristic functions of the blocks as its rows.

(1.3.2) Definition. A *block design* with parameters $(v, k; b, r, \lambda)$ is a $2-(v, k, \lambda)$ with $|\mathscr{B}| = b$. For every point there are r blocks containing that point. If $b = v$ then the block design is called *symmetric*.

(1.3.3) Definition. A *projective plane* of *order* n is a $2-(n^2+n+1, n+1, 1)$. In this case the blocks are called the lines of the plane. A projective plane of order n is denoted by $PG(2, n)$.

(1.3.4) Definition. The *affine geometry* of *dimension* m over the field $\mathbb{F}_q$ is the vector space $(\mathbb{F}_q)^m$ (we use the notation $AG(m, q)$ for the geometry). A k-dimensional *affine subspace* or a k-flat is a coset of a k-dimensional linear subspace (considered as a subgroup). If $k = m - 1$ we call the flat a *hyperplane*. The group generated by the linear transformations of $(\mathbb{F}_q)^m$ and the translations of the vector space is called the group of *affine transformations* and denoted by $AGL(m, q)$. The affine permutation group defined in Section 1.1 is the example with $m = 1$. The *projective geometry* of dimension m over $\mathbb{F}_q$ (notation $PG(m, q)$) consists of the linear subspaces of $AG(m+1, q)$. The subspaces of dimension 1 are called *points*, subspaces of dimension 2 are *lines*, etc.

We give one example. Consider $AG(3, 3)$. There are 27 points, $\frac{1}{2}(27-1) = 13$ lines through $(0, 0, 0)$ and also 13 planes through $(0, 0, 0)$. These 13 lines are the "points" of $PG(2, 3)$ and the 13 planes in $AG(3, 3)$ are the "lines" of the projective geometry. It is clear that this is a $2-(13, 4, 1)$. When speaking of the coordinates of a point in $PG(m, q)$ we mean the coordinates of any of the corresponding points different from $(0, 0, \dots, 0)$ in $AG(m+1, q)$. So, in the example of $PG(2, 3)$ the triples $(1, 2, 1)$ and $(2, 1, 2)$ are coordinates for the same point in $PG(2, 3)$.

In Chapter 10 we shall consider n-dimensional projective space $\mathbb{P}^n$ over a field k. A point will be denoted by $(a_0 : a_1 : \dots : a_n)$, not all $a_i = 0$, and $(a_0 : a_1 : \dots : a_n) = (b_0 : b_1 : \dots : b_n)$ if there is a $c \in k$, $c \neq 0$, such that $b_i = ca_i$ for $0 \le i \le n$.

(1.3.5) Definition. A square matrix H of order n with elements $+1$ and -1, such that $HH^T = nI$, is called a *Hadamard matrix*.

(1.3.6) Definition. A square matrix C of order n with elements 0 on the diagonal and $+1$ or -1 off the diagonal, such that $CC^T = (n-1)I$, is called a *conference matrix*.

There are several well known ways of constructing Hadamard matrices. One of these is based on the so-called *Kronecker product* of matrices which is defined as follows.

(1.3.7) Definition. If A is an $m \times m$ matrix with entries a_{ij} and B is an $n \times n$ matrix then the Kronecker product $A \otimes B$ is the $mn \times mn$ matrix given by

$$A \otimes B := \begin{bmatrix} a_{11}B & a_{12}B & \dots & a_{1m}B \\ a_{21}B & a_{22}B & \dots & a_{2m}B \\ \vdots & \vdots & \dots & \vdots \\ a_{m1}B & a_{m2}B & \dots & a_{mm}B \end{bmatrix}.$$

It is not difficult to show that the Kronecker product of Hadamard matrices is again a Hadamard matrix. Starting from $H_2 := \begin{pmatrix} 1 & 1 \\ 1 & -1 \end{pmatrix}$ we can find the sequence $H_2^{\otimes n}$, where $H_2^{\otimes 2} = H_2 \otimes H_2$, etc. These matrices appear in several places in the book (sometimes in disguised form).

One of the best known construction methods is due to R. E. A. C. Paley (see [93]). Let q be an odd prime power. We define the function χ on $\mathbb{F}_q$ by $\chi(0) := 0$, $\chi(x) := 1$ if x is a nonzero square, $\chi(x) = -1$ otherwise. Note that χ restricted to the multiplicative group of $\mathbb{F}_q$ is a character. Number the elements of $\mathbb{F}_q$ in any way as $a_0, a_1 \dots, a_{q-1}$, where $a_0 = 0$.

(1.3.8) Theorem. *The Paley matrix S of order q defined by $S_{ij} := \chi(a_i - a_j)$ has the properties:*

(i) $SJ = JS = 0$,
(ii) $SS^T = qI - J$,
(iii) $S^T = (-1)^{(q-1)/2}S$.

If we take such a matrix S and form the matrix C of order $q + 1$ as follows:

$$C := \begin{bmatrix} 0 & 1 & 1 & \dots & 1 \\ -1 & & & & \\ -1 & & S & & \\ \vdots & & & & \\ -1 & & & & \end{bmatrix},$$

then C is a conference matrix of order $q + 1$. If $q \equiv 3 \pmod 4$ we can then consider $H := I + C$. Since $C^T = -C$ because -1 is not a square in $\mathbb{F}_q$, we see that H is a Hadamard matrix of order $q + 1$.

§1.4. Probability Theory

Let x be a random variable which can take a finite number of values $x_1, x_2, \dots$. As usual, we denote the probability that x equals x_i, i.e. $P(\mathrm{x} = x_i)$, by p_i. The *mean* or *expected value* of x is $\mu = \mathscr{E}(\mathrm{x}) := \sum_i p_i x_i$.

If g is a function defined on the set of values of x then $\mathscr{E}(g(\mathrm{x})) = \sum_i p_i g(x_i)$. We shall use a number of well known facts such as

$$\mathscr{E}(a\mathbf{x} + b\mathbf{y}) = a\mathscr{E}(\mathbf{x}) + b\mathscr{E}(\mathbf{y}).$$

The *standard deviation* σ and the *variance* σ^2 are defined by: $\mu = \mathscr{E}(\mathbf{x})$,

$$\sigma^2 := \sum_i p_i x_i^2 - \mu^2 = \mathscr{E}(\mathbf{x} - \mu)^2, \qquad (\sigma > 0).$$

We also need a few facts about two-dimensional distributions. We use the notation $p_{ij} := P(\mathbf{x} = x_i \wedge \mathbf{y} = y_j)$, $p_{i.} := P(\mathbf{x} = x_i) = \sum_j p_{ij}$ and for the *conditional probability* $P(\mathbf{x} = x_i | \mathbf{y} = y_j) = p_{ij}/p_{.j}$. We say that $\mathbf{x}$ and $\mathbf{y}$ are *independent* if $p_{ij} = p_{i.}p_{.j}$ for all i and j. In that case we have

$$\mathscr{E}(\mathbf{xy}) = \sum_{i,j} p_{ij} x_i y_j = \mathscr{E}(\mathbf{x})\mathscr{E}(\mathbf{y}).$$

All these facts can be found in standard textbooks on probability theory (e.g. W. Feller [21]). The same is true for the following results that we shall use in Chapter 2.

(1.4.1) Theorem (Chebyshev's Inequality). *Let* $\mathbf{x}$ *be a random variable with mean* μ *and variance* σ^2. *Then for any* $k > 0$

$$P(|\mathbf{x} - \mu| \geq k\sigma) < k^{-2}.$$

The probability distribution which will play the most important role in the next chapter is the *binomial distribution.* Here, $\mathbf{x}$ takes the values $0, 1, \ldots, n$ and $P(\mathbf{x} = i) = \binom{n}{i} p^i q^{n-i}$, where $0 \leq p \leq 1$, $q := 1 - p$. For this distribution we have $\mu = np$ and $\sigma^2 = np(1 - p)$. An important tool used when estimating binomial coefficients is given in the following theorem

(1.4.2) Theorem (Stirling's Formula).

$$\begin{aligned}\log n! &= (n - \tfrac{1}{2}) \log n - n + \tfrac{1}{2}\log(2\pi) + o(1), && (n \to \infty)\\ &= n \log n - n + O(\log n), && (n \to \infty).\end{aligned}$$

Another useful lemma concerning binomial coefficients is Lemma 1.4.3.

(1.4.3) Lemma. *We have*

$$\binom{n}{m} \leq \frac{n^n}{m^m (n - m)^{n-m}}.$$

Proof.

$$n^n = \{m + (n - m)\}^n \geq \binom{n}{m} m^m (n - m)^{n-m}. \qquad \square$$

We shall now introduce a function that is very important in information theory. It is known as the *binary entropy* function and usually denoted by H. In (5.1.5) we generalize this to other q than 2. In the following the logarithms are to the base 2.

(1.4.4) Definition. The binary entropy function H is defined by

$$H(0) := 0,$$
$$H(x) := -x \log x - (1-x)\log(1-x), \qquad (0 < x \le \tfrac{1}{2}).$$

(1.4.5) Theorem. *Let* $0 \le \lambda \le \frac{1}{2}$. *Then we have*

(i) $\sum_{0 \le i \le \lambda n} \binom{n}{i} \le 2^{nH(\lambda)}$,

(ii) $\lim_{n\to\infty} n^{-1} \log \sum_{0 \le i \le \lambda n} \binom{n}{i} = H(\lambda)$.

PROOF.
(i)
$$1 = \{\lambda + (1-\lambda)\}^n \ge \sum_{0 \le i \le \lambda n} \binom{n}{i} \lambda^i (1-\lambda)^{n-i}$$
$$\ge \sum_{0 \le i \le \lambda n} \binom{n}{i} (1-\lambda)^n \left(\frac{\lambda}{1-\lambda}\right)^{\lambda n} = 2^{-nH(\lambda)} \sum_{0 \le i \le \lambda n} \binom{n}{i}.$$

(ii) Write $m := \lfloor \lambda n \rfloor$. Then $m = \lambda n + O(1)$ for $n \to \infty$. Therefore we find from Theorem 1.4.2:

$$\begin{aligned} n^{-1} \log \sum_{0 \le i \le \lambda n} \binom{n}{i} &\ge n^{-1} \log \binom{n}{m} \\ &= n^{-1}\{n \log n - m \log m - (n-m)\log(n-m) + o(n)\} \\ &= \log n - \lambda \log(\lambda n) - (1-\lambda)\log((1-\lambda)n) + o(1) \\ &= H(\lambda) + o(1) \qquad \text{for } n \to \infty. \end{aligned}$$

The result then follows from part (i). □

A probability distribution that plays an important role in information theory is the *normal* or *Gaussian* distribution. It is used to describe one of the common kinds of "*noise*" on communication channels. We say that a continuous random variable has Gaussian distribution with mean μ and variance σ^2 if it has density function

$$p(x) := \frac{1}{\sqrt{2\pi\sigma^2}} \exp\left(-\frac{(x-\mu)^2}{2\sigma^2}\right).$$

CHAPTER 2

Shannon's Theorem

§2.1. Introduction

This book will present an introduction to the mathematical aspects of the theory of *error-correcting codes.* This theory is applied in many situations which have as a common feature that information coming from some source is transmitted over a noisy communication channel to a receiver. Examples are telephone conversations, storage devices like magnetic tape units which feed some stored information to the computer, telegraph, etc. The following is a typical recent example. Many readers will have seen the excellent pictures which were taken of Mars, Saturn and other planets by satellites such as the Mariners, Voyagers, etc. In order to transmit these pictures to Earth a fine grid is placed on the picture and for each square of the grid the degree of blackness is measured, say in a scale of 0 to 63. These numbers are expressed in the binary system, i.e. each square produces a string of six 0s and 1s. The 0s and 1s are transmitted as two different signals to the receiver station on Earth (the Jet Propulsion Laboratory of the California Institute of Technology in Pasadena). On arrival the signal is very weak and it must be amplified. Due to the effect of thermal noise it happens occasionally that a signal which was transmitted as a 0 is interpreted by the receiver as a 1, and *vice versa.* If the 6-tuples of 0s and 1s that we mentioned above were transmitted as such, then the errors made by the receiver would have great effect on the pictures. In order to prevent this, so-called *redundancy* is built into the signal, i.e. the transmitted sequence consists of more than the necessary information. We are all familiar with the principle of redundancy from everyday language. The words of our language form a small part of all possible strings of letters (symbols). Consequently a misprint in a long(!) word is recognized because the word is changed into something that resembles the correct word more than it resembles any other word we know. This is the essence of the theory to be treated in this book. In the previous example the reader corrects the misprint. A more modest example of coding for noisy channels is the system used for the serial interface between a terminal and a computer or between a PC and the keyboard. In order to represent 128 distinct symbols, strings of seven 0s and 1s (i. e. the integers 0 to 127 in binary) are used. In practice one redundant *bit* (= binary digit) is added to the 7-tuple in such a way that the resulting 8-tuple (called a *byte*) has an even number of 1s. This is done for example in the ASCII character code. A failure in these interfaces occurs very rarely but it is possible that an occasional incorrect bit occurs. This results in incorrect parity of the 8-tuple (it will have an odd number of 1s). In this case, the 8-tuple is not accepted. This is an example of what is called a *single-error-detecting code.*

We mentioned above that the 6-tuples of 0s and 1s in picture transmission (e.g. Mariner 1969) are replaced by longer strings (which we shall always call *words*). In fact, in the case of Mariner 1969 the words consisted of 32 symbols (see [56]). At this point the reader should be satisfied with the knowledge that some device had been designed which changes the 64 possible information strings (6-tuples of 0s and 1s) into 64 possible *codewords* (32-tuples of 0s and 1s). This device is called the *encoder*. The codewords are transmitted. We consider the random noise, i.e. the *errors* as something that is added to the message (mod 2 addition).

At the receiving end, a device called the *decoder* changes a received 32-tuple, if it is not one of the 64 allowable codewords, into the *most likely* codeword and then determines the corresponding 6-tuple (the blackness of one square of the grid). The code which we have just described has the property that if not more than 7 of the 32 symbols are incorrect, then the decoder makes the right decision. Of course one should realize that we have paid a toll for this possibility of error correction. namely that the time available for the transmission of each bit is only 1/5 of what would be available with no coding, leading to increased error probability! We shall treat this example in more detail in §2.3.

In practice, the situation is more complicated because it is not the transmission time that changes, but the available energy per transmitted bit.

The most spectacular application of the theory of error-correcting codes is the Compact Disc Digital Audio system invented by Philips (Netherlands). Its success depends (among other things) on the use of Reed Solomon codes. These will be treated in Section 6.8. Figure 1 is a model of the situation described above.

In this book our main interest will be in the construction and the analysis of good codes. In a few cases we shall study the mathematical problems of decoding without considering the actual implementation. Even for a fixed code C there are many different ways to design an algorithm for a decoder. A *complete* decoding algorithm decodes every possible received word into some codeword. In some situations an *incomplete* decoding algorithm could be preferable, namely when a decoding error is very undesirable. In that case

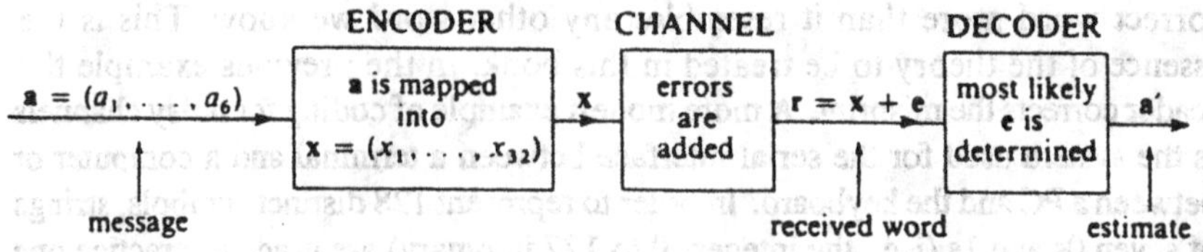

Figure 1

the algorithm will correct received messages that contain a few errors and for the other possible received messages there will be a decoding failure. In the latter case the receiver either ignores the message or, if possible, asks for a retransmission. Another distinction which is made is the one between so-called *hard decisions* and *soft decisions.* This regards the interpretation of received symbols. Most of them will resemble the signal for 0 or for 1 so much that the receiver has no doubt. In other cases however, this will not be true and then we could prefer putting a ? instead of deciding whether the symbol is 0 or it is 1. This is often referred to as an *erasure.* More complicated systems attach a probability to the symbol.

Introduction to Shannon's Theorem

In order to get a better idea about the origin of coding theory we consider the following imaginary experiment.

We are in a room where somebody is tossing a coin at a speed of t tosses per minute. The room is connected with another room by a telegraph wire. Let us assume that we can send two different symbols, which we call 0 and 1, over this communication channel. The channel is noisy and the effect is that there is a probability p that a transmitted 0 (resp. 1) is interpreted by the receiver as a 1 (resp. 0). Such a channel is called a *binary symmetric channel* (B.S.C.) Suppose furthermore that the channel can handle $2t$ symbols per minute and that we can use the channel for T minutes if the coin tossing also takes T minutes. Every time heads comes up we transmit a 0 and if tails comes up we transmit a 1. At the end of the transmission the receiver will have a fraction p of the received information which is incorrect. Now, if we did not have the time limitation specified above, we could achieve arbitrarily small error probability at the receiver as follows. Let N be odd. Instead of a 0 (resp. 1) we transmit N 0s (resp. 1s). The receiver considers a received N-tuple and decodes it into the symbol that occurs most often. The code which we are now using is called a *repetition code* of length N. It consists of two *code-words,* namely $\mathbf{0} = (0, 0, \ldots, 0)$ and $\mathbf{1} = (1, 1, \ldots, 1)$. As an example let us take $p = 0.001$. The probability that the decoder makes an error then is

$$(2.1.1)\qquad \sum_{0 \le k < N/2} \binom{N}{k} q^k p^{N-k} < (0.07)^N, \qquad \text{(here } q := 1 - p\text{)},$$

and this probability tends to 0 for $N \to \infty$ (the proof of (2.1.1) is Exercise 2.4.1).

Due to our time limitation we have a serious problem! We can only transmit two symbols for each toss of the coin. There is no point in sending each symbol twice instead of once. A most remarkable theorem, due to C. E. Shannon (cf. [62]), states that, in the situation described here, we can still achieve arbitrarily small error probability at the receiver for large T. The proof will be given in the next section. A first idea about the method of proof can be obtained in the following way. We transmit the result of two tosses of the coin as follows:

$$\text{heads, heads} \to 0\ 0\ 0\ 0,$$

$$\text{heads, tails} \to 0\ 1\ 1\ 1,$$

$$\text{tails, heads} \to 1\ 0\ 0\ 1,$$

$$\text{tails, tails} \to 1\ 1\ 1\ 0.$$

Observe that the first two transmitted symbols carry the actual information; the final two symbols are redundant. The decoder uses the following complete decoding algorithm. If a received 4-tuple is not one of the above, then assume that the fourth symbol is correct and that one of the first three symbols is incorrect. Any received 4-tuple can be uniquely decoded. The result is correct if the above assumptions are true. Without coding, the probability that two results are received correctly is $q^2 = 0.998$. With the code described above, this probability is $q^4 + 3q^3p = 0.999$. The second term on the left is the probability that the received word contains one error, but not in the fourth position. We thus have a nice improvement, achieved in a very easy way. The time requirement is fulfilled. We extend the idea used above by transmitting the coin tossing results three at a time. The information which we wish to transmit is then a 3-tuple of 0s and 1s, say (a_1, a_2, a_3). Instead of this 3-tuple, we transmit the 6-tuple $\mathbf{a} = (a_1, \ldots, a_6)$, where $a_4 := a_2 + a_3$, $a_5 := a_1 + a_3$, $a_6 := a_1 + a_2$ (the addition being addition mod 2). What we have done is to construct a *code* consisting of eight words, each with length 6. As stated before, we consider the noise as something added to the message, i.e. the received word $\mathbf{b}$ is $\mathbf{a} + \mathbf{e}$, where $\mathbf{e} = (e_1, e_2, \ldots, e_6)$ is called the *error pattern* (error vector). We have

$$e_2 + e_3 + e_4 = b_2 + b_3 + b_4 := s_1,$$

$$e_1 + e_3 + e_5 = b_1 + b_3 + b_5 := s_2,$$

$$e_1 + e_2 + e_6 = b_1 + b_2 + b_6 := s_3.$$

Since the receiver knows $\mathbf{b}$, he knows s_1, s_2, s_3. Given s_1, s_2, s_3 the decoder must choose the most likely error pattern $\mathbf{e}$ which satisfies the three equations. The most likely one is the one with the minimal number of symbols 1. One easily sees that if $(s_1, s_2, s_3) \neq (1, 1, 1)$ there is a unique choice for $\mathbf{e}$. If $(s_1, s_2, s_3) = (1, 1, 1)$ the decoder must choose one of the three possibilities (1,

0, 0, 1, 0, 0), (0, 1, 0, 0, 1, 0), (0, 0, 1, 0, 0, 1) for **e**. We see that an error pattern with one error is decoded correctly and among all other error patterns there is one with two errors that is decoded correctly. Hence, the probability that all three symbols a_1, a_2, a_3 are interpreted correctly after the decoding procedure, is

$$q^6 + 6q^5 p + q^4 p^2 = 0.999986.$$

This is already a tremendous improvement.

Through this introduction the reader will already have some idea of the following important concepts of coding theory.

(2.1.2) Definition. If a code C is used consisting of words of length n, then

$$R := n^{-1} \log|C|$$

is called the *information rate* (or just the rate) of the code.

The concept rate is connected with what was discussed above regarding the time needed for the transmission of information. In our example of the PC-keyboard interface, the rate is $\frac{7}{8}$. The Mariner 1969 used a code with rate $\frac{6}{32}$. The example given before the definition of rate had $R = \frac{1}{2}$.

We mentioned that the code used by Mariner 1969 had the property that the receiver is able to correct up to seven errors in a received word. The reason that this is possible is the fact that any two distinct codewords differ in at least 16 positions. Therefore a received word with less than eight errors resembles the intended codeword more than it resembles any other codeword. This leads to the following definition:

(2.1.3) Definition. If **x** and **y** are two n-tuples of 0s and 1s, then we shall say that their *Hamming-distance* (usually just distance) is

$$d(\mathbf{x}, \mathbf{y}) := |\{i | 1 \le i \le n, x_i \ne y_i\}|.$$

(Also see (3.1.1).)

The code C with eight words of length 6 which we treated above has the property that any two distinct codewords have distance at least 3. That is why any error-pattern with one error could be corrected. The code is a *single-error-correcting* code.

Our explanation of decoding rules was based on two assumptions. First of all we assumed that during communication all codewords are equally likely. Furthermore we used the fact that if $n_1 > n_2$ then an error pattern with n_1 errors is less likely than one with n_2 errors.

This means that if **y** is received we try to find a codeword **x** such that $d(\mathbf{x}, \mathbf{y})$ is minimal. This principle is called *maximum-likelihood-decoding.*

§2.2. Shannon's Theorem

We shall now state and prove Shannon's theorem for the case of the example given in Section 2.1. Let us state the problem. We have a binary symmetric channel with probability p that a symbol is received in error (again we write $q := 1 - p$). Suppose we use a code C consisting of M words of length n, each word occurring with equal probability. If $\mathbf{x}_1, \mathbf{x}_2, \ldots, \mathbf{x}_M$ are the codewords and we use maximum-likelihood-decoding, let P_i be the probability of making an incorrect decision given that $\mathbf{x}_i$ is transmitted. In that case the probability of incorrect decoding of a received word is:

(2.2.1) $$P_C := M^{-1} \sum_{i=1}^{M} P_i.$$

Now consider all possible codes C with the given parameters and define:

(2.2.2) $$P^*(M, n, p) := \text{minimal value of } P_C.$$

(2.2.3) Theorem (Shannon 1948). *If* $0 < R < 1 + p \log p + q \log q$ *and* $M_n := 2^{\lfloor Rn \rfloor}$ *then* $P^*(M_n, n, p) \to 0$ *if* $n \to \infty$.

(Here all logarithms have base 2.) We remark that in the example of the previous section $p = 0.001$, i.e. $1 + p \log p + q \log q$ is nearly 1. The requirement in the experiment was that the rate should be at least $\frac{1}{2}$. We see that for $\varepsilon > 0$ and n sufficiently large there is a code C of length n, with rate nearly 1 and such that $P_C < \varepsilon$. (Of course long codes cannot be used if T is too small.)

Before giving the proof of Theorem 2.2.3 we treat some technical details to be used later.

The probability of an error pattern with w errors is $p^w q^{n-w}$, i.e. it depends on w only.

The number of errors in a received word is a random variable with expected value np and variance $np(1 - p)$. If $b := (np(1 - p)/(\varepsilon/2))^{1/2}$, then by Chebyshev's inequality (Theorem 1.4.1) we have

(2.2.4) $$P(w > np + b) \le \tfrac{1}{2}\varepsilon.$$

Since $p < \frac{1}{2}$, the number $\rho := \lfloor np + b \rfloor$ is less than $\frac{1}{2}n$ for sufficiently large n. Let $B_\rho(\mathbf{x})$ be the set of words $\mathbf{y}$ with $d(\mathbf{x}, \mathbf{y}) \le \rho$. Then

(2.2.5) $$|B_\rho(\mathbf{x})| = \sum_{i \le \rho} \binom{n}{i} < \frac{1}{2} n \binom{n}{\rho} \le \frac{1}{2} n \cdot \frac{n^n}{\rho^\rho (n - \rho)^{n-\rho}}$$

(cf. Lemma 1.4.3). The set $B_\rho(\mathbf{x})$ is usually called the *sphere* with radius ρ and center $\mathbf{x}$ (although *ball* would have been more appropriate).

We shall use the following estimates:

(2.2.6) $$\frac{\rho}{n} \log \frac{\rho}{n} = \frac{1}{n} \lfloor np + b \rfloor \log \frac{\lfloor np + b \rfloor}{n} = p \log p + O(n^{-1/2}),$$

$$\left(1-\frac{\rho}{n}\right)\log\left(1-\frac{\rho}{n}\right) = q\log q + O(n^{-1/2}), \qquad (n\to\infty).$$

Finally we introduce two functions which play a role in the proof. Let

$$\mathbf{u} \in \{0, 1\}^n, \mathbf{v} \in \{0, 1\}^n.$$

Then

(2.2.7) $$f(\mathbf{u}, \mathbf{v}) := \begin{cases} 0, & \text{if } d(\mathbf{u}, \mathbf{v}) > \rho, \\ 1, & \text{if } d(\mathbf{u}, \mathbf{v}) \le \rho. \end{cases}$$

If $\mathbf{x}_i \in C$ and $\mathbf{y} \in \{0, 1\}^n$ then

(2.2.8) $$g_i(\mathbf{y}) := 1 - f(\mathbf{y}, \mathbf{x}_i) + \sum_{j\neq i} f(\mathbf{y}, \mathbf{x}_j).$$

Note that if $\mathbf{x}_i$ is the only codeword such that $d(\mathbf{x}_i, \mathbf{y}) \le \rho$, then $g_i(\mathbf{y}) = 0$ and that otherwise $g_i(\mathbf{y}) \ge 1$.

PROOF OF THEOREM 2.2.3. In the proof of Shannon's theorem we shall pick the codewords $\mathbf{x}_1, \mathbf{x}_2, \ldots, \mathbf{x}_M$ at random (independently). We decode as follows. If $\mathbf{y}$ is received and if there is exactly one codeword $\mathbf{x}_i$ such that $d(\mathbf{x}_i, \mathbf{y}) \le \rho$, then decode $\mathbf{y}$ as $\mathbf{x}_i$. Otherwise we declare an error (or if we must decode, then we always decode as $\mathbf{x}_1$).

Let P_i be as defined above. We have

$$P_i = \sum_{\mathbf{y}\in\{0,1\}^n} P(\mathbf{y}|\mathbf{x}_i)g_i(\mathbf{y})$$

$$= \sum_{\mathbf{y}} P(\mathbf{y}|\mathbf{x}_i)\{1 - f(\mathbf{y}, \mathbf{x}_i)\} + \sum_{\mathbf{y}}\sum_{j\neq i} P(\mathbf{y}|\mathbf{x}_i)f(\mathbf{y}, \mathbf{x}_j).$$

Here the first term on the right-hand side is the probability that the received word $\mathbf{y}$ is not in $B_\rho(\mathbf{x}_i)$. By (2.2.4) this probability is at most $\frac{1}{2}\varepsilon$.

Hence we have

$$P_C \le \frac{1}{2}\varepsilon + M^{-1}\sum_{i=1}^{M}\sum_{\mathbf{y}}\sum_{j\neq i} P(\mathbf{y}|\mathbf{x}_i)f(\mathbf{y}, \mathbf{x}_j).$$

The main principle of the proof is the fact that $P^*(M, n, p)$ is less than the expected value of P_C over all possible codes C picked at random. Therefore we have

$$P^*(M, n, p) \le \frac{1}{2}\varepsilon + M^{-1}\sum_{i=1}^{M}\sum_{\mathbf{y}}\sum_{j\neq i} \mathscr{E}(P(\mathbf{y}|\mathbf{x}_i))\mathscr{E}(f(\mathbf{y}, \mathbf{x}_j))$$

$$= \frac{1}{2}\varepsilon + M^{-1}\sum_{i=1}^{M}\sum_{\mathbf{y}}\sum_{j\neq i} \mathscr{E}(P(\mathbf{y}|\mathbf{x}_i))\cdot\frac{|B_\rho|}{2^n}$$

$$= \tfrac{1}{2}\varepsilon + (M-1)2^{-n}|B_\rho|.$$

We now take logarithms, apply (2.2.5) and (2.2.6), and then we divide by n.

The result is

$$\begin{aligned} n^{-1} &\log(P^*(M, n, p) - \tfrac{1}{2}\varepsilon) \\ &\leq n^{-1} \log M - (1 + p \log p + q \log q) + O(n^{-1/2}). \end{aligned}$$

Substituting $M = M_n$ on the right-hand side we find, using the restriction on R,

$$n^{-1} \log(P^*(M_n, n, p) - \tfrac{1}{2}\varepsilon) < -\beta < 0,$$

for $n > n_0$, i.e. $P^*(M, n, p) < \frac{1}{2}\varepsilon + 2^{-\beta n}$.

This proves the theorem. □

§2.3. On Coding Gain

In many practical applications, one has to choose B and W, where B equals the number of user bits per second that must be transmitted reliably through a noisy channel, using a power of at most W Watt. A well known example is in mobile telephony, where B determines the speech quality and W is related to the life time of the batteries. Another example is in deep space transmission, where B determines the number of pictures that can be transmitted in the fly by time, while W is the power that is available from the solar panels. In all these cases, the transmitter has an average energy of $E_b = W/B$ Joule per user bit available to generate signals to be sent to the receiver. Coding may influence the choices. The effect of coding is often expressed as "*coding gain*" which we now introduce. (Details from electrical engineering are not treated.)

If no coding is used, the energy E_b is available to the transmitter for mapping a user bit onto a signal with amplitude $s := \sqrt{E_b}$ if a 1 is transmitted, and $s = -\sqrt{E_b}$ for a 0. Often, the transmission channel is modeled as an Additive White Gaussian Noise (AWGN) channel. This means that the received signal amplitude r equals $r = s + n$, where the noise n is drawn from a Gaussian distribution having zero mean and variance σ^2. A receiver, making hard decisions, compares each received signal amplitude r with threshold 0 and decides for a 1 if $r > 0$, and for a 0 otherwise. Such a receiver makes an error if the noise n results in r having the wrong sign. Therefore, the error probability (per bit) p_e is

$$p_e = \int_{\sqrt{E}}^{\infty} \frac{1}{\sqrt{2\pi\sigma^2}} \exp\left(\frac{-y^2}{2\sigma^2}\right) dy = Q\left(\sqrt{\frac{E_b}{\sigma^2}}\right),$$

where

$$Q(x) := \frac{1}{\sqrt{2\pi}} \int_{x}^{\infty} \exp\left(\frac{-y^2}{2}\right) dy = \frac{1}{2}\operatorname{erfc}\left(\frac{x}{\sqrt{2}}\right).$$

The ratio E_b/σ^2 is called the *Signal to Noise Ratio* (SNR).

If we use a code C that maps k user bits onto n bits to be transmitted over the channel (*channel bits*), then we say that we are using a code with *rate* $R := k/n$

(see (3.1.3)). We have to send these channel bits $1/R$ times as fast to keep to our required speed of B user bits per second. Because of the power constraint of W Watt, we now only have available an energy of $E' := W/(B/R) = R \cdot E_b$ Joule per channel bit. Assuming a similar transmission scheme for the channel bits as before, we obtain an error rate p'_e (per channel bit) equal to

$$p'_e = Q\left(\sqrt{\frac{RE_b}{\sigma^2}}\right).$$

So, coding results in $p'_e > p_e$ because of the reduced energy budget. The whole idea of coding only makes sense if error correction more than makes up for this loss of energy per bit.

Let us look at the Mariner '69 again in more detail. Instead of calling the transmitted symbols 0 and 1, we denote them by $+1$ and -1. The 64 sequences of length 32 that the Mariner used were the rows of the matrices $H_2^{\otimes 5}$ and $-H_2^{\otimes 5}$ of §1.3. For a received signal (with hard decisions), the received row of 32 symbols ± 1 was taken and the inner product of this row and the 32 rows of $H_2^{\otimes 5}$ was calculated. If no error occurred, 31 of these products were 0 and one was ± 32 showing that the received signal was correct. In the case of one error, the inner products were ± 2, with one exception, where it was ± 30, yielding the correct signal. Note that, for up to seven errors, there is a unique inner product with absolute value greater than 16, pointing out the correct signal.

Let us now look at the effect of coding.

(2.3.1) Example. Consider the example of the Mariner code. Suppose that for a useful picture, each 6-tuple may be wrong with a probability P_E at most 10^{-4}. In case of no coding, we need $E_b/\sigma^2 \approx 17.22$ to achieve this, since $p_e = Q(\sqrt{17.22}) \approx 10^{-4}/6$, and $P_E = 1 - (1 - p_e)^6 \approx 10^{-4}$.

Next, suppose that we use the [32,6] code, correcting at most seven errors, at the same SNR of 17.22. Since $R = 6/32$, we obtain $p'_e \approx 0.036$. (Note that this error probability is 2000 times as large as before !) After decoding, we obtain erroneous 6-tuples with probability

$$P'_E = \sum_{i>7} \binom{32}{i} (p'_e)^i (1 - p'_e)^{32-i} \approx 1.4 \cdot 10^{-5},$$

which is almost an order of magnitude better than P_E.

When using soft decisions, the received waveform is not translated into a row of 1s and -1s, but correlated directly with the rows of $H_2^{\otimes 5}$. In that case, the probability that the signal we choose as most likely is indeed correct, is even larger.

We remark that if we had used soft decision decoding in Example 2.3.1, the error probability would have been reduced to $2 \cdot 10^{-11}$.

There is another way of looking at this situation. We could use coding to *need less energy*. We might choose to exploit C not for reducing the error rate, but for reducing the required SNR in the presence of coding.

In the Mariner example, we were satisfied with a probability of 10^{-4} of receiving an incorrect 6-tuple. To obtain $P'_E = 10^{-4}$, an SNR of 14.83 would suffice (by a calculation similar to the one above). This means that application of coding allows us to reduce the size of the solar panels by almost 15%. With soft decision decoding, the reduction would be more than 50%; (we need an SNR of 7.24 in that case).

(2.3.2) Definition. The ratio between SNR (uncoded) and SNR′ (coded) for equal error probability after decoding is called the *coding gain*.

The coding gain depends on the code, the decoding algorithm, the channel in question, and the required error probability after decoding. It is often expressed in "dB" (this is 10 times the logarithm to base 10 of the ratio in Definition 2.3.2). In engineering literature, the result of Example 2.3.1 will be described as a coding gain of 0.65 dB. We point out that for a given code, there will always be a signal to noise ratio at which the code becomes ineffective; it makes the situation worse than not using it.

We have only considered the energy aspect of transmission. One should realize that coding also increases the complexity of the process and in some way we pay for that too.

§2.4. Comments

C. E. Shannon's paper on the "Mathematical theory of communication" (1948) [62] marks the beginning of coding theory. Since the theorem shows that good codes exist, it was natural that one started to try to construct such codes. Since these codes had to be used with the aid of often very small electronic apparatus one was especially interested in codes with a lot of structure which would allow relatively simple decoding algorithms. In the following chapters we shall see that it is very difficult to obtain highly regular codes without losing the property promised by Theorem 2.2.3. We remark that one of the important areas where coding theory is applied is telephone communication. Many of the names which the reader will encounter in this book are names of (former) members of the staff of Bell Telephone Laboratories. Besides Shannon we mention Berlekamp, Gilbert, Hamming, Lloyd, MacWilliams, Slepian and Sloane. It is not surprising that much of the early literature on coding theory can be found in the *Bell System Technical Journal*. The author gratefully acknowledges that he acquired a large part of his knowledge of coding theory during his many visits to Bell Laboratories. The reader interested in more details about the code used in the Mariner 1969 is referred to [56]. For the coding in Compact Disc see [77], [78].

By consulting the references the reader can see that for many years now the most important results on coding theory have been published in *IEEE Transactions on Information Theory*.

§2.5. Problems

2.5.1. Prove (2.1.1).

2.5.2. Consider the code of length 6 which was described in the coin-tossing experiment in Section 2.2. We showed that the probability that a received word is decoded correctly is $q^6 + 6q^5p + q^4p^2$. Now suppose that after decoding we retain only the first three symbols of every decoded word (i.e. the information concerning the coin-tossing experiment). Determine the probability that a symbol in this sequence is incorrect; (this is called the *symbol error probability*, which without coding would be p).

2.5.3. Construct a code consisting of eight words of length 7 such that any two distinct codewords have distance at least 4. For a B.S.C. with error probability p, determine the probability that a received word is decoded correctly.

2.5.4. A binary channel has a probability $q = 0.9$ that a transmitted symbol is received correctly and a probability $p = 0.1$ that an erasure occurs (i.e. we receive ?). On this channel we wish to use a code with rate $\frac{1}{2}$. Does the probability of correct interpretation increase if we repeat each transmitted symbol ? Is it possible to construct a code with eight words of length 6 such that two erasures can do no harm? Compare the probabilities of correct interpretation for these two codes. (Assume that the receiver does not change the erasures by guessing a symbol.)

2.5.5. Consider the Mariner 1969 example. Suppose a row of 32 symbols is received with e_1 errors and e_2 erasures. Show that if $2e_1 + e_2 < 16$, the correct row can be retrieved.

2.5.6. Let C be a binary code of length 16 such that:

(i) Every codeword has weight 6;

(ii) Any two distinct codewords have distance 8.

Show that $|C| \leq 16$. Does such a code with $|C| = 16$ exist?

2.5.7. Let C be a binary single-error-correcting code of even length n. Show that $|C| \leq 2^n/(n+2)$.

Hint: Count pairs $(\mathbf{x}, \mathbf{c})$, where $\mathbf{x}$ is a word of length n and $\mathbf{c} \in C$, and $\mathbf{x}$ and $\mathbf{c}$ differ in two places.

CHAPTER 3

Linear Codes

§3.1. Block Codes

In this chapter we assume that information is coded using an *alphabet* Q with q distinct symbols. A code is called a *block code* if the coded information can be divided into blocks of n symbols which can be decoded independently. These blocks are the *codewords* and n is called the *block length* or *word length* (or just length). The examples in Chapter 2 were all block codes. In Chapter 13 we shall briefly discuss a completely different system, called *convolutional coding*, where an infinite sequence of information symbols $i_0, i_1, i_2, \ldots$ is coded into an infinite sequence of message symbols. For example, for rate $\frac{1}{2}$ one could have $i_0, i_1, i_2, \ldots \to i_0, i'_0, i_1, i'_1, \ldots$, where i'_n is a function of $i_0, i_1, \ldots, i_n$. For block codes we generalize (2.1.3) to arbitrary alphabets.

(3.1.1.) Definition. If $\mathbf{x} \in Q^n$, $\mathbf{y} \in Q^n$, then the *distance* $d(\mathbf{x}, \mathbf{y})$ of $\mathbf{x}$ and $\mathbf{y}$ is defined by

$$d(\mathbf{x}, \mathbf{y}) := |\{i \mid 1 \le i \le n, x_i \ne y_i\}|.$$

The *weight* $w(\mathbf{x})$ of $\mathbf{x}$ is defined by

$$w(\mathbf{x}) := d(\mathbf{x}, \mathbf{0}).$$

(We always denote $(0, 0, \ldots, 0)$ by $\mathbf{0}$ and $(1, 1, \ldots, 1)$ by $\mathbf{1}$.)

The distance defined in (3.1.1), again called *Hamming-distance*, is indeed a metric on Q^n. If we are using a channel with the property that an error in position i does not influence other positions and a symbol in error can be each of the remaining $q - 1$ symbols with equal probability, then Hamming-distance is a good way to measure the error content of a received message. In

Chapter 12 we shall see that in other situations a different distance function is preferable.

In the following a *code* C is a nonempty proper subset of Q^n. If $|C| = 1$ we call the code *trivial*. If $q = 2$ the code is called a *binary* code, for $q = 3$ a *ternary* code, etc. The following concepts play an essential rôle in this book (cf. Chapter 2).

(3.1.2) Definition. The *minimum distance* of a nontrivial code C is

$$\min\{d(\mathbf{x}, \mathbf{y}) | \mathbf{x} \in C, \mathbf{y} \in C, \mathbf{x} \neq \mathbf{y}\}.$$

The *minimum weight* of C is

$$\min\{w(\mathbf{x}) \mid \mathbf{x} \in C, \mathbf{x} \neq \mathbf{0}\}.$$

We also generalize the concept of rate.

(3.1.3) Definition. If $|Q| = q$ and $C \subset Q^n$ then

$$R := n^{-1} \log_q |C|$$

is called the *(information-) rate* of C.

Sometimes we shall be interested in knowing how far a received word can be from the closest codeword. For this purpose we introduce a counterpart of minimum distance.

(3.1.4) Definition. If $C \subset Q^n$ then the *covering radius* $\rho(C)$ of C is

$$\max\{\min\{d(\mathbf{x}, \mathbf{c}) | \mathbf{c} \in C\} | \mathbf{x} \in Q^n\}.$$

We remind the reader that in Chapter 2 the *sphere* $B_\rho(\mathbf{x})$ with radius ρ and center $\mathbf{x}$ was defined to be the set $\{\mathbf{y} \in Q^n | d(\mathbf{x}, \mathbf{y}) \leq \rho\}$. If ρ is the largest integer such that the spheres $B_\rho(\mathbf{c})$ with $\mathbf{c} \in C$ are disjoint, then $d = 2\rho + 1$ or $d = 2\rho + 2$. The covering radius is the smallest ρ such that the spheres $B_\rho(\mathbf{c})$ with $\mathbf{c} \in C$ cover the set Q^n. If these numbers are equal, then the code C is called *perfect*. This can be stated as follows.

(3.1.5) Definition. A code $C \subset Q^n$ with minimum distance $2e + 1$ is called a *perfect code* if every $\mathbf{x} \in Q^n$ has distance $\leq e$ to exactly one codeword.

The fact that the minimum distance is $2e + 1$ means that the code is e-error-correcting. The following is obvious.

(3.1.6) *Sphere-packing Condition*
If $C \subset Q^n$ is a perfect e-error-correcting code, then

$$|C| \sum_{i=0}^{e} \binom{n}{i} (q-1)^i = q^n.$$

Of course a trivial code is perfect even though one cannot speak of minimum distance for such a code. A simple example of a perfect code was treated in Chapter 2, namely the binary repetition code of odd length n consisting of the two words **0** and **1**.

§3.2. Linear Codes

We now turn to the problem of constructing codes which have some algebraic structure. The first idea is to take a group Q as alphabet and to take a subgroup C of Q^n as code. This is called a *group code*. In this section (in fact in most of this book) we shall require (a lot) more structure. In the following Q is the field $\mathbb{F}_q$, where $q = p^r$ (p prime). Then Q^n is an n-dimensional vector space, namely $\mathbb{F}_q^n$ (sometimes denoted by $\mathscr{R}$). In later chapters we sometimes use the fact that Q^n is isomorphic to the additive group of $\mathbb{F}_{q^n}$ (cf. Section 1.1).

(3.2.1) Definition. A *q-ary linear code C* is a linear subspace of $\mathbb{F}_q^n$. If C has dimension k then C is called an $[n, k]$ code.

From now on we shall use $[n, k, d]$ code as the notation for a k-dimensional linear code of length n with minimum distance d. An (n, M, d) code is any code with word length n, M codewords, and minimum distance d.

(3.2.2) Definition. A *generator matrix* G for a linear code C is a k by n matrix for which the rows are a basis of C.

If G is a generator matrix for C, then $C = \{\mathbf{a}G \mid \mathbf{a} \in Q^k\}$. We shall say that G is in *standard form* (often called reduced echelon form) if $G = (I_k \quad P)$, where I_k is the k by k identity matrix. The (6, 8, 3) code which we used in the example of Section 2.1 is a linear code with $G = (I \quad J - I)$. If G is in standard form, then the first k symbols of a codeword are called *information symbols*. These can be chosen arbitrarily and then the remaining symbols, which are called *parity check symbols*, are determined.

The code used on the PC-keyboard interface mentioned in the Introduction has one parity check bit (responsible for the name) and generator matrix

$$G = (I_7 \quad \mathbf{1}^T).$$

As far as error-correcting capability is concerned, two codes C_1 and C_2 are equally good if C_2 is obtained by applying a fixed permutation of the positions to all the codewords of C_1. We call such codes *equivalent*. Sometimes the definition of equivalence is extended by also allowing a permutation of the symbols of Q (for each position). It is well known from linear algebra that every linear code is equivalent to a code with a generator matrix in standard form.

In general a code C is called *systematic* on k positions (and the symbols in these positions are called information symbols) if $|C| = q^k$ and there is exactly one codeword for every possible choice of coordinates in the k positions. So we saw above that an $[n, k]$ code is systematic on at least one k-tuple of positions. Since one can separate information symbols and redundant symbols, these codes are also called *separable*. By (3.1.3) an $[n, k]$ code has rate k/n, in accordance with the fact that k out of n symbols carry information. The reader should check that the [6,3,3] code used in Section 2.1 is not systematic on four 3-tuples of positions.

The reader will have realized that if a code C has minimum distance $d = 2e + 1$, then it corrects up to e errors in a received word. If $d = 2e$ then an error pattern of weight e is always detected. In general if C has M words one must check $\binom{M}{2}$ pairs of codewords to find d. For linear codes the work is easier.

(3.2.3) Theorem. *For a linear code C the minimum distance is equal to the minimum weight.*

PROOF. $d(\mathbf{x}, \mathbf{y}) = d(\mathbf{x} - \mathbf{y}, \mathbf{0}) = w(\mathbf{x} - \mathbf{y})$ and if $\mathbf{x} \in C$, $\mathbf{y} \in C$ then $\mathbf{x} - \mathbf{y} \in C$. □

(3.2.4) Definition. If C is an $[n, k]$ code we define the *dual code* $C^\perp$ by

$$C^\perp := \{\mathbf{y} \in \mathscr{R} \mid \forall_{\mathbf{x} \in C}[\langle \mathbf{x}, \mathbf{y} \rangle = 0]\}.$$

The dual code $C^\perp$ is clearly a linear code, namely an $[n, n - k]$ code. The reader should be careful not to think of $C^\perp$ as an orthogonal complement in the sense of vector spaces over $\mathbb{R}$. In the case of a finite field Q, the subspaces C and $C^\perp$ can have an intersection larger than $\{\mathbf{0}\}$ and in fact they can even be equal. If $C = C^\perp$ then C is called a *self-dual* code.

If $G = (I_k \quad P)$ is a generator matrix in standard form of the code C, then $H = (-P^\top \quad I_{n-k})$ is a generator matrix for $C^\perp$. This follows from the fact that H has the right size and rank and that $GH^\top = 0$ implies that every codeword $\mathbf{a}G$ has inner product 0 with every row of H. In other words we have

$$\mathbf{x} \in C \Leftrightarrow \mathbf{x}H^\top = \mathbf{0}. \tag{3.2.5}$$

In (3.2.5) we have $n - k$ linear equations which every codeword must satisfy.

If $\mathbf{y} \in C^\perp$ then the equation $\langle \mathbf{x}, \mathbf{y} \rangle = 0$ which holds for every $\mathbf{x} \in C$, is called a *parity check* (equation). H is called a *parity check matrix* of C. For the [6, 3] code used in Section 2.1 the equation $a_4 = a_2 + a_3$ is one of the parity checks. (The code is not systematic on positions 2,3, and 4.)

(3.2.6) Definition. If C is a linear code with parity check matrix H then for every $\mathbf{x} \in Q^n$ we call $\mathbf{x}H^\top$ the *syndrome* of $\mathbf{x}$. Observe that the covering radius $\rho(C)$ of an $[n, k]$ code (cf. (3.14)) is the smallest integer ρ such that any (column-)vector in Q^{n-k} can be written as the sum of at most ρ columns of H.

In (3.2.5) we saw that codewords are characterized by syndrome $\mathbf{0}$. The syndrome is an important aid in decoding received vectors $\mathbf{x}$. Once again this idea was introduced using the [6, 3] code in Section 2.1. Since C is a subgroup of Q^n we can partition Q^n into cosets of C. Two vectors $\mathbf{x}$ and $\mathbf{y}$ are in the same coset iff they have the same syndrome ($\mathbf{x}H^\top = \mathbf{y}H^\top \Leftrightarrow \mathbf{x} - \mathbf{y} \in C$). Therefore if a vector $\mathbf{x}$ is received for which the error pattern is $\mathbf{e}$ then $\mathbf{x}$ and $\mathbf{e}$ have the same syndrome. It follows that for maximum likelihood decoding of $\mathbf{x}$ one must choose a vector $\mathbf{e}$ of minimal weight in the coset which contains $\mathbf{x}$ and then decode $\mathbf{x}$ as $\mathbf{x} - \mathbf{e}$. The vector $\mathbf{e}$ is called the *coset leader*. How this works in practice was demonstrated in Section 2.1 for the [6, 3]-code. For seven of the eight cosets there was a unique coset leader. Only for the syndrome $(s_1, s_2, s_3) = (1, 1, 1)$ did we have to pick one out of three possible coset leaders.

Here we see the first great advantage of introducing algebraic structure. For an $[n, k]$ code over $\mathbb{F}_q$ there are q^k codewords and q^n possible received messages. Let us assume that the rate is reasonably high. The receiver needs to know the q^{n-k} coset leaders corresponding to all possible syndromes. Now q^{n-k} is much smaller than q^n. If the code had no structure, then for every possible received word $\mathbf{x}$ we would have to list the most likely transmitted word.

It is clear that if C has minimum distance $d = 2e + 1$, then every error pattern of weight $\leq e$ is the unique coset leader of some coset because two vectors with weight $\leq e$ have distance $\leq 2e$ and are therefore in different cosets. If C is perfect then there are no other coset leaders. If a code C has minimum distance $2e + 1$ and all coset leaders have weight $\leq e + 1$ then the code is called *quasi-perfect*. The [6, 3] code of Section 2.1 is an example. The covering radius is the weight of a coset leader with maximum weight.

We give one other example of a very simple decoding procedure (cf. [3]). Let C be a $[2k, k]$ binary self-dual code with generator matrix $G = (I_k \quad P)$. The decoding algorithm works if C can correct 3 errors and if the probability that more than 3 errors occur in a received vector is very small. We have the parity check matrix $H = (P^\top \quad I_k)$ but G is also a parity check matrix because C is self-dual. Let $\mathbf{y} = \mathbf{c} + \mathbf{e}$ be the received vector. We write $\mathbf{e}$ as $(\mathbf{e}_1; \mathbf{e}_2)$ where $\mathbf{e}_1$ corresponds to the first k places, $\mathbf{e}_2$ to the last k places. We calculate the two syndromes

$$\mathbf{s}^{(1)} := \mathbf{y}H^\top = \mathbf{e}_1 P + \mathbf{e}_2,$$

$$\mathbf{s}^{(2)} := \mathbf{y}G^\top = \mathbf{e}_1 + \mathbf{e}_2 P^\top.$$

If the $t \leq 3$ errors all occur in the first or last half of $\mathbf{y}$, i.e. $\mathbf{e}_1 = \mathbf{0}$ or $\mathbf{e}_2 = \mathbf{0}$, then one of the syndromes will have weight ≤ 3 and we immediately have $\mathbf{e}$. If this is not the case then the assumption $t \leq 3$ implies that $\mathbf{e}_1$ or $\mathbf{e}_2$ has weight 1. We consider $2k$ vectors $\mathbf{y}^{(i)}$ obtained by changing the ith coordinate of $\mathbf{y}$ $(1 \leq i \leq 2k)$. For each of these vectors we calculate $\mathbf{s}_1$ (for $i \leq k$) resp. $\mathbf{s}_2$ (if $i > k$). If we find a syndrome with weight ≤ 2, we can correct the remaining errors. If we find a syndrome with weight 3, we have detected four errors if C is a code with distance 8 and if C has distance ≥ 10 we can correct this pattern of four errors.

It will often turn out to be useful to adjoin one extra symbol to every codeword of a code C according to some natural rule. The most common of these is given in the following definition.

(3.2.7) Definition. If C is a code of length n over the alphabet $\mathbb{F}_q$ we define the *extended code* $\overline{C}$ by

$$\overline{C} := \left\{(c_1, c_2, \ldots, c_n, c_{n+1}) | (c_1, \ldots, c_n) \in C, \sum_{i=1}^{n+1} c_i = 0\right\}.$$

If C is a linear code with generator matrix G and parity check matrix H then $\overline{C}$ has generator matrix $\overline{G}$ and parity check matrix $\overline{H}$, where $\overline{G}$ is obtained by adding a column to G in such a way that the sum of the columns of $\overline{G}$ is 0 and where

$$\overline{H} := \begin{bmatrix} 1 & 1 & 1\ldots & 1 \\ & & & 0 \\ & H & & 0 \\ & & & \vdots \\ & & & 0 \end{bmatrix}.$$

If C is a binary code with an odd minimum distance d, then $\overline{C}$ has minimum distance $d + 1$ since all weights and distances for $\overline{C}$ are even.

§3.3. Hamming Codes

Let G be the k by n generator matrix of an $[n, k]$ code C over $\mathbb{F}_q$. If any two columns of G are linearly independent, i.e. the columns represent distinct points of $\mathrm{PG}(k - 1, q)$, then C is called a *projective code*. The dual code $C^\perp$ has G as parity check matrix. If $\mathbf{c} \in C^\perp$ and if $\mathbf{e}$ is an error vector of weight 1, then the syndrome $(\mathbf{c} + \mathbf{e})G^\top$ is a multiple of a column of G. Since this uniquely determines the column of G it follows that $C^\perp$ is a code which corrects at least one error. We now look at the case in which n is maximal (given k).

(3.3.1) Definition. Let $n := (q^k - 1)/(q - 1)$. The $[n, n - k]$ *Hamming code* over $\mathbb{F}_q$ is a code for which the parity check matrix has columns that are pairwise linearly independent (over $\mathbb{F}_q$), i.e. the columns are a maximal set of pairwise linearly independent vectors.

Here we obviously do not distinguish between equivalent codes. Clearly the minimum distance of a Hamming code is equal to 3.

(3.3.2) Theorem. *Hamming codes are perfect codes.*

PROOF. Let C be the $[n, n-k]$ Hamming code over $\mathbb{F}_q$, where $n = (q^k - 1)/(q - 1)$. If $\mathbf{x} \in C$ then

$$|B_1(\mathbf{x})| = 1 + n(q-1) = q^k.$$

Therefore the q^{n-k} disjoint spheres of radius 1 around the codewords of C contain $|C| \cdot q^k = q^n$ words, i.e. all possible words. Hence C is perfect (cf. (3.1.5) and (3.1.6)).

(3.3.3) EXAMPLE. The [7, 4] binary Hamming code C has parity check matrix

$$H = \begin{bmatrix} 0 & 0 & 0 & 1 & 1 & 1 & 1 \\ 0 & 1 & 1 & 0 & 0 & 1 & 1 \\ 1 & 0 & 1 & 0 & 1 & 0 & 1 \end{bmatrix}.$$

If we consider two columns of H and the sum of these two (e.g. the first three columns of H), then there is a word of weight 3 in C with 1s in the positions corresponding to these columns (e.g. (1110000)). Therefore C has seven words of weight 3 which, when listed as rows of a matrix, form PG(2, 2). The words of even weight in C are a solution to Problem 2.4.3. By inspection of H we see that the extended code $\bar{C}$ is self-dual.

(3.3.4) EXAMPLE. Suppose that we use an extended Hamming code of length $n = 2^m$ on a B.S.C. with bit error probability p; $(q := 1 - p)$. The expected number of errors per block before decoding is np. If one error occurs, it is corrected. If two errors occur, then we have error detection but no correction. So, the two errors remain. Otherwise, it is possible that the decoder introduces an extra error by changing a received word with ≥ 3 errors into the closest codeword. Therefore, the expected number of errors per block after decoding is at most

$$\begin{aligned} 2\binom{n}{2} p^2 q^{n-2} &+ \sum_{i=3}^{n} (i+1) \binom{n}{i} p^i q^{n-i} \\ &= 2\binom{n}{2} p^2 \left\{ q^{n-2} + \sum_{i=3}^{n} \frac{(i+1)\binom{n}{i}}{2\binom{n}{2}} p^{i-2} q^{n-i} \right\} \\ &\leq 2\binom{n}{2} p^2 \left\{ q^{n-2} + \sum_{i=3}^{n} \binom{n-2}{i-2} p^{i-2} q^{n-i} \right\} \\ &= n(n-1)p^2 < (np)^2. \end{aligned}$$

If p is small enough, this is a considerable improvement. We shall use this estimate in §4.4.

§3.4. Majority Logic Decoding

In this section we shall briefly sketch a decoding method which is used with many linear codes. Generalizations will occur in later chapters. The method is simple and it has the advantage that in some cases more errors are corrected than one expects to be able to correct.

(3.4.1) Definition. A system of parity check equations $\langle \mathbf{x}, \mathbf{y}^{(\nu)} \rangle = 0$ $(1 \le \nu \le r)$, is said to be *orthogonal with respect to position* i (for the code C; $\mathbf{y}^{(\nu)} \in C^{\perp}$) if

(i) $y_i^{(\nu)} = 1$ $(1 \le \nu \le r)$,
(ii) if $j \ne i$ then $y_j^{(\nu)} \ne 0$ for at most one value of ν.

Now suppose $\mathbf{x}$ is a received word that contains t errors, where $t \le \frac{1}{2}r$. Then

$$\langle \mathbf{x}, \mathbf{y}^{(\nu)} \rangle \ne 0 \qquad \text{for} \begin{cases} \le t \text{ values of } \nu, & \text{if } x_i \text{ is correct,} \\ \ge r - (t-1) \text{ values of } \nu, & \text{if } x_i \text{ is incorrect.} \end{cases}$$

Since $r - (t-1) > t$, the majority of the values of $\langle \mathbf{x}, \mathbf{y}^{(\nu)} \rangle$ (i.e. 0, resp. not 0) decides for us whether x_i is correct or not. In the case of a binary code we can subsequently correct the error. If we have such orthogonal check sets for every i, we can correct the different positions one by one.

As an example we consider the dual of the [7, 4] Hamming code (cf. (3.3.3)). The parity check equations

$$x_1 + x_2 + x_3 = 0,$$
$$x_1 + x_4 + x_5 = 0,$$
$$x_1 + x_6 + x_7 = 0,$$

are othogonal with respect to position 1. If $\mathbf{x}$ contains one error, then the three equations yield 1, 1, 1 if x_1 is incorrect, respectively two 0s and one 1 if x_1 is correct. If two outcomes are 1, we see that more than one error has been made (the code is two-error detecting).

Consider the [6, 3, 3] code with generator matrix $G := (I \quad J - I)$ and adjoin two symbols $a_7 = a_8 = a_1$. The reader should check that we still have $d = 3$ but that the new parity check matrix has four rows that are orthogonal with respect to position 1. So, even if two errors occur, position 1 is correct after decoding.

§3.5. Weight Enumerators

The minimum distance of a linear code tells us how many errors a received word may contain and still be decoded correctly. Often it is necessary to have more detailed information about the distances in the code. For this purpose we introduce the so-called *weight enumerator* of the code.

(3.5.1) Definition. Let C be a linear code of length n and let A_i be the number of codewords of weight i.

Then

$$A(z) := \sum_{i=0}^{n} A_i z^i$$

is called the *weight enumerator* of C. The sequence $(A_i)_{i=0}^n$ is called the *weight distribution* of C.

If C is linear and $\mathbf{c} \in C$, then the number of codewords at distance i from $\mathbf{c}$ equals A_i. For a nonlinear code this is generally not true. A code that does have this property (for all codewords and all i) is called *distance invariant*. (Also see Definition 5.3.2.)

As an example we calculate the weight enumerator of the binary Hamming code of length n. Consider $i-1$ columns of the parity check matrix of this code. There are three possibilities:

(i) the sum of these columns is $\mathbf{0}$;
(ii) the sum of these columns is one of the chosen columns;
(iii) the sum of these columns is one of the remaining columns.

We can choose the $i-1$ columns in $\binom{n}{i-1}$ ways. Possibility (i) occurs A_{i-1} times, possibility (ii) occurs $(n-(i-2))A_{i-2}$ times, and possibility (iii) occurs iA_i times. Therefore

$$iA_i = \binom{n}{i-1} - A_{i-1} - (n-i+2)A_{i-2},$$

which is trivially correct if $i > n+1$. If we multiply both sides by z^{i-1} and then sum over i we find

$$A'(z) = (1+z)^n - A(z) - nzA(z) + z^2 A'(z).$$

Since $A(0) = 1$, this differential equation has the unique solution

$$A(z) = \frac{1}{n+1}(1+z)^n + \frac{n}{n+1}(1+z)^{(n-1)/2}(1-z)^{(n+1)/2}. \tag{3.5.2}$$

One of the most fundamental results in coding theory is a theorem due to F. J. MacWilliams (1963) which gives a relation between the weight enumerators of a linear code and its dual.

(3.5.3) Theorem. *Let C be an $[n, k]$ code over $\mathbb{F}_q$ with weight enumerator $A(z)$ and let $B(z)$ be the weight enumerator of $C^\perp$. Then*

$$B(z) = q^{-k}(1+(q-1)z)^n A\left(\frac{1-z}{1+(q-1)z}\right).$$

Proof. Let χ be any nontrivial character of $(\mathbb{F}_q, +)$. As usual let $\mathcal{R} = \mathbb{F}_q^n$. We define

$$g(\mathbf{u}) := \sum_{\mathbf{v} \in \mathcal{R}} \chi(\langle \mathbf{u}, \mathbf{v} \rangle) z^{w(\mathbf{v})}.$$

Then we have

$$\sum_{\mathbf{u}\in C} g(\mathbf{u}) = \sum_{\mathbf{u}\in C}\sum_{\mathbf{v}\in \mathscr{R}} \chi(\langle \mathbf{u},\mathbf{v}\rangle) z^{w(\mathbf{v})} = \sum_{\mathbf{v}\in \mathscr{R}} z^{w(\mathbf{v})} \sum_{\mathbf{u}\in C} \chi(\langle \mathbf{u},\mathbf{v}\rangle).$$

Here, if $\mathbf{v}\in C^\perp$ the inner sum is $|C|$. If $\mathbf{v}\notin C^\perp$ then in the inner sum $\langle \mathbf{u},\mathbf{v}\rangle$ takes every value in $\mathbb{F}_q$ the same number of times, i.e. the inner sum is 0. Therefore

(3.5.4) $$\sum_{\mathbf{u}\in C} g(\mathbf{u}) = |C|\cdot B(z).$$

Extend the weight function to $\mathbb{F}_q$ by writing $w(v)=0$ if $v=0$ and $w(v)=1$ otherwise. Then, writing $\mathbf{u}=(u_1,u_2,\dots,u_n)$ and $\mathbf{v}=(v_1,v_2,\dots,v_n)$, we have from the definition of $g(\mathbf{u})$:

$$\begin{aligned} g(\mathbf{u}) &= \sum_{(v_1,v_2,\dots,v_n)\in\mathscr{R}} z^{w(v_1)+\cdots+w(v_n)}\chi(u_1v_1+\cdots+u_nv_n) \\ &= \sum_{(v_1,v_2,\dots,v_n)\in\mathscr{R}} z^{w(v_1)}\chi(u_1v_1)z^{w(v_2)}\chi(u_2v_2)\cdots z^{w(v_n)}\chi(u_nv_n) \\ &= \prod_{i=1}^{n}\sum_{v\in\mathbb{F}_q} z^{w(v)}\chi(u_iv). \end{aligned}$$

In the last expression the inner sum is equal to $1+(q-1)z$ if $u_i=0$ and it is equal to

$$1+z\sum_{\alpha\in\mathbb{F}_q\setminus\{0\}}\chi(\alpha)=1-z, \qquad \text{if } u_i\neq 0.$$

Therefore

(3.5.5) $$g(\mathbf{u}) = (1-z)^{w(\mathbf{u})}(1+(q-1)z)^{n-w(\mathbf{u})}.$$

Since $|C|=q^k$ the theorem now follows by substituting (3.5.5) in (3.5.4). □

For a generalization we refer to Section 7.2.

Sometimes the weight enumerator of a code C is given in homogeneous form as

$$Ham_C(x,y) := \sum_{c\in C} x^{n-w(c)}y^{w(c)}.$$

In this notation, the MacWilliams relation for a binary code C and its dual $C^\perp$ is given by

$$Ham_{C^\perp}(x,y) = \frac{1}{|C|}Ham_C(x+y,x-y).$$

This follows directly from Theorem 3.5.3.

§3.6. The Lee Metric

In many communication schemes used in practice, one can model the alphabet as a set of points regularly spaced on a circle. Take as example an alphabet of this kind with seven symbols. We identify these symbols (still on a circle) with the elements of $\mathbb{Z}_7$. In these channels, the effect of additive Gaussian noise does not make all errors equally likely. It is much more likely that a transmitted symbol is received as a symbol close to it. In our terminology, this means that if a 4 is sent and an error occurs, it is more likely that a 3 or a 5 is received than a 2 or a 6, etc.

So, for these channels Hamming distance is not a natural metric for measuring errors. Instead, one uses the so-called *Lee weight* and *Lee distance*.

(3.6.1) Definition. Consider $\mathbb{Z}_m$ as alphabet. The Lee weight of an integer i $(0 \le i < m)$ is defined by

$$w_L(i) := \min\{i, m - i\}.$$

The Lee metric on $\mathbb{Z}_m^n$ is defined by

$$w_L(\mathbf{a}) := \sum_{i=1}^{n} w_L(a_i),$$

where the sum is defined in $\mathbb{N}_0$. We define Lee distance by

$$d_L(\mathbf{x}, \mathbf{y}) := w_L(\mathbf{x} - \mathbf{y}).$$

It is not difficult to see that this is indeed a distance function.

In a later chapter, we shall be especially interested in the alphabet $\mathbb{Z}_4$. We treat this in more detail now. In $\mathbb{Z}_4$, the Lee weights of 0,1, and 2 are 0,1, and 2 respectively, but the Lee weight of 3 is 1.

For a code $C \subseteq \mathbb{Z}_4^n$ (see (8.1.1)), we define two weight enumerators, the *symmetrized weight enumerator* and the *Lee weight enumerator*.

(3.6.2) Definition. The symmetrized weight enumerator of a code $C \subseteq \mathbb{Z}_4^n$ is given by

$$\mathrm{swe}_C(w, x, y) := \sum_{\mathbf{c} \in C} w^{n_0(\mathbf{c})} x^{n_1(\mathbf{c}) + n_3(\mathbf{c})} y^{n_2(\mathbf{c})},$$

where $n_i(\mathbf{c})$ denotes the number of coordinates of $\mathbf{c}$ equal to i.

(3.6.3) Definition. The Lee weight enumerator of a code $C \subseteq \mathbb{Z}_4^n$ is defined by

$$\mathrm{Lee}_C(x, y) := \sum_{\mathbf{c} \in C} x^{2n - w_L(\mathbf{c})} y^{w_L(\mathbf{c})}.$$

Note that

(3.6.4) $$\mathrm{Lee}_C(x, y) = \mathrm{swe}_C(x^2, xy, y^2).$$

Let us see if a slight modification of the proof of Theorem 3.5.3 can yield a generalization of the MacWilliams relation to codes over $\mathbb{Z}_4$. We take χ to be a character on $(\mathbb{Z}_4, +)$; below, we will take

$$\chi(a) := i^a, \quad \text{where} \quad i^2 = -1 \text{ in } \mathbb{C}.$$

We consider a function f defined on $\mathcal{R} := \mathbb{Z}_4^n$ and define

$$g(\mathbf{u}) := \sum_{\mathbf{v} \in \mathcal{R}} \chi(\langle \mathbf{u}, \mathbf{v} \rangle) f(\mathbf{v}).$$

In the same way as for (3.5.4), we find

(3.6.5) $$\sum_{\mathbf{u}\in C} g(\mathbf{u}) = |C| \sum_{\mathbf{v}\in C^\perp} f(\mathbf{v}).$$

In the next part of the proof, we choose

$$f(\mathbf{v}) := w^{n_0(\mathbf{v})} x^{n_1(\mathbf{v})+n_3(\mathbf{v})} y^{n_2(\mathbf{v})}.$$

Continuing exactly as in the proof of (3.5.3), we find

$$g(\mathbf{u}) = \prod_{i=1}^{n} \sum_{v\in \mathbb{Z}_4} \chi(u_i v) w^{n_0(v)} x^{n_1(v)+n_3(v)} y^{n_2(v)}.$$

To calculate the inner sum, we must distinguish between $u_i = 0$, $u_i = 1$ or 3, and $u_i = 2$. In the three cases, we find $(w + 2x + y)$, $(w - y)$, and $(w - 2x + y)$ respectively. Hence

(3.6.6) $$g(\mathbf{u}) = (w + 2x + y)^{n_0(\mathbf{u})} (w - y)^{n_1(\mathbf{u})+n_3(\mathbf{u})} (w - 2x + y)^{n_2(\mathbf{u})}.$$

Substituting (3.6.6) in (3.6.5) yields

(3.6.7) $$\mathrm{swe}_{C^\perp}(w, x, y) = \frac{1}{|C|} \mathrm{swe}_C(w + 2x + y, w - y, w - 2x + y).$$

We find the following generalization of Theorem 3.5.3.

(3.6.8) Theorem. *If C is a quaternary code and $C^\perp$ its dual, then*

$$\mathrm{Lee}_{C^\perp}(x, y) = \frac{1}{|C|} \mathrm{Lee}_C(x + y, x - y).$$

Proof. Apply (3.6.4) to (3.6.7). □

§3.7. Comments

The subject of linear codes was greatly influenced by papers by D. E. Slepian and R. W. Hamming written in the 1950s. The reader interested in knowing more about majority logic decoding should consult the book by J. L. Massey [47]. There are several generalizations of MacWilliams' theorem even to nonlinear codes. An extensive treatment can be found in Chapter 5 of [46]. For an application of (3.5.2) see Chapter 2 of [42].

§3.8. Problems

3.8.1. Let C be a binary perfect code of length n with minimum distance 7. Show that $n = 7$ or $n = 23$.

3.8.2. Let C be an $[n, k]$ code over $\mathbb{F}_q$ which is systematic on any set of k positions. Show that C has minimum distance $d = n - k + 1$.

3.8.3. Let C be a $[2k + 1, k]$ binary code such that $C \subset C^\perp$. Describe $C^\perp \backslash C$.

3.8.4. Let $\mathcal{R} = \mathbb{F}_2^6$ and let $\mathbf{x} \in \mathcal{R}$. Determine $|B_1(\mathbf{x})|$. Is it possible to find a set $C \subset \mathcal{R}$ with $|C| = 9$ such that for all $\mathbf{x} \in C$, $\mathbf{y} \in C$, $\mathbf{x} \neq \mathbf{y}$ the distance $d(\mathbf{x}, \mathbf{y})$ is at least 3?

3.8.5. Let C be an $[n, k]$ code over $\mathbb{F}_q$ with generator matrix G. If G does not have a column of 0s then the sum of the weights of the codewords of C is $n(q - 1)q^{k-1}$. Prove this.

3.8.6. Let C be a binary $[n, k]$ code. If C has word of odd weight then the words of even weight in C form an $[n, k - 1]$ code. Prove this.

3.8.7. Let C be a binary code with generator matrix

$$\begin{bmatrix} 1 & 0 & 0 & 0 & 1 & 0 & 1 \\ 0 & 1 & 0 & 0 & 1 & 0 & 1 \\ 0 & 0 & 1 & 0 & 0 & 1 & 1 \\ 0 & 0 & 0 & 1 & 0 & 1 & 1 \end{bmatrix}.$$

Decode the following received words:

(a) $(1\ \ 1\ \ 0\ \ 1\ \ 0\ \ 1\ \ 1)$;
(b) $(0\ \ 1\ \ 1\ \ 0\ \ 1\ \ 1\ \ 1)$;
(c) $(0\ \ 1\ \ 1\ \ 1\ \ 0\ \ 0\ \ 0)$.

3.8.8. Let p be a prime. Is there an $[8, 4]$ self-dual code over $\mathbb{F}_p$?

3.8.9. For $q = 2$ let R_k denote the rate of the Hamming code defined in (3.3.1). Determine $\lim_{k\to\infty} R_k$.

3.8.10. Let C be a binary code with weight enumerator $A(z)$. What is the weight enumerator of $\bar{C}$? What is the weight enumerator of the dual of the extended binary Hamming code of length 2^k?

3.8.11. Let C be a binary $[n, k]$ code with weight enumerator $A(z)$. We use C on a binary symmetric channel with error probability p. Our purpose is error detection only. What is the probability that an incorrect word is received and the error is not detected?

3.8.12. The n_2 by n_1 matrices over $\mathbb{F}_2$ clearly form a vector space $\mathcal{R}$ of dimension $n_1 n_2$. Let C_i be an $[n_i, k_i]$ binary code with minimum distance d_i $(i = 1, 2)$. Let C be the subset of $\mathcal{R}$ consisting of those matrices for which every column, respectively row, is a codeword in C_1, respectively C_2. Show that C is an $[n_1 n_2, k_1 k_2]$ code with minimum distance $d_1 d_2$. This code is called *direct product* of C_1 and C_2.

3.8.13. Let C be the binary $[10, 5]$ code with generator matrix

$$G = \begin{bmatrix} 1 & 0 & 0 & 0 & 0 & 0 & 0 & 0 & 1 & 1 \\ 0 & 1 & 0 & 0 & 0 & 0 & 1 & 1 & 0 & 0 \\ 0 & 0 & 1 & 0 & 0 & 1 & 0 & 1 & 0 & 0 \\ 0 & 0 & 0 & 1 & 0 & 1 & 1 & 0 & 0 & 0 \\ 0 & 0 & 0 & 0 & 1 & 1 & 1 & 1 & 0 & 0 \end{bmatrix}.$$

Show that C is *uniquely decodable* in the following sense: For every received word $\mathbf{x}$ there is a unique code word $\mathbf{c}$ such that $d(\mathbf{x}, \mathbf{c})$ is minimal.

3.8.14. Let us define *lexicographically least binary codes* with distance d as follows. The word length is not specified at first. Start with $\mathbf{c}_0 = \mathbf{0}$ and $\mathbf{c}_1 = (1, 1, \dots, 1, 0, 0, 0, \dots, 0)$ of weight d. If $\mathbf{c}_0, \mathbf{c}_1, \dots, \mathbf{c}_{l-1}$ have been chosen then $\mathbf{c}_l$ is chosen as the word which comes first in the lexicographic ordering (with 1s as far to the left as possible) such that $d(\mathbf{c}_i, \mathbf{c}_l) \geq d (0 \leq i \leq l-1)$. After l steps the length of the code is defined to be the length of the part where coordinates 1 occur.

(i) Show that after 2^k vectors have been chosen the lexicographically least code is linear!

(ii) For $d = 3$ the Hamming codes occur among the lexicographically least codes. Prove this.

3.8.15. Show that a $[15,8,5]$ code does not exist.
Hint: Show that such a code would have a generator matrix with a row of weight 5 and consider the subcode generated by the other rows.

CHAPTER 4

Some Good Codes

§4.1. Hadamard Codes and Generalizations

Let H_n be a Hadamard matrix of order n (see (1.3.5)). In H_n and $-H_n$ we replace -1 by 0. In this way we find $2n$ rows which are words in $\mathbb{F}_2^n$. Since any two rows of a Hadamard matrix differ in half of the positions we have constructed an $(n, 2n, \frac{1}{2}n)$ code. For $n = 8$ this is an extended Hamming code. For $n = 32$ the code is the one used by Mariner 1969 which was mentioned in Section 2.1. In general these codes are called *Hadamard codes*.

A similar construction starts from a Paley matrix S of order n (see (1.3.8)). We construct a code C with codewords $\mathbf{0}$, $\mathbf{1}$, the rows of $\frac{1}{2}(S + I + J)$ and $\frac{1}{2}(-S + I + J)$. From Theorem 1.3.8 it follows that C is an $(n, 2(n + 1), d)$ code, where $d = \frac{1}{2}(n - 1)$ if $n \equiv 1 \pmod 4$ and $d = \frac{1}{2}(n - 3)$ if $n \equiv 3 \pmod 4$. In the case $n = 9$ the code consists of the rows of the matrix

$$(4.1.1)\qquad \begin{bmatrix} 0\,0\,0 & 0\,0\,0 & 0\,0\,0 \\ J & P^2 & P \\ P & J & P^2 \\ P^2 & P & J \\ I & J - P^2 & J - P \\ J - P & I & J - P^2 \\ J - P^2 & J - P & I \\ 1\,1\,1 & 1\,1\,1 & 1\,1\,1 \end{bmatrix}$$

where I and J are 3 by 3 and

$$P = \begin{bmatrix} 0 & 1 & 0 \\ 0 & 0 & 1 \\ 1 & 0 & 0 \end{bmatrix}$$

§4.2. The Binary Golay Code

The most famous of all (binary) codes is the so-called binary *Golay code* $\mathscr{G}_{23}$. There are very many constructions of this code, some of them quite elegant and with short proofs of the properties of this code. We shall prove that $\mathscr{G}_{24}$, the extended binary Golay code, is unique and treat a few constructions. From these it follows that the automorphism group of the extended code is transitive and hence $\mathscr{G}_{23}$ is also unique.

We consider the incidence matrix N of a 2-(11, 6, 3) design. It is easy to show (by hand) that this design is unique. We have $NN^{\mathrm{T}} = 3I + 3J$. Consider N as a matrix with entries in $\mathbb{F}_2$. Then $NN^{\mathrm{T}} = I + J$. So N has rank 10 and the only nonzero vector $\mathbf{x}$ with $\mathbf{x}N = \mathbf{0}$ is $\mathbf{1}$. The design properties imply trivially that the rows of N all have weight 6, and that the sum of any two distinct rows of N also has weight 6. Furthermore, we know that the sum of three or four rows of N is not $\mathbf{0}$.

Next, let G be the 12 by 24 matrix (over $\mathbb{F}_2$) given by $G := (I_{12}P)$, where

$$(4.2.1) \qquad P := \begin{bmatrix} 0 & 1 & \cdots & 1 \\ 1 & & & \\ \vdots & & N & \\ 1 & & & \end{bmatrix}.$$

Every row of G has a weight $\equiv 0 \pmod 4$. Any two rows of G have inner product 0. This implies that the weight of any linear combination of the rows of G is $\equiv 0 \pmod 4$ (proof by induction). The observations made about N then show that a linear combination of any number of rows of G has weight at least 8. Consider the binary code generated by G and call it $\mathscr{G}_{24}$. Delete any coordinate to find a binary [23, 12] code with minimum distance at least 7. The distance cannot be larger, since (3.1.6) is satisfied with $e = 3$, which shows that in fact this [23, 12, 7] code is a *perfect code*! We denote this code by $\mathscr{G}_{23}$; (as mentioned above, we shall prove its uniqueness, justifying the notation).

(4.2.2) Theorem. *The codewords of weight* 8 *in* $\mathscr{G}_{24}$ *form a* 5-(24, 8, 1) *design.*

PROOF. By an easy counting argument, one can show that the weight enumerator of a perfect code containing $\mathbf{0}$ is uniquely determined. In fact, we have $A_0 = A_{23} = 1$, $A_7 = A_{16} = 253$, $A_8 = A_{15} = 506$, $A_{11} = A_{12} = 1288$. So, $\mathscr{G}_{24}$ has 759 words of weight 8, no two overlapping in more than four positions. Hence, these words together cover $759 \cdot \binom{8}{5} = \binom{24}{5}$ fivetuples. □

(4.2.3) Theorem. *If C is a binary code of length* 24, *with* $|C| = 2^{12}$, *minimum distance* 8, *and if* $\mathbf{0} \in C$, *then C is equivalent to* $\mathscr{G}_{24}$.

PROOF. (i) The difficult part of the proof is to show that C must be a *linear* code. To see this, observe that deleting any coordinate produces a code C' of length 23 and distance 7 with $|C'| = 2^{12}$. So, this code is perfect and its weight enumerator is as in the proof of the previous theorem. From the fact that this is the case, no matter which of the 24 positions is deleted, it follows that all codewords in C have weight 0, 8, 12, 16, or 24. Furthermore, a change of origin, obtained by adding a fixed codeword of C to all the words of C, shows that we can conclude that the distance of any two words of C is also 0, 8, 12, 16, or 24. Since all weights and all distances are $\equiv 0 \pmod 4$, any two codewords have inner product 0. Therefore the codewords of C span a linear code that is selforthogonal. This span clearly has dimension at most 12, i.e. at most 2^{12} codewords. It follows that this span must be C itself. In other words, C is a selfdual linear code!

(ii) We form a generator matrix G of C, taking as first row any word of weight 12. After a permutation of positions we have

$$G = \begin{pmatrix} 1 & \dots & 1 & 0 & \dots & 0 \\ & A & & & B & \end{pmatrix}.$$

We know that any linear combination of the rows of B must have even weight $\neq 0$. So, B has rank 11. This implies that the code generated by B is the [12, 11, 2] even weight code. We may therefore assume that B is the matrix I_{11} bordered by a column of 1's. A second permutation of the columns of G yields a generator matrix G' of the form $(I_{12}P)$, where P has the same form as in (4.2.1). What do we know about the matrix N in this case? Clearly any row of N must have weight 6 (look at the first row of G'). In the same way we see that the sum of any two rows of N has weight 6. This implies that N is the incidence matrix of the (unique!) 2-(11, 6, 3) design. Hence C is equivalent to $\mathscr{G}_{24}$. □

The following construction of $\mathscr{G}_{24}$ is due to R. J. Turyn. We consider the [7, 4] Hamming code H in the following representation. Take $\mathbf{0}$ and the seven cyclic shifts of (1 1 0 1 0 0 0); (note that these seven vectors form the incidence matrix of $PG(2, 2)$). Then take the eight complements of these words. Together these form H. Let H^* be obtained by reversing the order of the symbols in the words of H. By inspection we see that $\overline{H}$ and $\overline{H^*}$ are [8, 4] codes with the property $\overline{H} \cap \overline{H^*} = \{\mathbf{0}, \mathbf{1}\}$. We know that both $\overline{H}$ and $\overline{H^*}$ are selfdual codes with minimum distance 4.

We now form a code C with word length 24 by concatenating as follows:

$$C := \{(\mathbf{a} + \mathbf{x}, \mathbf{b} + \mathbf{x}, \mathbf{a} + \mathbf{b} + \mathbf{x}) | \mathbf{a} \in \overline{H}, \mathbf{b} \in \overline{H}, \mathbf{x} \in \overline{H^*}\}.$$

By letting $\mathbf{a}$ and $\mathbf{b}$ run through a basis of $\overline{H}$ and $\mathbf{x}$ run through a basis of $\overline{H^*}$, we see that the words $(\mathbf{a}, \mathbf{0}, \mathbf{a})$, $(\mathbf{0}, \mathbf{b}, \mathbf{b})$, $(\mathbf{x}, \mathbf{x}, \mathbf{x})$ form a basis for the code C. Hence C is a [24, 12] code. Any two (not necessarily distinct) basis vectors of C are orthogonal, i. e. C is selfdual. Since all the basis vectors have a weight divisible by 4, this holds for every word in C. Can a word $\mathbf{c} \in C$ have weight less than 8? Since the three components $\mathbf{a} + \mathbf{x}$, $\mathbf{b} + \mathbf{x}$, $\mathbf{a} + \mathbf{b} + \mathbf{x}$ all obviously have even weight, one of them must be $\mathbf{0}$. Our observation on the intersection of $\overline{H}$ and $\overline{H^*}$ then leads to the conclusion that $\mathbf{x} = \mathbf{0}$ or $\mathbf{1}$. Without loss of generality we assume that $\mathbf{x} = \mathbf{0}$. Since the words of H have weight 0, 4, or 8, it follows that $\mathbf{c} = \mathbf{0}$.

We have shown that C is a [24, 12, 8] code; hence $C = \mathscr{G}_{24}$.

The next construction is due to J. H. Conway. Let $\mathbb{F}_4 = \{0, 1, \omega, \bar{\omega}\}$. Let C be the [6, 3] code over $\mathbb{F}_4$ with codewords $(a, b, c, f(1), f(\omega), f(\bar{\omega}))$, where $f(x) := ax^2 + bx + c$. It is an easy exercise to show that C has minimum weight 4 and no words of weight 5. C is known as the *hexacode*.

Now, let G be a binary code of length 24 for which the words are represented as 4 by 6 binary matrices A. Denote the four rows of such a matrix A by $\mathbf{a}_0$, $\mathbf{a}_1$, $\mathbf{a}_\omega$ and $\mathbf{a}_{\bar{\omega}}$. A matrix A belongs to G iff the following two conditions are satisfied

(1) Every column of A has the same parity as its first row $\mathbf{a}_0$;
(2) $\mathbf{a}_1 + \omega\mathbf{a}_\omega + \bar{\omega}\mathbf{a}_{\bar{\omega}} \in C$.

These conditions obviously define a linear code.

If the first row of A has even parity and the codeword in (2) is not $\mathbf{0}$, then A has at least four columns of weight ≥ 2, i. e. weight ≥ 8. If, on the other hand, the codeword in (2) is $\mathbf{0}$, then either A is the zero word or A has at least two columns of weight 4, again a total weight at least 8. If A has a first row of odd parity, then all the columns of A have odd weight. These weights cannot all be 1, because this would imply that the word of C in condition (2) has odd weight. We have shown that G has minimum distance 8. We leave it as an exercise for the reader to show that conditions (1) and (2) and the fact that C has dimension 3 imply that G has dimension 12. Therefore, the matrices A form $\mathscr{G}_{24}$.

In Section 6.9 we shall find yet another construction of $\mathscr{G}_{23}$ as a cyclic code, i. e. a code with an automorphism of order 23. All these constructions together show that the automorphism group of $\mathscr{G}_{24}$ is transitive (in fact 5-transitive; it is the *Mathieu group* M_{24}). Therefore $\mathscr{G}_{23}$ is also unique.

We mention that the construction of Exercise 3.8.14 with $d = 8$ and $k = 12$ also produces the extended binary Golay code $\mathscr{G}_{24}$.

The following decoding algorithm for $\mathscr{G}_{24}$ is a generalization of Section 3.4 based on Theorem 4.2.1. Let $\mathbf{y}_i$ $(1 \leq i \leq 253)$ be the 253 code words of weight 8 of $\mathscr{G}_{24}$ with a 1 in a given position, say position 1. Consider the parity checks $\langle \mathbf{x}, \mathbf{y}_i \rangle$ $(1 \leq i \leq 253)$; here we use the fact that $\mathscr{G}_{24}$ is self-dual. Suppose $\mathbf{x}$ is received and contains $t \leq 4$ errors. Theorem 4.2.1 implies that the number of parity checks which fail is given by the following table.

	x_1 correct	x_1 incorrect
$t = 1$	77	253
2	112	176
3	125	141
4	128	128

So in case $t \leq 3$ we can correct the symbol x_1. The line corresponding to $t = 4$ demonstrates that $\mathscr{G}_{24}$ is 4-error-detecting but not 4-error-correcting.

We remark that the procedure for self-dual codes that we described in Section 3.2, when applied to the extended binary Golay code, involves the calculation of at most $26 \times 12 = 312$ parity checks and produces all the coordinates of the error vector (if $t \leq 3$).

§4.3. The Ternary Golay Code

Let S_5 be the Paley matrix of size 5 defined in (1.3.8), i.e.

$$S_5 = \begin{bmatrix} 0 & + & - & - & + \\ + & 0 & + & - & - \\ - & + & 0 & + & - \\ - & - & + & 0 & + \\ + & - & - & + & 0 \end{bmatrix}.$$

Consider the [11, 6] ternary code C defined by the generator matrix

$$G = \left[\begin{array}{c|c} & 1\ 1\ 1\ 1\ 1 \\ I_6 & \\ & S_5 \end{array}\right]$$

The code C is an [11, 6] code. From (1.3.8) it follows that $\overline{C}$ is self-dual. Therefore all the words of $\overline{C}$ have weight divisible by 3. The generator $\overline{G}$ for $\overline{C}$ is obtained by adding the column $(0, -1, -1, -1, -1, -1)^T$ to G. Every row of $\overline{G}$ has weight 6. A linear combination of two rows of $\overline{G}$ has weight at least $2 + 2$, hence it has weight 6. Therefore a linear combination of two rows of $\overline{G}$ has exactly two zeros in the last six positions and this implies that a linear combination of three rows of $\overline{G}$ has weight at least $3 + 1$, i.e. weight at least 6. Therefore $\overline{C}$ has minimum distance 6. It follows that C is an $(11, 3^6, 5)$ code. From $|B_2(\mathbf{x})| = \sum_{i=0}^{2} \binom{11}{i} 2^i = 3^5$ it then follows that C is a perfect code. This code is known as the *ternary Golay code*. It has been shown that any $(11, 3^6, 5)$ code is equivalent to C (cf. [46]). A simple uniqueness proof such as we gave for the binary Golay code has not been found yet.

§4.4. Constructing Codes from Other Codes

Many good codes have been constructed by modifying (in various ways) previously constructed codes. In this section we shall give several examples. The first method was introduced in (3.2.7), namely *extending* a code by adding an extra symbol called the overall parity check. The inverse process, which we used in Section 4.2 to obtain the binary Golay code from its extension is called *puncturing* a code. If we consider as another example the (9, 20, 4) code of (4.1.1) and puncture it, i.e. delete the last symbol in every word, we obtain an (8, 20, 3) code. In the next chapter we shall see that this is indeed a good code. We remark that an equivalent code can also be obtained by taking all cyclic permutations of the words (1 1 0 1 0 0 0 0), (1 1 1 0 0 1 0 0), and (1 0 1 0 1 0 1 0) together with **0** and **1**.

A third procedure is *shortening* a code C. Here, one takes all codewords in C that end in the same symbol and subsequently deletes this symbol. This procedure decreases the length and the number of codewords but it does not lower the minimum distance. Note that if the deleted symbol is not 0, this procedure changes a linear code to a nonlinear code (generally).

Let us now look at a slightly more complicated method. From one of the constructions of the extended binary Golay code $\mathscr{G}_{24}$ in Section 4.2 one immediately sees that $\mathscr{G}_{24}$ has a subcode consisting of 32 words with 0 in the first eight positions. Similarly, if we take $c_8 = 1$ and exactly one of the symbols c_1 to c_7 equal to 1, then we find a subcode of 32 words $(c_1, c_2, \ldots, c_{24})$. Doing this in all possible ways, we have a subset of 256 words of $\mathscr{G}_{24}$ with the property that any two of them differ in at most two positions among the first eight. Now we delete the first eight symbols from these words. The result is a binary (16, 256, 6) code which is nonlinear. This code is called the *Nordstrom-Robinson code*. This code is the first one in an infinite sequence that we discuss in Section 7.4. If we shorten this code twice and then puncture once the result is a (13, 64, 5) code, which we denote by Y. This will be an important example in the next chapter. It is known that Y is unique and that if we shorten Y there are two possible results (J.-M. Goethals 1977; cf. [26]). The two codes are: a code known as the *Nadler code* and the code of Problem 4.8.7.

A construction similar to our construction of the extended binary Golay code is known as the $(\mathbf{u}, \mathbf{u} + \mathbf{v})$-construction. Let C_i be an (n, M_i, d_i) binary code $(i = 1, 2)$. Define

$$C := \{(\mathbf{u}, \mathbf{u} + \mathbf{v}) | \mathbf{u} \in C_1, \mathbf{v} \in C_2\}. \tag{4.4.1}$$

Then C is a $(2n, M_1 M_2, d)$ code, where $d := \min\{2d_1, d_2\}$. To show this we consider two codewords $(\mathbf{u}_1, \mathbf{u}_1 + \mathbf{v}_1)$ and $(\mathbf{u}_2, \mathbf{u}_2 + \mathbf{v}_2)$. If $\mathbf{v}_1 = \mathbf{v}_2$ and $\mathbf{u}_1 \neq \mathbf{u}_2$, their distance is at least $2d_1$. If $\mathbf{v}_1 \neq \mathbf{v}_2$ the distance is $w(\mathbf{u}_1 - \mathbf{u}_2) + w(\mathbf{u}_1 - \mathbf{u}_2 + \mathbf{v}_1 - \mathbf{v}_2)$ which clearly exceeds $w(\mathbf{v}_1 - \mathbf{v}_2)$, i.e. it is at least d_2. As an example we take for C_2 the (8, 20, 3) code constructed above, and for C_1 we take the [8, 7] even weight code. The construction yields a $(16, 5 \cdot 2^9, 3)$ code. There is no $(16, M, 3)$ code known at present with $M > 5 \cdot 2^9$.

Many good codes were constructed using the following idea due to H. J. Helgert and R. D. Stinaff (1973; cf. [34]). Let C be an $[n, k]$ binary code with minimum distance d. We may assume that C has a generator matrix G with a word of weight d as its first row, say

$$G = \left[\begin{array}{cccc|cccc} 1 & 1 & \dots & 1 & 0 & 0 & \dots & 0 \\ & & G_1 & & & & G_2 & \end{array}\right].$$

Let d' be the minimum distance of the $[n-d, k-1]$ code generated by G_2 which we call the *residual code* w.r.t. the first row of G. From G we see that to each codeword of the residual code there correspond two codewords of C, at least one of which has weight $\le \frac{1}{2}d$ on the first d positions. Hence $d' \ge \frac{1}{2}d$. To illustrate this method we now show that a linear code with the parameters of the Nadler code does not exist. If there were such a code it would have a generator matrix G as above where G_2 generates a $[7, 4]$ code with distance $d' \ge 3$. Therefore the residual code is a Hamming code. W.l.o.g. we can take G_2 to have four rows of weight 3 and then G_1 must have (w.l.o.g.) four rows of weight 2. There are only a few possibilities to try and these do not yield a code with $d = 5$. Even for small parameter values it is often quite difficult to find good codes. For example, a rather complicated construction (cf. [46], Chapter 2, Section 7) produced a $(10, M, 4)$ code with $M = 38$ and for a long time it was believed that this could not be improved. M. R. Best (1978; cf. [8]) found a $(10, 40, 4)$ code which we describe below. In the next chapter we shall see that for $n = 10$, $d = 4$ this is indeed the *Best code*! Consider the $[5, 3]$ code C_1 with generator $\begin{bmatrix} 1&0&0&0&1 \\ 0&1&0&1&1 \\ 0&0&1&1&0 \end{bmatrix}$. By doubling all the codewords we have a $[10, 3]$ code C_2 with minimum distance $d = 4$. Now add (10000 00100) to all the words of C_2. The new code is no longer linear and does not contain $\mathbf{0}$. Numbering the positions from 1 to 10 we subsequently permute the positions of the codewords by elements of the subgroup of S_{10} generated by $(1\ 2\ 3\ 4\ 5)(6\ 7\ 8\ 9\ 10)$. This yields 40 codewords which turn out to have minimum distance 4.

In many technical applications (such as the compact disc) two codes are used. These codes collaborate in some way. Sometimes the goal is to combat burst errors. Quite often, more errors can be corrected than one would expect from the minimum distance.

We saw an example of collaborating codes in Problem 3.7.12, namely a direct product code. Let us have another look at such a code. Consider the product of an [8,4,4] extended Hamming code with a [16,11,4] extended Hamming code. The product can correct up to 7 errors. Now suppose a received word (i. e. a 16 by 8 matrix) has five rows with no errors, eight rows with one error, and three rows with two errors. We have 14 errors, twice the number we expect to be able to handle. However, when we decode the rows, thirteen are corrected and the three bad ones are recognized. We now declare these rows to be *erasures*! When we decode the columns, we will not encounter words with errors, but all of them have three erasures. Since the column code has distance 4, we can handle these erasures. At the end, all 14 errors have been corrected.

The codes used in practice apply variations of this idea. In the compact disc, two codes, each with distance 5, collaborate. For one of them, the decoder only corrects if at most one error occurs; otherwise, the word is declared an erasure. In the end, this turns out to increase the efficiency of the collaborating pair.

We extend the example treated above, to introduce a sequence of codes defined by P. Elias in 1954. We start with an extended Hamming code C_1 of length $n_1 = 2^m$. Assume that the codes are to be used on a B. S. C. with bit error probability p, where $n_1 p < \frac{1}{2}$. For C_2 we take the extended Hamming code of length 2^{m+1}. Define $V_1 := C_1$ and define V_2 to be the direct product of C_1 and C_2. We continue in this way: if V_i has been defined, then V_{i+1} is the direct product of V_i and the extended Hamming code C_{i+1} of length 2^{m+i}. Denote the length of V_i by n_i and its dimension by k_i. Finally, let E_i be the expected number of errors per block in words of V_i after decoding.

From the definition, we have:

$$\begin{aligned} n_{i+1} &= n_i \cdot 2^{m+i}, \\ k_{i+1} &= k_i \cdot (2^{m+i} - m - i - 1), \end{aligned}$$

and from Example 3.3.4 it follows that $E_{i+1} \le E_i^2$ and $E_1 \le (n_1 p)^2 \le \frac{1}{4}$. So these codes have the property that E_i tends to zero as $i \to \infty$.

From the recurrence relations for n_i and k_i, we find

$$n_i = 2^{mi+\frac{1}{2}i(i-1)} \quad ; \quad k_i = n_i \prod_{j=0}^{i-1} \left(1 - \frac{m+j+1}{2^{m+j}}\right).$$

So, if R_i denotes the rate of V_i, then

$$R_i \to \prod_{i=0}^{\infty} \left(1 - \frac{m+j+1}{2^{m+j}}\right) > 0$$

for $i \to \infty$. So we have a sequence of codes for which the length tends to ∞, the rate *does not* tend to 0, and nevertheless the error probability tends to 0. This is close to what Shannon's theorem promises us. Note that these codes, called *Elias codes*, have minimum distance $d_i = 4^i$ and hence $d_i/n_i \to 0$ as $i \to \infty$.

§4.5. Reed–Muller Codes

We shall now describe a class of binary codes connected with finite geometries. The codes were first treated by D. E. Muller (1954) and I. S. Reed (1954). The codes are not as good as some of the codes that will be treated in later chapters but in practice they have the advantage that they are easy to decode. The method is a generalization of majority logic decoding (see Section 3.4).

There are several ways of representing the codewords of a Reed–Muller code. We shall try to give a unified treatment which shows how the different points of view are related. As preparation we need a theorem from number theory that is a century old (Lucas (1878)).

(4.5.1) Theorem. *Let p be a prime and let*

$$n = \sum_{i=0}^{l} n_i p^i \quad \text{and} \quad k = \sum_{i=0}^{l} k_i p^i$$

be representations of n and k in base p (i.e. $0 \le n_i \le p-1$, $0 \le k_i \le p-1$*). Then*

$$\binom{n}{k} \equiv \prod_{i=0}^{l} \binom{n_i}{k_i} \pmod{p}.$$

Proof. We use the fact that $(1+x)^p \equiv 1 + x^p \pmod{p}$. If $0 \le r < p$ then

$$(1+x)^{ap+r} \equiv (1+x^p)^a(1+x)^r \pmod{p}.$$

Comparing coefficients of x^{bp+s} (where $0 \le s < p$) on both sides yields

$$\binom{ap+r}{bp+s} \equiv \binom{a}{b}\binom{r}{s} \pmod{p}.$$

The result now follows by induction. □

The following theorem on weights of polynomials is also a preparation. Let $q = 2^r$. For a polynomial $P(x) \in \mathbb{F}_q[x]$ we define the Hamming weight $w(P(x))$ to be the number of nonzero coefficients in the expansion of $P(x)$. Let $c \in \mathbb{F}_q$, $c \ne 0$. The polynomials $(x+c)^i$, $i \ge 0$, are a basis of $\mathbb{F}_q[x]$.

(4.5.2) Theorem (Massey *et al.* 1973; cf. [49]). *Let* $P(x) = \sum_{i=0}^{l} b_i(x+c)^i$, *where* $b_l \ne 0$ *and let* i_0 *be the smallest index i for which* $b_i \ne 0$. *Then*

$$w(P(x)) \ge w((x+c)^{i_0}).$$

Proof. For $l = 0$ the assertion is obvious. We use induction. Assume the theorem is true for $l < 2^n$. Now let $2^n \le l < 2^{n+1}$. Then we have

$$\begin{aligned} P(x) &= \sum_{i=0}^{2^n-1} b_i(x+c)^i + \sum_{i=2^n}^{l} b_i(x+c)^i \\ &= P_1(x) + (x+c)^{2^n}P_2(x) = (P_1(x) + c^{2^n}P_2(x)) + x^{2^n}P_2(x), \end{aligned}$$

where $P_1(x)$ and $P_2(x)$ are polynomials for which the theorem holds. We distinguish two cases.

(i) If $P_1(x) = 0$ then $w(P(x)) = 2w(P_2(x))$ and since $i_0 \ge 2^n$

$$w((x+c)^{i_0}) = w((x^{2^n} + c^{2^n})(x+c)^{i_0-2^n}) = 2w((x+c)^{i_0-2^n}),$$

from which the assertion follows.

(ii) If $P_1(x) \ne 0$ then for every term in $c^{2^n}P_2(x)$ that cancels a term in $P_1(x)$ we have a term in $x^{2^n}P_2(x)$ that does not cancel. Hence $w(P(x)) \ge w(P_1(x))$ and the result follows from the induction hypothesis. □

The three representations of codewords in Reed–Muller codes which we now introduce are: (i) characteristic functions of subsets in AG(m, 2); (ii) coefficients of binary expansions of polynomials; and (iii) lists of values which are taken by a Boolean function on $\mathbb{F}_2^m$.

First some notations and definitions. We consider the points of AG(m, 2), i.e. $\mathbb{F}_2^m$ as column vectors and denote the standard basis by $\mathbf{u}_0, \mathbf{u}_1, \ldots, \mathbf{u}_{m-1}$. Let the binary representation of j be $j = \sum_{i=0}^{m-1} \xi_{ij} 2^i$ $(0 \le j < 2^m)$.

We define $\mathbf{x}_j := \sum_{i=0}^{m-1} \xi_{ij} \mathbf{u}_i$. This represents a point of AG(m, 2) and all points are obtained in this way. Let E be the matrix with columns $\mathbf{x}_j$ $(0 \le j < 2^m)$. Write $n := 2^m$. The m by n matrix E is a list of the points of AG(m, 2), written as column vectors.

(4.5.3) Definitions.

(i) $A_i := \{\mathbf{x}_j \in \mathrm{AG}(m, 2) | \xi_{ij} = 1\}$, i.e. A_i is an $(m-1)$-dimensional affine subspace (a hyperplane), for $0 \le i < m$;

(ii) $\mathbf{v}_i :=$ the ith row of E, i.e. the characteristic function of A_i. The vector $\mathbf{v}_i$ is a word in $\mathbb{F}_2^n$; as usual we write $\mathbf{1} := (1, 1, \ldots, 1)$ for the characteristic function of AG(m, 2);

(iii) if $\mathbf{a} = (a_0, a_1, \ldots, a_{n-1})$ and $\mathbf{b} = (b_0, b_1, \ldots, b_{n-1})$ are words in $\mathbb{F}_2^n$, we define

$$\mathbf{ab} := (a_0 b_0, a_1 b_1, \ldots, a_{n-1} b_{n-1});$$

(iv) if $S \subset \{0, 1, \ldots, m-1\}$ we define

$$C(S) := \left\{ j = \sum_{i=0}^{m-1} \xi_{ij} 2^i | i \notin S \Rightarrow \xi_{ij} = 0 \; (0 \le i < m) \right\}.$$

(4.5.4) Lemma. *Let $l = \sum_{i=0}^{m-1} \xi_{il} 2^i$ and let $i_1, \ldots, i_s$ be the values of i for which $\xi_{il} = 0$. If*

$$\mathbf{v}_{i_1} \mathbf{v}_{i_2} \cdots \mathbf{v}_{i_s} = (a_{l,0}, a_{l,1}, \ldots, a_{l,n-1}),$$

then

$$(x+1)^l = \sum_{j=0}^{n-1} a_{l,j} x^{n-1-j}.$$

(Here, as usual, a product with no factors ($s = 0$) is defined to be $\mathbf{1}$.)

Proof. By Theorem 4.5.1 the binomial coefficient $\binom{l}{n-1-j}$ is 1 iff $\xi_{ij} = 1$ for every i for which $\xi_{il} = 0$. By (4.5.3) (i), (ii) and (iii) we also have $a_{l,j} = 1$ iff $\xi_{ij} = 1$ for $i = i_1, \ldots, i_s$. □

The following shows how to interpret the products $\mathbf{v}_{i_1} \ldots \mathbf{v}_{i_s}$ geometrically.

(4.5.5) Lemma. *If $i_1, i_2, \ldots, i_s$ are different then*

(i) $\mathbf{v}_{i_1} \mathbf{v}_{i_2} \cdots \mathbf{v}_{i_s}$ *is the characteristic function of the $(m-s)$-flat*

$$A_{i_1} \cap A_{i_2} \cap \cdots \cap A_{i_s},$$

(ii) *the weight* $w(\mathbf{v}_{i_1} \ldots \mathbf{v}_{i_s})$ *of the vector* $\mathbf{v}_{i_1} \ldots \mathbf{v}_{i_s}$ *in* $\mathbb{F}_2^n$ *is* 2^{m-s},
(iii) *the characteristic function of* $\{\mathbf{x}_j\}$, *i.e. the jth basis vector of* $\mathbb{F}_2^n$ *is*

$$\mathbf{e}_j = \prod_{i=0}^{m-1} \{\mathbf{v}_i + (1 + \xi_{ij})\mathbf{1}\},$$

(iv) *the products* $\mathbf{v}_{i_1} \ldots \mathbf{v}_{i_s}$ $(0 \le s \le m)$ *are a basis of* $\mathbb{F}_2^n$.

PROOF.

(i) This is a consequence of (4.5.3)(i)–(iii).
(ii) By (i) the weight is the cardinality of an $(m - s)$-flat.
(iii) Consider the matrix E. For every i such that $\xi_{ij} = 0$ we replace the ith row of E, i.e. $\mathbf{v}_i$, by its complement $\mathbf{1} + \mathbf{v}_i$. If we then multiply the rows of the new matrix, the product vector will have entry 1 only in position j, since all possible columns occur only once. As an example consider $\{\mathbf{x}_{14}\}$ in the following table. Since $14 = 0 + 2 + 2^2 + 2^3$ we see that $1 + \xi_{ij} = 1$ only if $i = 0$ (here $j = 14$). So in the table we complement the row corresponding to $\mathbf{v}_0$ and then multiply to find $(\mathbf{v}_0 + \mathbf{1})\ \mathbf{v}_1\ \mathbf{v}_2\ \mathbf{v}_3$ which is a row vector which has a 1 only in the fourteenth position.
(iv) There are $\sum_{s=0}^{m} \binom{m}{s} = 2^m = n$ products $\mathbf{v}_{i_1} \ldots \mathbf{v}_{i_s}$. The result follows from (iii). Since the polynomials $(x + 1)^l$ are independent we could also have used Lemma 4.5.4. □

The following table illustrates Lemmas 4.5.4 and 4.5.5. For example, $\mathbf{v}_0\ \mathbf{v}_2$ corresponds to $l = 15 - 2^0 - 2^2 = 10$ and hence $(x + 1)^{10} = x^{10} + x^8 + x^2 + 1$.

$\mathbf{v}_{i_1}\mathbf{v}_{i_2}\ldots\mathbf{v}_{i_s}$	Coordinates = coefficients of $(x + 1)^l$	$l = n - 1 - \sum 2^{i_s}$
$\mathbf{1}$	1 1 1 1 1 1 1 1 1 1 1 1 1 1 1 1	15 = 1111
$\mathbf{v}_0$	0 1 0 1 0 1 0 1 0 1 0 1 0 1 0 1	14 = 1110
$\mathbf{v}_1$	0 0 1 1 0 0 1 1 0 0 1 1 0 0 1 1	13 = 1101
$\mathbf{v}_2$	0 0 0 0 1 1 1 1 0 0 0 0 1 1 1 1	11 = 1011
$\mathbf{v}_3$	0 0 0 0 0 0 0 0 1 1 1 1 1 1 1 1	7 = 0111
$\mathbf{v}_0\ \mathbf{v}_1$	0 0 0 1 0 0 0 1 0 0 0 1 0 0 0 1	12 = 1100
$\mathbf{v}_0\ \mathbf{v}_2$	0 0 0 0 0 1 0 1 0 0 0 0 0 1 0 1	10 = 1010
$\mathbf{v}_0\ \mathbf{v}_3$	0 0 0 0 0 0 0 0 0 1 0 1 0 1 0 1	6 = 0110
$\mathbf{v}_1\ \mathbf{v}_2$	0 0 0 0 0 0 1 1 0 0 0 0 0 0 1 1	9 = 1001
$\mathbf{v}_1\ \mathbf{v}_3$	0 0 0 0 0 0 0 0 0 0 1 1 0 0 1 1	5 = 0101
$\mathbf{v}_2\ \mathbf{v}_3$	0 0 0 0 0 0 0 0 0 0 0 0 1 1 1 1	3 = 0011
$\mathbf{v}_0\ \mathbf{v}_1\ \mathbf{v}_2$	0 0 0 0 0 0 0 1 0 0 0 0 0 0 0 1	8 = 1000
$\mathbf{v}_0\ \mathbf{v}_1\ \mathbf{v}_3$	0 0 0 0 0 0 0 0 0 0 0 1 0 0 0 1	4 = 0100
$\mathbf{v}_0\ \mathbf{v}_2\ \mathbf{v}_3$	0 0 0 0 0 0 0 0 0 0 0 0 0 1 0 1	2 = 0010
$\mathbf{v}_1\ \mathbf{v}_2\ \mathbf{v}_3$	0 0 0 0 0 0 0 0 0 0 0 0 0 0 1 1	1 = 0001
$\mathbf{v}_0\ \mathbf{v}_1\ \mathbf{v}_2\ \mathbf{v}_3$	0 0 0 0 0 0 0 0 0 0 0 0 0 0 0 1	0 = 0000

(4.5.6) Definition. Let $0 \le r < m$. The linear code of length $n = 2^m$ which has the products $\mathbf{v}_{i_1} \ldots \mathbf{v}_{i_s}$ with $s \le r$ factors as basis is called the rth *order binary Reed–Muller code* (RM code; notation $\mathscr{R}(r, m)$).

The special case $\mathscr{R}(0, m)$ is the repetition code. From Lemma 4.5.5(i) we see that the Boolean function $x_{i_1} x_{i_2} \ldots x_{i_s}$ where $\mathbf{x} = (x_0, \ldots, x_{m-1})$ runs through $\mathbb{F}_2^m$ has value 1 iff $\mathbf{x} \in A_{i_1} \cap \cdots \cap A_{i_s}$. Hence $\mathscr{R}(r, m)$ consists of the sequences of values taken by polynomials in $x_0, \ldots, x_{m-1}$ of degree at most r.

(4.5.7) Theorem. *$\mathscr{R}(r, m)$ has minimum distance 2^{m-r}.*

Proof. By the definition and Lemma 4.5.5(ii) the minimum distance is at most 2^{m-r} and by Lemma 4.5.4 and Theorem 4.5.2 it is at least 2^{m-r}. (Also see Problem 4.7.9.) □

(4.5.8) Theorem. *The dual of $\mathscr{R}(r, m)$ is $\mathscr{R}(m - r - 1, m)$.*

Proof.

(a) By the definition and the independence of the products $\mathbf{v}_{i_1} \ldots \mathbf{v}_{i_s}$ the dimension of $\mathscr{R}(r, m)$ is $1 + \binom{m}{1} + \cdots + \binom{m}{r}$. So $\dim \mathscr{R}(r, m) + \dim \mathscr{R}(m - r - 1, m) = n$.

(b) Let $\mathbf{v}_{i_1} \ldots \mathbf{v}_{i_s}$ and $\mathbf{v}_{j_1} \ldots \mathbf{v}_{j_t}$ be basis vectors of $\mathscr{R}(r, m)$ and $\mathscr{R}(m - r - 1, m)$ respectively. Then $s + t < m$. Hence the product of these two basis vectors has the form $\mathbf{v}_{k_1} \ldots \mathbf{v}_{k_u}$ where $u < m$. By Lemma 4.5.5(ii) this product has even weight, i.e. the original two basis vectors are orthogonal. □

Corollary. $\mathscr{R}(m - 2, m)$ is the $[n, n - m - 1]$ extended Hamming code.

We have chosen the characteristic functions of certain flats as basis for an RM-code. We shall now show that for every flat of suitable dimension the characteristic function is in certain RM codes.

(4.5.9) Theorem. *Let $C = \mathscr{R}(m - l, m)$ and let A be an l-flat in* AG$(m, 2)$. *Then the characteristic function of A is in C.*

Proof. Let $\mathbf{f} = \sum_{j=0}^{n-1} f_j \mathbf{e}_j$ be the characteristic function of A. By Definition 4.5.3(iv) and Lemma 4.5.5(iii) we have

$$\mathbf{e}_j = \sum_{s=0}^{m} \sum_{\substack{(i_1, \ldots, i_s) \\ j \in C(i_1, \ldots, i_s)}} \mathbf{v}_{i_1} \mathbf{v}_{i_2} \ldots \mathbf{v}_{i_s}$$

and therefore

$$\mathbf{f} = \sum_{s=0}^{m} \sum_{(i_1, \ldots, i_s)} \left(\sum_{j \in C(i_1, \ldots, i_s)} f_j \right) \mathbf{v}_{i_1} \ldots \mathbf{v}_{i_s}.$$

Here the inner sum counts the number of points in the intersection of A and the s-flat

$$L = \{\mathbf{x}_j \in AG(m, 2) \mid j \in C(i_i, \ldots, i_s)\}.$$

If $s > m - l$ then $L \cap A$ is either empty or an affine subspace of positive dimension. In both cases $|L \cap A|$ is even, i.e. the inner sum is 0. □

This theorem and the definition show that a word is in $\mathscr{R}(r, m)$ iff it is the sum of characteristic functions of affine subspaces of dimension $\geq m - r$. In the terminology of Boolean functions $\mathscr{R}(r, m)$ is the set of polynomials in $x_0, x_1, \ldots, x_{m-1}$ of degree $\leq r$.

In Section 3.2 we defined the notion of equivalence of codes using permutations acting on the positions of the codewords. Let us now consider a code C of length n and the permutations $\pi \in S_n$ which map every word in C to a word in C. These permutations form a group, called the *automorphism group* of C (Notation: Aut(C)). For example, if C is the repetition code then Aut(C) $= S_n$.

(4.5.10) Theorem. $AGL(m, 2) \subset \text{Aut}(\mathscr{R}(r, m))$.

Proof. This is an immediate consequence of Theorem 4.5.9 and the fact that $AGL(m, 2)$ maps a k-flat onto a k-flat (for every k). □

Remark. The reader should realize that we consider $AGL(m, 2)$ acting on $AG(m, 2)$ as a group of permutations of the n positions, which have been numbered by the elements of $AG(m, 2)$.

Without going into details we briefly describe a decoding procedure for RM codes which is a generalization of majority decoding. Let $C = \mathscr{R}(r, m)$. By Theorems 4.5.8 and 4.5.9 the characteristic function of any $(r + 1)$-flat in $AG(m, 2)$ is a parity check vector for C. Given an r-flat A there are $2^{m-r} - 1$ distinct $(r + 1)$-flats which contain A. A point not in A is in exactly one of these $(r + 1)$-flats. Each of these $(r + 1)$-flats contains the points of A and exactly as many points not in A.

Now let us look at the result of the parity checks. Let a received word contain less than 2^{m-r-1} errors (see Theorem 4.5.7). Let t parity checks fail. These are two possible explanations:

(i) This was caused by an odd number of errors in the positions of A, compensated $2^{m-r} - 1 - t$ times by an odd number of errors in the remaining positions of the check set.
(ii) The number of errors in the positions of A is even but in t of the parity check equations there is an odd number of errors in the remaining positions.

By maximum likelihood (ii) is more probable than (i) if $t < 2^{m-r-1}$ and otherwise (i) is more probable. This means that it is possible to determine the parity of the number of errors in the positions of any r-flat. Then, using a similar procedure, the same thing is done for $(r-1)$-flats, etc. After $r+1$ steps the errors have been located. This procedure is called *multistep majority decoding.*

§4.6. Kerdock Codes

We shall briefly discuss a class of nonlinear codes known as *Kerdock codes,* (cf. [75], [11]). A Kerdock code is a subcode of a second order Reed-Muller code consisting of a number of cosets of the corresponding first order Reed-Muller code. Note that $\mathscr{R}(2, m)$ is itself a union of cosets of $\mathscr{R}(1, m)$, each coset corresponding to some quadratic form

(4.6.1) $$Q(\mathbf{v}) := \sum_{0 \le i < j < m} q_{ij} \mathbf{v}_i \mathbf{v}_j.$$

Corresponding to Q, there is an alternating bilinear form B defined by

$$B(\mathbf{v}, \mathbf{w}) := Q(\mathbf{v} + \mathbf{w}) - Q(\mathbf{v}) - Q(\mathbf{w}) = \mathbf{v}B\mathbf{w}^{\mathrm{T}},$$

where B is a symplectic matrix (zero on the diagonal and $B = -B^{\mathrm{T}}$). By an easy induction proof one can show that, by a suitable affine transformation, Q can be put into the form

(4.6.2) $$\sum_{i=0}^{h-1} \mathbf{v}_{2i} \mathbf{v}_{2i+1} + L(\mathbf{v}),$$

where L is linear and $2h$ is the rank of B. In fact, one can see to it that $L(\mathbf{v}) = 0$, 1 or $\mathbf{v}_{2h}$.

(4.6.3) Lemma. *The number of points* $(x_0, x_1, \ldots, x_{2h-1}) \in \mathbb{F}_2^{2h}$ *for which* $\sum_{i=0}^{h-1} x_{2i} x_{2i+1} = 0$ *is* $2^{2h-1} + 2^{h-1}$.

Proof. If $x_0 = x_2 = \cdots = x_{2h-2} = 0$, then there are 2^h choices for $(x_1, \ldots, x_{2h-1})$. Otherwise there are 2^{h-1} choices. So, the number of zeros is $2^h + (2^h - 1)2^{h-1}$. □

From (4.6.2) and (4.6.3) we find the following lemma.

(4.6.4) Lemma. *Let m be even. If $Q(\mathbf{v})$ is a quadratic form corresponding to a symplectic form of rank m, then the coset of $\mathscr{R}(1, m)$ determined by $Q(\mathbf{v})$ has 2^m words of weight $2^{m-1} - 2^{m/2-1}$ and 2^m words of weight $2^{m-1} + 2^{m/2-1}$.*

(Note that this implies that if Q has rank smaller than m, the corresponding coset has smaller minimum weight).

Clearly, a union of cosets of $\mathscr{R}(1, m)$ will be a code with minimum distance at most $2^{m-1} - 2^{m/2-1}$. We wish to form a code C by taking the union of cosets corresponding to certain quadratic forms $Q_1, \ldots, Q_l$ (with associated symplectic forms $B_1, \ldots, B_l$). To find the minimum distance of this code, we must consider codes corresponding to cosets defined by the forms $Q_i - Q_j$ $(i \neq j)$ and find their minimum weight. The best that we can achieve is that each difference $Q_i - Q_j$ corresponds to a symplectic form of maximal rank, that is a *nonsingular* symplectic form. Since the symplectic forms correspond to skew-symmetric matrices with zero diagonal and no two of these can have the same first row, it follows that $l \leq 2^{m-1}$ if the minimum distance d of C is to be $2^{m-1} - 2^{m/2-1}$.

(4.6.5) Definition. Let m be even. A set of 2^{m-1} symplectic matrices of size m such that the difference of any two distinct elements is nonsingular, is called a *Kerdock set.*

(4.6.6) Definition. Let m be even. Let $l = 2^{m-1}$ and let $Q_1, \ldots, Q_l$ be a Kerdock set. The nonlinear code $\mathcal{K}(m)$ of length $n = 2^m$ consisting of cosets of $\mathcal{R}(1, m)$, corresponding to the forms Q_i $(1 \leq i \leq l)$, is called a *Kerdock code.*

To show that such codes actually exist is a nontrivial problem related to the geometry of $\mathbb{F}_2^m$ (cf. [11]). We only give one example. Let $m = 4$. If Q is a quadratic form $\sum_{i=0}^{3} q_{ij}x_i x_j$, we represent Q by a graph on the vertices $x_0, \ldots, x_3$ with an edge $\{x_i, x_j\}$ if and only if $q_{ij} = 1$.

If Q corresponds to a nonsingular sympletic form, then the graph must be isomorphic to one of the following graphs:

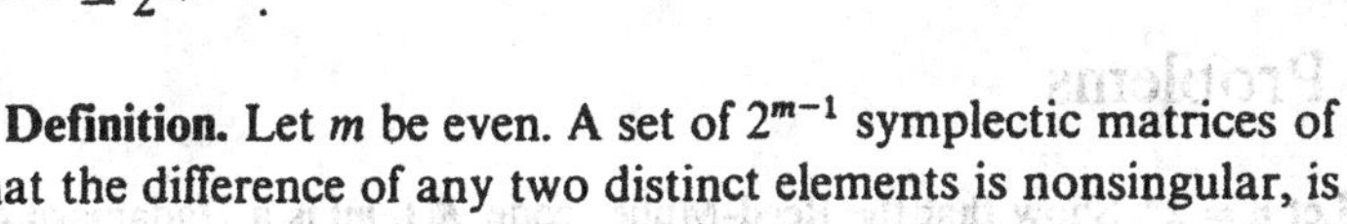

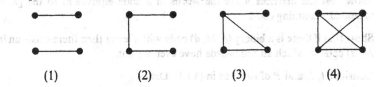

(1) (2) (3) (4)

Order the partitions (12)(34), (13)(24), and (14)(23) cyclically. Form six graphs of type (2) by taking two sides from one pair of these partitions and one from the following one. It is easily seen that these six graphs, the empty graph and the graph of type (4) have the property that the sum (or difference) of any two corresponds to a nonsingular symplectic form. In this way we find the $8 \cdot 2^5 = 2^8$ words of a $(16, 2^8, 6)$ code, which is in fact the Nordstrom-Robinson code of §4.4.

In the general case $\mathcal{K}(m)$ is a $(2^m, 2^{2m}, 2^{m-1} - 2^{m/2-1})$ code. So, the number of words is considerably larger than for $\mathcal{R}(1, m)$ although the minimum distance is only slightly smaller.

§4.7. Comments

For details about the application of Hadamard codes in the Mariner expeditions we refer to reference [56].

The Golay codes were constructed by M. J. E. Golay in 1949 in a different way from our treatment. For more about these codes and several connections to combinatorial theory we refer to a book by P. J. Cameron and J. H. van Lint [11] or to [46]; also see [19].

The reader who is interested in more material related to Section 4.4 is referred to references [64] and [65]. For more about encoding and decoding of RM codes see [2] or [46].

§4.8. Problems

4.8.1. Let $n = 2^m$. Show that the Reed-Muller code $\mathscr{R}(1, m)$ is a Hadamard code of length n.

4.8.2. Show that the ternary Golay code has 132 words of weight 5. For each pair $\{\mathbf{x}, 2\mathbf{x}\}$ of codewords of weight 5 consider the subset of positions where $x_i \neq 0$. Show that these 66 sets form a $4-(11, 5, 1)$ design.

4.8.3. Let S be the Paley matrix of order 11 and $A = \frac{1}{2}(S + I + J)$. Consider the rows of A, all 55 sums of two distinct rows of A, and the complements of these vectors. Show that this is an (11, 132, 3) code.

4.8.4. Construct a (17, 36, 8) code.

4.8.5. Consider Conway's construction of $\mathscr{G}_{24}$. Then consider the subcode consisting of the matrices A that have the form (B, B, B), where each B is a 4 by 2 matrix. Show that the matrices B are the words of a code equivalent to the [8, 4] extended Hamming code.

4.8.6. Show that if there is a binary (n, M, d) code with d even then there exists an (n, M, d) code in which all codewords have even weight.

4.8.7. Consider I, J, and P of size 3 as in (4.1.1). Define

$$A := \begin{bmatrix} J-I & I & I & I \\ I & J-I & I & I \\ I & I & J-I & I \\ I & I & I & J-I \end{bmatrix}, \qquad B := \begin{bmatrix} J & P & I & P^2 \\ P & J & P^2 & I \\ I & P^2 & J & P \\ P^2 & I & P & J \end{bmatrix},$$

$$C := (J-I \quad J-I \quad J-I \quad J-I), \qquad D := \begin{bmatrix} 000 & 111 & 111 & 111 \\ 111 & 000 & 111 & 111 \\ 111 & 111 & 000 & 111 \\ 111 & 111 & 111 & 000 \end{bmatrix}.$$

Show that $\mathbf{0}$ and the rows of A, B, C and D are the words of a (12, 32, 5) code.

4.8.8. Let H be the Hadamard matrix H_{12} of (1.3.9) and let $A := H - I$, $G := (I \quad A)$. Show that G is the generator matrix of a ternary [24, 12] code with minimum distance 9.

4.8.9. Show that the $(\mathbf{u}, \mathbf{u} + \mathbf{v})$-construction of (4.4.1) with $C_1 = \mathscr{R}(r + 1, m)$, $C_2 = \mathscr{R}(r, m)$ yields $C = \mathscr{R}(r + 1, m + 1)$. Use this to give a second proof of Theorem 4.5.7.

4.8.10. (i) Let $n = 2^m$. For $\mathbf{x} \in \mathbb{F}_2^n$ we define $\mathbf{x}^* \in \{1, -1\}^n$ as the vector obtained by replacing the 0s in $\mathbf{x}$ by -1. In Problem 4.7.1 we saw that this mapping applied to $\mathscr{R}(1, m)$ yields vectors $\pm\mathbf{a}_1, \pm\mathbf{a}_2, \ldots, \pm\mathbf{a}_n$ where the $\mathbf{a}_i$ are the rows of a Hadamard matrix. By using this show that if $\mathbf{x} \in \mathbb{F}_2^n$ then there exists a codeword $\mathbf{c} \in \mathscr{R}(1, m)$ such that $d(\mathbf{x}, \mathbf{c}) \le (n - \sqrt{n})/2$.
If $m = 2k$ and $\mathbf{x}$ is the word $\in \mathscr{R}(2, m)$ corresponding to the Boolean function $x_1x_2 + x_3x_4 + \cdots + x_{2k-1}x_{2k}$ show that $d(\mathbf{x}, \mathbf{c}) \ge (n - \sqrt{n})/2$ for all $\mathbf{c} \in \mathscr{R}(1, m)$. (In other words: the covering radius of (1, 2k) is $2^{2k-1} - 2^{k-1}$.)

4.8.11. Let H be a parity check matrix for the [4, 2] ternary Hamming code and let I and J be the 4 by 4 identity resp. all one matrix. Show that

$$G := \begin{bmatrix} J+I & I & I \\ 0 & H & -H \end{bmatrix}$$

generates a [12, 6] code C with $d = 6$, i.e. a code equivalent to the extended ternary Golay code.

CHAPTER 5
Bounds on Codes

§5.1. Introduction; The Gilbert Bound

In this chapter we shall be interested in codes that have as many codewords as possible, given their length and minimum distance. We shall not be interested in questions like usefulness in practice, encoding or decoding of such codes. We again consider as alphabet a set Q of q symbols and we define $\theta := (q-1)/q$. Notation is as in Section 3.1. We assume q has been chosen and then define an $(n, *, d)$ code as a code with length n and minimum distance d. We are interested in the maximal number of codewords (i.e. the largest M which can be put in place of the $*$). An (n, M, d) code which is not contained in any $(n, M+1, d)$ code is called *maximal*.

(5.1.1) Definition. $A(n, d) := \max\{M \mid \text{an } (n, M, d) \text{ code exists}\}$. A code C such that $|C| = A(n, d)$ is called *optimal*.

Some authors use the term "optimal" for $[n, k]$ codes with $d = n - k + 1$ (see Problem 3.8.2). Such codes are optimal in the sense of (5.1.1) (cf. (5.2.2)). Usually $[n, k, n-k+1]$ codes are called *maximum distance separable* codes (MDS codes).

The study of the numbers $A(n, d)$ is considered to be the central problem in combinatorial coding theory. In Chapter 2 we learned that good codes are long, or more precisely, given a channel with a certain error probability p, we can reduce the probability of error by looking at a sequence of codes with increasing length n. Clearly the average number of errors in a received word is np and hence d must grow at least as fast as $2np$ if we wish to correct these errors. This explains the importance of the number $\alpha(\delta)$ which we define as follows.

(5.1.2) Definition.

$$\alpha(\delta) := \limsup_{n\to\infty} n^{-1} \log_q A(n, \delta n).$$

In Chapter 2 we studied good codes with a given rate R. In that case we should ask how large d/n is (as a function of n). By (5.1.2) this means that we are interested in the inverse function $\alpha^{-}(R)$.

The functions A and α are not known in general. We shall study upper and lower bounds for both of them and special values of $A(n, d)$. The techniques of extending, shortening, or puncturing (see Section 4.4) will often come in handy. These immediately yield the following theorem.

(5.1.3) Theorem. *For binary codes we have*

$$A(n, 2l - 1) = A(n + 1, 2l).$$

We remind the reader of the definition of a sphere $B_r(\mathbf{x})$, given in Section 3.1 and we define

(5.1.4) $$V_q(n, r) := |B_r(\mathbf{x})| = \sum_{i=0}^{r} \binom{n}{i}(q-1)^i$$

(cf. (3.1.6)).

In order to study the function α, we need a generalization of the entropy function defined in (1.4.4). We define the *entropy function* H_q on $[0, \theta]$, where $\theta := (q-1)/q$, by

(5.1.5) $$H_q(0) := 0,$$

$$H_q(x) := x \log_q(q-1) - x \log_q x - (1-x) \log_q(1-x) \quad \text{for } 0 < x \le \theta.$$

Note that $H_q(x)$ increases from 0 to 1 as x runs from 0 to θ.

(5.1.6) Lemma. *Let* $0 \le \lambda \le \theta$, $q \ge 2$. *Then*

$$\lim_{n\to\infty} n^{-1} \log_q V_q(n, \lfloor \lambda n \rfloor) = H_q(\lambda).$$

Proof. For $r = \lfloor \lambda n \rfloor$ the last term of the sum of the right-hand side of (5.1.4) is the largest. Hence

$$\binom{n}{\lfloor \lambda n \rfloor}(q-1)^{\lfloor \lambda n \rfloor} \le V_q(n, \lfloor \lambda n \rfloor) \le (1 + \lfloor \lambda n \rfloor)\binom{n}{\lfloor \lambda n \rfloor}(q-1)^{\lfloor \lambda n \rfloor}.$$

By taking logarithms, dividing by n, and then proceeding as in the proof of Theorem 1.4.5 the result follows. □

To finish this section we discuss a lower found for $A(n, d)$ and the corresponding bound for $\alpha(\delta)$. Although the result is nearly trivial, it was thought for a long time that $\alpha(\delta)$ would be equal to this lower bound. In 1982, Tsfasman, Vlăduţ, and Zink [81] improved the lower bound (for $q \geq 49$) using methods from algebraic geometry.

(5.1.7) Theorem. *For $n \in \mathbb{N}$, $d \in \mathbb{N}$, $d \leq n$, we have*

$$A(n, d) \geq q^n / V_q(n, d-1).$$

Proof. Let the (n, M, d) code C be maximal. This implies that there is no word in Q^n with distance d or more to all the words of C. In other words: the spheres $B_{d-1}(\mathbf{c})$, with $\mathbf{c} \in C$, cover Q^n. Therefore the sum of their "volumes", i.e. $|C|\, V_q(n, d-1)$ exceeds $q^n = |Q|^n$. □

The proof shows that a code which has at least $q^n/V_q(n, d-1)$ codewords can be constructed by simply starting with any word $\mathbf{c}_0$ and then consecutively adding new words that have distance at least d to the words which have been chosen before, until the code is maximal. Such a code has no structure. Surprisingly enough, the requirement that C is linear is not an essential restriction as the following theorem shows.

(5.1.8) Theorem. *If $n \in \mathbb{N}$, $d \in \mathbb{N}$, $k \in \mathbb{N}$ satisfy $V_q(n, d-1) < q^{n-k+1}$, then an $[n, k, d]$ code exists.*

Proof. For $k = 0$ this is trivial. Let C_{k-1} be an $[n, k-1, d]$ code. Since $|C_{k-1}|\, V_q(n, d-1) < q^n$, this code is not maximal. Hence there is a word $\mathbf{x} \in Q^n$ with distance $\geq d$ to all words of C_{k-1}. Let C_k be the code spanned by C_{k-1} and $\{\mathbf{x}\}$. Let $\mathbf{z} = a\mathbf{x} + \mathbf{y}$ (where $0 \neq a \in Q$, $\mathbf{y} \in C_{k-1}$) be a codeword in C_k. Then

$$w(\mathbf{z}) = w(a^{-1}\mathbf{z}) = w(\mathbf{x} + a^{-1}\mathbf{y}) = d(\mathbf{x}, -a^{-1}\mathbf{y}) \geq d. \qquad \square$$

The codes of Problem 3.8.14 are an example of Theorem 5.1.8.

Example. Let $q = 2$, $n = 13$, $d = 5$. Then from (5.1.4) we find $V_2(13, 4) = 1093$ and hence $A(13, 5) \geq \lfloor 8192/1093 \rfloor = 8$. In fact Theorem 5.1.8 guarantees the existence of a $[13, 3, 5]$ code. Clearly this is not a very good code since by Theorem 4.5.7 puncturing $\mathscr{R}(1, 4)$ three times yields a $[13, 5, 5]$ code and in fact the code Y of Section 4.4 is an even better nonlinear code, namely a $(13, 64, 5)$ code. This example shows one way of finding bounds for $A(n, d)$, namely by constructing good codes. We know that $A(13, 5) \geq 64$.

The bound of Theorem 5.1.7 is known as the *Gilbert bound* (or *Gilbert-Varshamov* bound). Let us now look at the corresponding bound for α.

(5.1.9) Theorem (Asymptotic Gilbert Bound). *If $0 \leq \delta \leq \theta$ then*

$$\alpha(\delta) \geq 1 - H_q(\delta).$$

PROOF. By (5.1.7) and (5.1.6) we have

$$\alpha(\delta) = \limsup_{n\to\infty} n^{-1} \log_q A(n, \delta n) \geq \lim_{n\to\infty} \{1 - n^{-1} \log_q V_q(n, \delta n)\}$$
$$= 1 - H_q(\delta).$$ □

§5.2. Upper Bounds

In this section we treat a number of upper bounds for $A(n, d)$ that are fairly easy to derive. In the seventies more complicated methods produced better bounds, which we shall discuss in Section 5.3.

By puncturing an (n, M, d) code $d - 1$ times we obtain an $(n - d + 1, M, 1)$ code, i.e. the M punctured words are different. Hence $M \leq q^{n-d+1}$. We have proved the following theorem, known as the *Singleton bound.*

(5.2.1) Theorem. *For $q, n, d \in \mathbb{N}$, $q \geq 2$ we have*

$$A(n, d) \leq q^{n-d+1}.$$

(5.2.2) Corollary. *For an $[n, k]$ code over $\mathbb{F}_q$ we have $k \leq n - d + 1$.*

A code achieving this bound is called an MDS code (see Problem 3.8.2).

EXAMPLE. Let $q = 2$, $n = 13$, $d = 5$. Then we have $A(13, 5) \leq 512$.

The asymptotic form of Theorem 5.2.1 is as follows.

(5.2.3) Theorem. *For $0 \leq \delta \leq 1$ we have $\alpha(\delta) \leq 1 - \delta$.*

Our next bound is obtained by calculating the maximal possible value of the average distance between two distinct codewords. Suppose C is an (n, M, d) code. We make a list of words of C. Consider a column in this list. Let the jth symbol of Q $(0 \leq j \leq q - 1)$ occur m_j times in this column. The contribution of this column to the sum of the distances between all ordered pairs of distinct codewords is $\sum_{j=0}^{q-1} m_j(M - m_j)$. Since $\sum_{j=0}^{q-1} m_j = M$ we have from the Cauchy-Schwarz inequality

$$\sum_{j=0}^{q-1} m_j(M - m_j) = M^2 - \sum_{j=0}^{q-1} m_j^2 \leq M^2 - q^{-1}\left(\sum_{j=0}^{q-1} m_j\right)^2 = \theta M^2.$$

Since our list has n columns and since there are $M(M - 1)$ ordered pairs of codewords, we find

$$M(M - 1)d \leq n\theta M^2.$$

We have proved the so-called *Plotkin bound.*

(5.2.4) Theorem. *For $q, n, d \in \mathbb{N}$, $q \geq 2$, $\theta = 1 - q^{-1}$ we have*

$$A(n, d) \leq \frac{d}{d - \theta n}, \qquad \text{if } d > \theta n.$$

EXAMPLES. (a) Let $q = 2$, $n = 13$, $d = 5$. Then $\theta = \frac{1}{2}$. In order to be able to apply Theorem 5.2.4 we consider a $(13, M, 5)$ code and shorten it four times to obtain a $(9, M', 5)$ code with $M' \geq 2^{-4}M$. By the Plotkin bound $M' \leq 5/(5 - 4\frac{1}{2}) = 10$. So $M \leq 160$, i.e. $A(13, 5) \leq 160$. A better bound can be obtained by first applying Theorem 5.1.3 to get $A(13, 5) = A(14, 6)$ and then repeating the above argument to get $A(14, 6) \leq 2^3 \cdot 6/(6 - 5\frac{1}{2}) = 96$.

(b) Let $q = 3$, $n = 13$, $d = 9$. Then $\theta = \frac{2}{3}$ and the Plotkin bound yields $A(13, 9) \leq 27$ for ternary codes. Consider the dual of the ternary Hamming code (see (3.3.1)). This code has generator matrix

$$G = \begin{bmatrix} 0 & 0 & 0 & 0 & 1 & 1 & 1 & 1 & 1 & 1 & 1 & 1 & 1 \\ 0 & 1 & 1 & 1 & 0 & 0 & 0 & 1 & 1 & 1 & 2 & 2 & 2 \\ 1 & 0 & 1 & 2 & 0 & 1 & 2 & 0 & 1 & 2 & 0 & 1 & 2 \end{bmatrix}.$$

This matrix has the points of PG(2, 3) as columns. The positions where $(a_1, a_2, a_3)G$ has a zero correspond to the points of PG(2, 3) on the projective line with equation $a_1x_1 + a_2x_2 + a_3x_3 = 0$, i.e. if $\mathbf{a} \neq \mathbf{0}$ there are exactly four such positions. Hence every codeword $\neq \mathbf{0}$ has weight 9 and hence any two distinct codewords have distance 9. So this is a linear code satisfying Theorem 5.2.4 with equality.

From the proof of Theorem 5.2.4 we can see that equality is possible only if all pairs of distinct codewords indeed have the same distance. Such a code is called an *equidistant code*.

Again we derive an asymptotic result.

(5.2.5) Theorem (Asymptotic Plotkin Bound). *We have*

$$\alpha(\delta) = 0, \qquad \text{if } \theta \leq \delta \leq 1,$$

$$\alpha(\delta) \leq 1 - \delta/\theta, \qquad \text{if } 0 \leq \delta < \theta.$$

PROOF. The first assertion is a trivial consequence of Theorem 5.2.4. For the second assertion we define $n' := \lfloor (d-1)/\theta \rfloor$. Then $1 \leq d - \theta n' \leq 1 + \theta$. Shorten an (n, M, d) code to an (n', M', d) code. Then $M' \geq q^{n'-n}M$ and by Theorem 5.2.4 we have $M' \leq d/(d - \theta n') \leq d$. So $M \leq dq^{n-n'}$. From this and $n'/n \to \delta/\theta$ if $n \to \infty$ and $d = \delta n$ we find $\alpha(\delta) \leq 1 - \delta/\theta$. □

The following bound, found by J. H. Griesmer (1960), is a bound for linear codes which is asymptotically equivalent to the Plotkin bound but in some cases it is better. Even though the proof is elementary, it turns out that the bound is sharp quite often. The proof is based on the same ideas as the method of Helgert and Stinaff treated in Section 4.4. Let G be the generator matrix of an $[n, k, d]$ code. We may assume that the first row of G has weight d, in fact we may assume w.l.o.g. that it is $(111\cdots 10\cdots 0)$ with d ones. Every other row has at least $\lceil d/q \rceil$ coordinates in the first d positions that are the same. Therefore the residual code with respect to the first row is an $[n-d, k-1, d']$ code with $d' \geq \lceil d/q \rceil$. Using induction we then find the following theorem.

(5.2.6) Theorem (Griesmer Bound). *For an* $[n, k, d]$ *code over* $\mathbb{F}_q$ *we have*

$$n \geq \sum_{i=0}^{k-1} \lceil d/q^i \rceil.$$

EXAMPLES. (a) Let $q = 2$, $n = 13$, $d = 5$. Since $\sum_{i=0}^{5} \lceil 5/2^i \rceil = 13$ we see that a $[13, k, 5]$ code must have $k \leq 6$. The code Y of Section 4.4 has 64 words but it is not linear. In fact a $[13, 6, 5]$ code cannot exist because it would imply the existence of a $[12, 5, 5]$ code contradicting the analysis given in Section 4.4. So in this case the Griesmer bound is not sharp.

(b) Let $q = 3$, $n = 14$, $d = 9$. From $\sum_{i=0}^{3} \lceil 9/3^i \rceil = 14$ it follows that a $[14, k, 9]$ ternary code has $k \leq 4$. A shortened version of such a code would be like Example (b) following Theorem 5.2.4. Suppose such a code exists. As before we can assume w.l.o.g. that $(11\ldots 100000)$ of weight 9 is the first row of the generator matrix. Then, as in the proof of the Griesmer bound, the residual code is a $[5, 3, 3]$ ternary code. W.l.o.g. the generator of such a code would be

$$G = \begin{bmatrix} 1 & 0 & 0 & 1 & 1 \\ 0 & 1 & 0 & a & b \\ 0 & 0 & 1 & c & d \end{bmatrix}, \qquad \text{where } a, b, c, d \text{ are not } 0.$$

Clearly $a \neq b$ and $c \neq d$ and hence there is a combination of rows 2 and 3 with weight 2, a contradiction. Again the Griesmer bound is not sharp.

One of the easiest bounds to understand generalizes (3.1.6). It is known as the *Hamming bound* or *sphere packing bound*.

(5.2.7) Theorem. *If* $q, n, e \in \mathbb{N}$, $q \geq 2$, $d = 2e + 1$, *then*

$$A(n, d) \leq q^n / V_q(n, e).$$

PROOF. The spheres $B_e(\mathbf{c})$, where $\mathbf{c}$ runs through an $(n, M, 2e+1)$ code, are disjoint. Therefore $M \cdot V_q(n, e) \leq q^n$. □

EXAMPLE. Let $q = 2$, $n = 13$, $d = 5$. Then from $V_2(13, 2) = 1 + 13 + 78 = 92$ we find $A(13, 5) \leq \lfloor 2^{13}/92 \rfloor = 89$.

We have defined a perfect code to be a code that satisfies (5.2.7) with equality. We return to this question in Chapter 7.

(5.2.8) Theorem (Asymptotic Hamming Bound). *We have*

$$\alpha(\delta) \le 1 - H_q(\tfrac{1}{2}\delta).$$

PROOF. $A(n, \lceil \delta n \rceil) \le A(n, 2\lceil \tfrac{1}{2}\delta n \rceil - 1) \le q^n / V_q(n, \lceil \tfrac{1}{2}\delta n \rceil - 1)$.

The result follows from Lemma 5.1.6. □

We now come to an upper bound which is somewhat more difficult to prove. For a long time it was the best known upper bound. From the proof of the Plotkin bound it should be clear that that bound cannot be good if the distances between codewords are not all close to the average distance. The following idea, due to P. Elias, gives a stronger result. Apply the method of proof of the Plotkin bound to the set of codewords in a suitably chosen sphere in Q^n. The following lemma shows how to choose the sphere. W.l.o.g. we take $Q = \mathbb{Z}/q\mathbb{Z}$.

(5.2.9) Lemma. *If A and C are subsets of Q^n then there is an $\mathbf{x} \in Q^n$ such that*

$$\frac{|(\mathbf{x} + A) \cap C|}{|A|} \ge \frac{|C|}{q^n}.$$

PROOF. Choose $\mathbf{x}_0$ such that $|(\mathbf{x}_0 + A) \cap C|$ is maximal. Then

$$\begin{aligned} |(\mathbf{x}_0 + A) \cap C| &\ge q^{-n} \sum_{\mathbf{x} \in Q^n} |(\mathbf{x} + A) \cap C| \\ &= q^{-n} \sum_{\mathbf{x} \in Q^n} \sum_{\mathbf{a} \in A} \sum_{\mathbf{c} \in C} |\{\mathbf{x} + \mathbf{a}\} \cap \{\mathbf{c}\}| \\ &= q^{-n} \sum_{\mathbf{a} \in A} \sum_{\mathbf{c} \in C} 1 = q^{-n} |A| \cdot |C|. \end{aligned}$$ □

Now let C be an (n, M, d) code and let A be $B_r(\mathbf{0})$. We may assume w.l.o.g. that the point $\mathbf{x}_0$ of the lemma is $\mathbf{0}$. Consider the code $A \cap C$. This is an (n, K, d) code with $K \ge MV_q(n, r)/q^n$. We list the words of this code as rows of a K by n matrix. Let m_{ij} denote the number of occurrences of the symbol j in the ith column of this matrix. We know

(i) $\sum_{j=0}^{q-1} m_{ij} = K$

and

(ii) $\sum_{i=1}^{n} m_{i0} =: S \ge K(n - r)$

because every row of the matrix has weight at most r.

Therefore:

(iii) $\sum_{j=1}^{q-1} m_{ij}^2 \ge (q-1)^{-1}(\sum_{j=1}^{q-1} m_{ij})^2 = (q-1)^{-1}(K - m_{i0})^2$

and

(iv) $\sum_{i=1}^{n} m_{i0}^2 \ge n^{-1}(\sum_{i=1}^{n} m_{i0})^2 = n^{-1}S^2$.

We again calculate the sum of the distances of all ordered pairs of rows of the matrix. We find from (i) to (iv):

$$\sum_{i=1}^{n}\sum_{j=0}^{q-1} m_{ij}(K - m_{ij}) = nK^2 - \sum_{i=1}^{n}\left(m_{i0}^2 + \sum_{j=1}^{q-1} m_{ij}^2\right)$$

$$\le nK^2 - (q-1)^{-1}\sum_{i=1}^{n}(qm_{i0}^2 + K^2 - 2Km_{i0})$$

$$\le nK^2 - (q-1)^{-1}(qn^{-1}S^2 + nK^2 - 2KS).$$

In this inequality, we substitute $S \ge K(n - r)$, where we now pick $r \le \theta n$, and hence $S \ge q^{-1}nK$. We find $\sum_{i=1}^{n}\sum_{j=0}^{q-1} m_{ij}(K - m_{ij}) \le K^2 r(2 - (r/\theta n))$. Since the number of pairs of rows is $K(K-1)$, we have

$$K(K-1)d \le K^2 r(2 - r\theta^{-1}n^{-1}).$$

Therefore we have proved the following lemma.

(5.2.10) Lemma. *If the words of an (n, K, d) code all have weight $\le r \le \theta n$, then*

$$d \le \frac{Kr}{K-1}\left(2 - \frac{r}{\theta n}\right).$$

(5.2.11) Theorem (Elias Bound). *Let q, n, d, $r \in \mathbb{N}$, $q \ge 2$, $\theta = 1 - q^{-1}$ and assume that $r \le \theta n$ and $r^2 - 2\theta nr + \theta nd > 0$. Then*

$$A(n, d) \le \frac{\theta nd}{r^2 - 2\theta nr + \theta nd}\cdot\frac{q^n}{V_q(n, r)}.$$

Proof. From Lemma 5.2.9 we saw that an (n, M, d) code has a subcode with $K \ge MV_q(n, r)/q^n$ words which are all in some $B_r(\mathbf{x})$. So we may apply Lemma 5.2.10. This yields

$$q^{-n}MV_q(n, r) \le K \le \frac{\theta nd}{r^2 - 2\theta nr + \theta nd}. \qquad \square$$

Note that $r = \theta n$, $d > \theta n$ yields the Plotkin bound.

Example. Let $q = 2$, $n = 13$, $d = 5$. Then $\theta = \frac{1}{2}$. The best result is obtained if we estimate $A(14, 6)$ in (5.2.11). The result is

$$A(13, 5) = A(14, 6) \le \frac{42}{r^2 - 14r + 42}\cdot\frac{2^{14}}{\sum_{i\le r}\binom{14}{i}}$$

and then the best choice is $r = 3$ which yields $A(13, 5) \le 162$.

The result in the example is not as good as earlier estimates. However, asymptotically the Elias bound is the best result of this section.

(5.2.12) Theorem (Asymptotic Elias Bound). *We have*

$$\alpha(\delta) \le 1 - H_q(\theta - \sqrt{\theta(\theta - \delta)}), \qquad \text{if } 0 \le \delta < \theta,$$
$$\alpha(\delta) = 0, \qquad \text{if } \theta \le \delta < 1.$$

PROOF. The second part follows from Theorem 5.2.5. So let $0 < \delta \le \theta$. Choose $0 \le \lambda < \theta - \sqrt{\theta(\theta - \delta)}$ and take $r = \lfloor \lambda n \rfloor$. Then $\theta\delta - 2\theta\lambda + \lambda^2 > 0$. From Theorem 5.2.11 we find, with $d = \lfloor \delta n \rfloor$

$$\begin{aligned} n^{-1} \log_q A(n, \delta n) &\le n^{-1} \log_q \left(\frac{\theta n d}{r^2 - 2\theta n r + \theta n d} \cdot \frac{q^n}{V_q(n, r)} \right) \\ &\sim n^{-1} \left\{ \log_q \left(\frac{\theta\delta}{\lambda^2 - 2\theta\lambda + \theta\delta} \right) + n - nH_q(\lambda) \right\} \\ &\sim 1 - H_q(\lambda), \qquad (n \to \infty). \end{aligned}$$

Therefore $\alpha(\delta) \le 1 - H_q(\lambda)$. Since this is true for every λ with $\lambda < \theta - \sqrt{\theta(\theta - \delta)}$ the result follows. □

The next bound is also based on the idea of looking at a subset of the codewords. In this case we consider codewords with a fixed weight w.

We must first study certain numbers similar to $A(n, d)$. We restrict ourselves to the case $q = 2$.

(5.2.13) Definition. We denote by $A(n, d, w)$ the maximal number of codewords in a binary code of length n and minimum distance $\ge d$ for which all codewords have weight w.

(5.2.14) Lemma. *We have*

$$A(n, 2k-1, w) = A(n, 2k, w) \le \left\lfloor \frac{n}{w} \left\lfloor \frac{n-1}{w-1} \left\lfloor \cdots \left\lfloor \frac{n-w+k}{k} \right\rfloor \cdots \right\rfloor \right\rfloor \right\rfloor.$$

PROOF. Since words with the same weight have even distance $A(n, 2k - 1, w) = A(n, 2k, w)$. Suppose we have a code C with $|C| = K$ satisfying our conditions. Write the words of C as rows of a matrix. Every column of this matrix has at most $A(n - 1, 2k, w - 1)$ ones. Hence $Kw \le nA(n - 1, 2k, w - 1)$, i.e.

$$A(n, 2k, w) \le \left\lfloor \frac{n}{w} A(n-1, 2k, w-1) \right\rfloor.$$

Since $A(n, 2k, k - 1) = 1$ the result follows by induction. □

This lemma shows how to estimate the numbers $A(n, d, w)$. The numbers can be used to estimate $A(n, d)$ as is done in the following generalization of the Hamming bound, which is known as the *Johnson bound.*

(5.2.15) Theorem. *Let $q = 2$, $n, e \in \mathbb{N}$, $d = 2e + 1$. Then*

$$A(n, d) \leq \frac{2^n}{\displaystyle\sum_{i=0}^{e} \binom{n}{i} + \frac{\binom{n}{e+1} - \binom{d}{e} A(n, d, d)}{\left\lfloor \dfrac{n}{e+1} \right\rfloor}}.$$

PROOF. The idea is the same as the proof of the Hamming bound. Let there be N_{e+1} words in $\{0, 1\}^n$ which have distance $e + 1$ to the (n, M, d) code C. Then

$$M \sum_{i=0}^{e} \binom{n}{i} + N_{e+1} \leq 2^n.$$

In order to estimate N_{e+1} we consider an arbitrary codeword $\mathbf{c}$ which we can take to be $\mathbf{0}$ (w.l.o.g.). Then the number of words in C with weight d is clearly at most $A(n, d, d)$. Each of these words has distance e to $\binom{d}{e}$ words of weight $e + 1$. Since there are $\binom{n}{e+1}$ words of weight $e + 1$ there must be at least $\binom{n}{e+1} - \binom{d}{e} A(n, d, d)$ among them that have distance $e + 1$ to C. By varying $\mathbf{c}$ we thus count $M\left\{\binom{n}{e+1} - \binom{d}{e} A(n, d, d)\right\}$ words in $\{0, 1\}^n$ that have distance $e + 1$ to the code. How often has each of these words been counted? Take one of them; again w.l.o.g. we call it $\mathbf{0}$. The codewords with distance $e + 1$ to $\mathbf{0}$ have mutual distances $\geq 2e + 1$ iff they have 1s in different positions. Hence there are at most $\lfloor n/(e + 1) \rfloor$ such codewords. This gives us the desired estimate for N_{e+1}. □

From Lemma 5.2.14 we find, taking $k = e + 1$, $w = 2e + 1$

$$\binom{d}{e} A(n, d, d) \leq \binom{n}{e} \left\lfloor \frac{n-e}{e+1} \right\rfloor.$$

Substitution in Theorem 5.2.15 shows that a code C satisfies

$$\text{(5.2.16)} \qquad |C| \left\{ \sum_{i=0}^{e} \binom{n}{i} + \frac{\binom{n}{e}}{\left\lfloor \dfrac{n}{e+1} \right\rfloor} \left(\frac{n-e}{e+1} - \left\lfloor \frac{n-e}{e+1} \right\rfloor \right) \right\} \leq 2^n,$$

which is the original form of the Johnson bound.

EXAMPLE. Let $q = 2$, $n = 13$, $d = 5$ (i.e. $e = 2$). Then $A(13, 5, 5) \leq \lfloor \frac{13}{5} \lfloor \frac{12}{4} \lfloor \frac{11}{3} \rfloor \rfloor \rfloor = 23$ and the Johnson bound yields

$$A(13,5) \le \left\lfloor \frac{2^{13}}{1+13+78+\dfrac{286-10\cdot 23}{4}} \right\rfloor = 77.$$

For $n = 13$, $q = 2$, $d = 5$ this is the best result up to now. Only the powerful methods of the next section are enough to produce the true value of $A(13, 5)$.

§5.3. The Linear Programming Bound

Many of the best known bounds for the numbers $A(n, d)$ known at present are based on a method which was developed by P. Delsarte (1973). The idea is to derive inequalities that have a close connection to the MacWilliams identity (Theorem 3.5.3) and then to use linear programming techniques to analyze these inequalities. In this section we shall have to rely heavily on properties of the so-called *Krawtchouk polynomials.*

In order to avoid cumbersome notation, we assume that q and n have been chosen and are fixed. Then we define

$$K_k(x) := \sum_{j=0}^{k} (-1)^j \binom{x}{j}\binom{n-x}{k-j}(q-1)^{k-j},$$

where

$$\binom{x}{j} := \frac{x(x-1)\cdots(x-j+1)}{j!}, \qquad (x \in \mathbb{R}).$$

For a discussion of these polynomials and the properties which we need, we refer to Section 1.2.

In the following we assume that the alphabet Q is the ring $\mathbb{Z}/q\mathbb{Z}$ (which we may do w.l.o.g). Then $\langle \mathbf{x}, \mathbf{y} \rangle$ denotes the usual inner product $\sum_{i=1}^{n} x_i y_i$ for $\mathbf{x}$, $\mathbf{y} \in Q^n$.

(5.3.1) Lemma. *Let ω be a primitive qth root of unity in $\mathbb{C}$ and let $\mathbf{x} \in Q^n$ be a fixed word of weight i. Then*

$$\sum_{\substack{\mathbf{y} \in Q^n \\ w(\mathbf{y}) = k}} \omega^{\langle \mathbf{x}, \mathbf{y} \rangle} = K_k(i).$$

PROOF. We may assume that $\mathbf{x} = (x_1, x_2, \ldots, x_i, 0, 0, \ldots, 0)$ where the coordinates x_1 to x_i are not 0. Choose k positions, $h_1, h_2, \ldots, h_k$ such that $0 < h_1 < h_2 < \cdots < h_j \le i < h_{j+1} < \cdots < h_k \le n$. Let D be the set of all words (of weight k) that have their nonzero coordinates in these positions. Then by Lemma 1.1.32

$$\sum_{y \in D} \omega^{\langle \mathbf{x}, \mathbf{y} \rangle} = \sum_{y_{h_1} \in Q\setminus\{0\}} \cdots \sum_{y_{h_k} \in Q\setminus\{0\}} \omega^{x_{h_1}y_{h_1} + \cdots + x_{h_k}y_{h_k}}$$

$$= (q-1)^{k-j} \prod_{l=1}^{j} \sum_{y \in Q\setminus\{0\}} \omega^{x_{h_l}y} = (-1)^j (q-1)^{k-j}.$$

Since there are $\binom{i}{j}\binom{n-i}{k-j}$ choices for D the result follows. □

In order to be able to treat arbitrary (i.e. not necessarily linear) codes we generalize (3.5.1).

(5.3.2) Definition. Let $C \subseteq Q^n$ be a code with M words. We define

$$A_i := M^{-1} |\{(\mathbf{x}, \mathbf{y}) | \mathbf{x} \in C, \mathbf{y} \in C, d(\mathbf{x}, \mathbf{y}) = i\}|.$$

The sequence $(A_i)_{i=0}^n$ is called the *distance distribution* or *inner distribution* of C.

Note that if C is linear or distance invariant, the distance distribution is the weight distribution.

The following lemma is the basis of the *linear programming bound* (Theorem 5.3.4).

(5.3.3) Lemma. *Let $(A_i)_{i=0}^n$ be the distance distribution of a code $C \subseteq Q^n$. Then*

$$\sum_{i=0}^{n} A_i K_k(i) \geq 0,$$

for $k \in \{0, 1, \ldots, n\}$.

Proof. By Lemma 5.3.1 we have

$$M \sum_{i=0}^{n} A_i K_k(i) = \sum_{i=0}^{n} \sum_{\substack{(\mathbf{x}, \mathbf{y}) \in C^2 \\ d(\mathbf{x}, \mathbf{y}) = i}} \sum_{\substack{\mathbf{z} \in Q^n \\ w(\mathbf{z}) = k}} \omega^{\langle \mathbf{x} - \mathbf{y}, \mathbf{z} \rangle}$$

$$= \sum_{\substack{\mathbf{z} \in Q^n \\ w(\mathbf{z}) = k}} \left| \sum_{\mathbf{x} \in C} \omega^{\langle \mathbf{x}, \mathbf{z} \rangle} \right|^2 \geq 0. \qquad \square$$

(5.3.4) Theorem. *Let $q, n, d \in \mathbb{N}$, $q \geq 2$. Then*

$$A(n, d) \leq \max \left\{ \sum_{i=0}^{n} A_i \,\middle|\, A_0 = 1, A_i = 0 \text{ for } 1 \leq i < d, \right.$$

$$\left. A_i \geq 0, \sum_{i=0}^{n} A_i K_k(i) \geq 0 \text{ for } k \in \{0, 1, \ldots, n\} \right\}.$$

If $q = 2$ and d is even we may take $A_i = 0$ for i odd.

PROOF. By Lemma 5.3.3, the distance distribution of an (n, M, d) code satisfies the inequalities $\sum_{i=0}^{n} A_i K_k(i) \geq 0$. Clearly the A_i are nonnegative and $A_0 = 1$, $A_i = 0$ for $1 \leq i < d$. Furthermore by (5.3.2) we have $\sum_{i=0}^{n} A_i = M^{-1}|C^2| = M$.
The final assertion is Problem 4.8.6. □

EXAMPLE. As in several previous examples we wish to estimate $A(13, 5) = A(14, 6)$ for $q = 2$. For the distance distribution of a $(14, M, 6)$ code we may assume

$$A_0 = 1, A_1 = A_2 = A_3 = A_4 = A_5 = A_7 = A_9 = A_{11} = A_{13} = 0,$$

$$A_6 \geq 0, A_8 \geq 0, A_{10} \geq 0, A_{12} \geq 0, A_{14} \geq 0.$$

For these we have the following inequalities from Lemma 5.3.3. (The values of $K_k(i)$ are found by using (1.2.10).)

$$\begin{aligned}
14 + 2A_6 - 2A_8 - 6A_{10} - 10A_{12} - 14A_{14} &\geq 0,\\
91 - 5A_6 - 5A_8 + 11A_{10} + 43A_{12} + 91A_{14} &\geq 0,\\
364 - 12A_6 + 12A_8 + 4A_{10} - 100A_{12} - 364A_{14} &\geq 0,\\
1001 + 9A_6 + 9A_8 - 39A_{10} + 121A_{12} + 1001A_{14} &\geq 0,\\
2002 + 30A_6 - 30A_8 + 38A_{10} - 22A_{12} - 2002A_{14} &\geq 0,\\
3003 - 5A_6 - 5A_8 + 27A_{10} - 165A_{12} + 3003A_{14} &\geq 0,\\
3432 - 40A_6 + 40A_8 - 72A_{10} + 264A_{12} - 3432A_{14} &\geq 0.
\end{aligned}$$

We must find an upper bound for $M = 1 + A_6 + A_8 + A_{10} + A_{12} + A_{14}$. This linear programming problem turns out to have a unique solution, namely

$$A_6 = 42, A_8 = 7, A_{10} = 14, A_{12} = A_{14} = 0.$$

Hence $M \leq 64$. In Section 4.4 we constructed a $(13, 64, 5)$ code Y. Therefore we have now proved that $A(13, 5) = 64$.

We shall now put Theorem 5.3.4 into another form which often has advantages over the original form. The reader familiar with linear programming will realize that we are applying the duality theorem (cf. [32]).

(5.3.5) Theorem. *Let $\beta(x) = 1 + \sum_{k=1}^{n} \beta_k K_k(x)$ be any polynomial with $\beta_k \geq 0$ $(1 \leq k \leq n)$ such that $\beta(j) \leq 0$ for $j = d, d + 1, \ldots, n$. Then $A(n, d) \leq \beta(0)$.*

PROOF. Suppose $A_0, A_1, \ldots, A_n$ satisfy the conditions of Theorem 5.3.4, i.e. $K_k(0) + \sum_{i=d}^{n} A_i K_k(i) \geq 0$ $(k = 0, 1, \ldots, n;\ A_i \geq 0$ for $i = d, d + 1, \ldots, n)$. Then the condition on β yields $\sum_{i=d}^{n} A_i \beta(i) \leq 0$ i.e.

$$-\sum_{i=d}^{n} A_i \geq \sum_{k=1}^{n} \beta_k \sum_{i=d}^{n} A_i K_k(i) \geq -\sum_{k=1}^{n} \beta_k K_k(0) = 1 - \beta(0)$$

and hence

$$1 + \sum_{i=d}^{n} A_i \le \beta(0). \qquad \square$$

The advantage of Theorem 5.3.5 is that any polynomial β satisfying the conditions of the theorem yields a bound for $A(n, d)$ whereas in Theorem 5.3.4 one has to find the optimal solution to the system of inequalities.

EXAMPLE. Let $q = 2$, $n = 2l + 1$, $d = l + 1$. We try to find a bound for $A(n, d)$ by taking $\beta(x) = 1 + \beta_1 K_1(x) + \beta_2 K_2(x) = 1 + \beta_1(n - 2x) + \beta_2(2x^2 - 2nx + \frac{1}{2}n(n-1))$. Choose β_1 and β_2 in such a way that $\beta(d) = \beta(n) = 0$. We find $\beta_1 = (n+1)/2n$, $\beta_2 = 1/n$ and hence the conditions of Theorem 5.3.5 are satisfied. So we have $A(2l+1, l+1) \le \beta(0) = 1 + \beta_1 n + \beta_2 \binom{n}{2} = 2l + 2$. This is the same as the Plotkin bound (5.2.4).

The best bound for $\alpha(\delta)$ that is known at present is due to R. J. McEliece, E. R. Rodemich, H. C. Rumsey and L. R. Welch (1977; cf. [50]). We shall not treat this best bound but we give a slightly weaker result (actually equal for $\delta > 0.273$), also due to these authors. It is based on an application of Theorem 5.3.5.

(5.3.6) Theorem. *Let $q = 2$. Then*

$$\alpha(\delta) \le H_2(\tfrac{1}{2} - \sqrt{\delta(1-\delta)}).$$

PROOF. We consider an integer t with $1 \le t \le \frac{1}{2}n$ and a real number a in the interval $[0, n]$. Define the polynomial $\alpha(x)$ by

$$\alpha(x) := (a - x)^{-1}\{K_t(a)K_{t+1}(x) - K_{t+1}(a)K_t(x)\}^2.$$

By applying (1.2.12) we find

$$\text{(5.3.7)} \quad \alpha(x) = \frac{2}{t+1}\binom{n}{t}\{K_t(a)K_{t+1}(x) - K_{t+1}(a)K_t(x)\} \sum_{k=0}^{t} \frac{K_k(a)K_k(x)}{\binom{n}{k}}.$$

Let $\alpha(x) = \sum_{k=0}^{2t+1} \alpha_k K_k(x)$ be the Krawtchouk expansion of $\alpha(x)$. We wish to choose a and t in such a way that $\beta(x) := \alpha(x)/\alpha_0$ satisfies the conditions of Theorem 5.3.5. If we take $a \le d$ then the only thing we have to check is whether $\alpha_i \ge 0$ $(i = 1, \ldots, n)$, $\alpha_0 > 0$. If $x_1^{(k)}$ denotes the smallest zero of K_k then we know that $0 < x_1^{(t+1)} < x_1^{(t)}$ (cf. (1.2.13)).

In order to simplify the following calculations, we choose t in such a way that $x_1^{(t)} < d$ and then choose a between $x_1^{(t+1)}$ and $x_1^{(t)}$ in such a way that $K_t(a) = -K_{t+1}(a) > 0$. It follows that (5.3.7) expresses $\alpha(x)$ in the form $\sum c_{kl} K_k(x) K_l(x)$, where all coefficients c_{kl} are nonnegative. Then it follows from (1.2.14) that all α_i are nonnegative. Furthermore, $\alpha_0 = -[2/(t+1)] \times$

$\binom{n}{t} K_t(a)K_{t+1}(a) > 0$. Hence we can indeed apply Theorem 5.3.5. We find

$$(5.3.8) \qquad A(n, d) \le \beta(0) = \frac{\alpha(0)}{\alpha_0} = \frac{(n+1)^2}{2a(t+1)}\binom{n}{t}.$$

To finish the proof, we need to know more about the location of the zero $x_1^{(t)}$. It is known that if $0 < \tau < \frac{1}{2}$, $n \to \infty$ and $t/n \to \tau$ then $x_1^{(t)}/n \to \frac{1}{2} - \sqrt{\tau(1-\tau)}$. It follows that we can apply (5.3.8) to the situation $n \to \infty$, $d/n \to \delta$ with a sequence of values of t such that $t/n \to \frac{1}{2} - \sqrt{\delta(1-\delta)}$. Taking logarithms in (5.3.8) and dividing by n, the assertion of the theorem follows. (For a proof of the statement about $x_1^{(t)}$, we refer the reader to one of the references on orthogonal polynomials or [46] or [50].) □

§5.4. Comments

For a treatment of the codes defined using algebraic geometry, and for the improvement of the Gilbert bound, we refer to [73]. For an example see §6.8.

In Figure 2 we compare the asymptotic bounds derived in this chapter. We

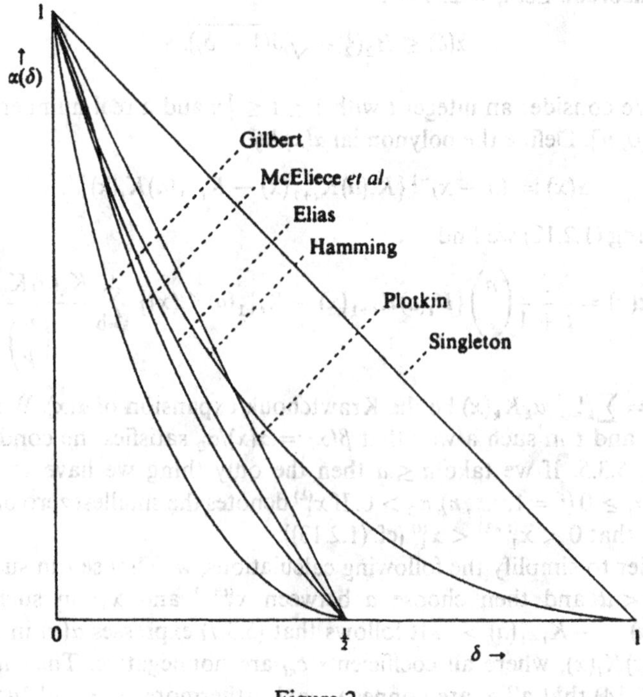

Figure 2

have not included bounds given by V. I. Levenshtein (1975; cf. [40]) and V. M. Sidelnikov (1975; cf. [63]) because these are not as good as the results by McEliece *et al.* [50] and rather difficult to derive. The best known bound mentioned above is

$$\alpha(\delta) \le \min\{1 + g(u^2) - g(u^2 + 2\delta u + 2\delta) | 0 \le u \le 1 - 2\delta\}, \tag{5.4.1}$$

where

$$g(x) := H_2\left(\frac{1 - \sqrt{1 - x}}{2}\right).$$

For a proof we refer to [50] or [52]. For very small values of δ, the Elias bound is better than (5.3.6) but not as good as (5.4.1).

In a paper by M. R. Best *et al.* (1977; cf. [6]) (5.3.3) and (5.3.4) are generalized:

(i) by observing that if $|C|$ is odd, then the inequalities of (5.3.3) are much stronger; and
(ii) by adding inequalities (such as the obvious inequality $A_{n-1} + A_n \le 1$) to (5.3.4). This yields several very good bounds (see Problem 5.5.12).

§5.5. Problems

5.5.1. Use the fact that a linear code can be defined by its parity check matrix to show that an $[n, k, d]$ code over $\mathbb{F}_q$ exists if $V_q(n - 1, d - 2) < q^{n-k}$. Compare this with Theorem 5.1.8.

5.5.2. Determine $A(10, 5)$ for $q = 2$.

5.5.3. Let $q = 2$. Show that if in Theorem 5.2.4 the right-hand side is an odd integer l then $A(n, d) \le l - 1$.

5.5.4. Determine bounds for $A(17, 8)$ if $q = 2$.

5.5.5. Consider a generator matrix for the [31, 5] dual binary Hamming code. Show that it is possible to leave out a number of columns of this matrix in such a way that the resulting code has $d = 10$ and meets the Griesmer bound.

5.5.6. Let C be a binary code of length n with minimum distance $d = 2k$ and let all codewords of C have weight w. Suppose $|C| = [n(n - 1)/w(w - 1)] \times A(n - 2, 2k, w - 2)$. Show that the words of C are the blocks of a 2-design.

5.5.7. Show that a shortened binary Hamming code is optimal.

5.5.8. Let $w \in \mathbb{N}$, $w > 4$. Let C_l be the binary code of length n defined by

$$C_l := \left\{(c_0, c_1, \ldots, c_{n-1}) \middle| \sum_{i=0}^{n-1} c_i = w, \sum_{i=0}^{n} ic_i \equiv l \pmod{n}\right\},$$

where the summations are in $\mathbb{Z}$. Show that

$$A(n, 4, w) \sim \frac{n^{w-1}}{w!}, \qquad (n \to \infty).$$

5.5.9. Let $q = 2$. Show that $\binom{n}{w} A(n, 2k) \leq 2^n A(n, 2k, w)$.

5.5.10. (i) Show that $A(n, 2k, w) \leq \left(1 - \frac{w}{k}\left(1 - \frac{w}{n}\right)\right)^{-1}$, if the right-hand side is positive.
(ii) Using this and Problem 5.5.9 derive the Elias bound.

5.5.11. Let C be a binary (n, M, d) code with $n - \sqrt{n} < 2d \leq n$. Suppose C has the property that if $\mathbf{x} \in C$ then also $\mathbf{x} + \mathbf{1} \in C$. Show that $k = 2$ in (5.3.3) yields the bound

$$M \leq \frac{8d(n-d)}{n - (n-2d)^2}.$$

(This is known as the *Grey bound.*)

5.5.12. Show that the (8, 20, 3) code of Section 4.4 is optimal. (This is difficult. See Section 5.4.)

CHAPTER 6
Cyclic Codes

§6.1. Definitions

In Section 4.5 we defined the automorphism group Aut(C) of a code C. Corresponding to this group there is a group of permutation matrices. Sometimes the definition of Aut(C) is extended by replacing permutation matrices by monomial matrices, i.e. matrices for which the nonzero entries correspond to a permutation matrix. In both cases we are interested in the group of permutations. In this chapter we shall study linear codes for which the automorphism group contains the cyclic group of order n, where n is the word length.

(6.1.1) Definition. A linear code C is called *cyclic* if

$$\forall_{(c_0, c_1, \ldots, c_{n-1}) \in C}[(c_{n-1}, c_0, c_1, \ldots, c_{n-2}) \in C].$$

This definition is extended by using monomial matrices instead of permutations as follows. If for every codeword $(c_0, c_1, \ldots, c_{n-1})$, the word $(\lambda c_{n-1}, c_0, c_1, \ldots, c_{n-2})$ is also in C (here λ is fixed), the code is called *constacyclic* (and *negacyclic* if $\lambda = -1$). We shall present the theory for cyclic codes; the generalization to constacyclic codes is an easy exercise for the reader.

The most important tool in our description of cyclic codes is the following isomorphism between $\mathbb{F}_q^n$ and a group of polynomials. The multiples of $x^n - 1$ form a principal ideal in the polynomial ring $\mathbb{F}_q[x]$. The residue class ring $\mathbb{F}_q[x]/(x^n - 1)$ has the set of polynomials

$$\{a_0 + a_1 x + \cdots + a_{n-1}x^{n-1} \mid a_i \in \mathbb{F}_q, 0 \le i < n\}$$

as a system of representatives. Clearly $\mathbb{F}_q^n$ is isomorphic to this ring (considered only as an additive group). In the following we shall also use the multiplicative structure which we have now introduced, namely multiplication of polynomials $\mathrm{mod}(x^n - 1)$. From now on we make the following identification:

$$(6.1.2) \qquad (a_0, a_1, \ldots, a_{n-1}) \in \mathbb{F}_q^n \rightleftarrows a_0 + a_1 x + \cdots + a_{n-1}x^{n-1} \in \mathbb{F}_q[x]/(x^n - 1)$$

and we shall often speak of a codeword **c** as the codeword $c(x)$, using (6.1.2). Extending this, we interpret a linear code as a subset of $\mathbb{F}_q[x]/(x^n - 1)$.

(6.1.3) Theorem. *A linear code C in $\mathbb{F}_q^n$ is cyclic if and only if C is an ideal in $\mathbb{F}_q[x]/(x^n - 1)$.*

PROOF.

(i) If C is an ideal in $\mathbb{F}_q[x]/(x^n - 1)$ and $c(x) = c_0 + c_1 x + \cdots + c_{n-1}x^{n-1}$ is any codeword, then $xc(x)$ is also a codeword, i.e.

$$(c_{n-1}, c_0, c_1, \ldots, c_{n-2}) \in C.$$

(ii) Conversely, if C is cyclic, then for every codeword $c(x)$ the word $xc(x)$ is also in C. Therefore $x^i c(x)$ is in C for every i, and since C is linear $a(x)c(x)$ is in C for every polynomial $a(x)$. Hence C is an ideal. □

(6.1.4) Convention. From now on we only consider cyclic codes of length n over $\mathbb{F}_q$ with $(n, q) = 1$. For some theory of binary cyclic codes with even length see §6.10.

Since $\mathbb{F}_q[x]/(x^n - 1)$ is a principal ideal ring every cyclic code C consists of the multiples of a polynomial $g(x)$ which is the monic polynomial of lowest degree (i.e. not the zero polynomial) in the ideal (cf. Section 1.1).

This polynomial $g(x)$ is called the *generator polynomial* of the cyclic code. The generator polynomial is a divisor of $x^n - 1$ (since otherwise the g.c.d. of $x^n - 1$ and $g(x)$ would be a polynomial in C of degree lower than the degree of $g(x)$). Let $x^n - 1 = f_1(x)f_2(x)\ldots f_t(x)$ be the decomposition of $x^n - 1$ into irreducible factors. Because of (6.1.4), these factors are different. We can now find all cyclic codes of length n by picking (in all possible ways) one of the 2^t factors of $x^n - 1$ as generator polynomial $g(x)$ and defining the corresponding code to be the set of multiples of $g(x)\ \mathrm{mod}(x^n - 1)$.

(6.1.5) EXAMPLE. Over $\mathbb{F}_2$ we have

$$x^7 - 1 = (x - 1)(x^3 + x + 1)(x^3 + x^2 + 1).$$

There are altogether eight cyclic codes of length 7. One of these has **0** as the only codeword and one contains all possible words. The code with generator $x - 1$ contains all words of even weight. The $[7, 1]$ cyclic code has **0** and **1** as codewords. The remaining four codes have dimension 3, 3, 4, and 4 respectively. For example, taking $g(x) := (x - 1)(x^3 + x + 1) = x^4 + x^3 + x^2 + 1$, we find a [7, 3] cyclic code. This code is an example of the irreducible cyclic codes defined below.

(6.1.6) Definition. The cyclic code generated by $f_i(x)$ is called a *maximal cyclic code* (since it is a maximal ideal) and denoted by M_i^+. The code generated by $(x^n - 1)/f_i(x)$ is called a *minimal cyclic code* and denoted by M_i^-. Minimal cyclic codes are also called *irreducible* cyclic codes.

Our definition (6.1.1) guarantees that the automorphism group of a cyclic code C contains the cyclic group generated by the permutation

$$i \mapsto i + 1 \pmod{n}.$$

However, since $a(x^q) = a(x)^q$ is in the same cyclic codes as $a(x)$, we see that the permutation π_q defined by $\pi_q(i) = qi \pmod{n}$ (i.e. $x \mapsto x^q$) also maps a cyclic code onto itself. If m is the order of $q \pmod{n}$ then the two permutations $i \mapsto i + 1$ and π_q generate a group of order nm contained in $\mathrm{Aut}(C)$.

§6.2. Generator Matrix and Check Polynomial

Let $g(x)$ be the generator polynomial of a cyclic code C of length n. If $g(x)$ has degree $n - k$, then the codewords $g(x), xg(x), \ldots, x^{k-1}g(x)$ clearly form a basis for C, i.e. C is an $[n, k]$ code. Hence, if $g(x) = g_0 + g_1 x + \cdots + g_{n-k}x^{n-k}$, then

$$G = \begin{bmatrix} g_0 & g_1 & \cdots & g_{n-k} & 0 & 0 & \cdots & 0 \\ 0 & g_0 & \cdots & g_{n-k-1} & g_{n-k} & 0 & \cdots & 0 \\ 0 & 0 & \cdots & & & & \cdots & 0 \\ 0 & 0 & \cdots & & g_0 & g_1 & \cdots & g_{n-k} \end{bmatrix}$$

is a generator matrix for C. This means that we encode an information sequence $(a_0, a_1, \ldots, a_{k-1})$ as $\mathbf{a}G$ which is the polynomial

$$(a_0 + a_1 x + \cdots + a_{k-1}x^{k-1})g(x).$$

A more convenient form of the generator matrix is obtained by defining (for $i \geq n - k$), $x^i = g(x)q_i(x) + r_i(x)$, where $r_i(x)$ is a polynomial of degree $< n - k$. The polynomials $x^i - r_i(x)$ are codewords of C and form a basis for the code, which yields a generator matrix of C in standard form (with I_k in back). In this case $(a_0, a_1, \ldots, a_{k-1})$ is encoded as follows: divide $(a_0 + a_1 x + \cdots + a_{k-1}x^{k-1})x^{n-k}$ by $g(x)$ and subtract the remainder from $(a_0 + a_1 x + \cdots + a_{k-1}x^{k-1})x^{n-k}$, thus obtaining a codeword.

Technically this is a very easy way to encode information because the division by a fixed polynomial can be realized by a simple shift register (for a definition see Chapter 13).

Since $g(x)$ is a divisor of $x^n - 1$, there is a polynomial $h(x) = h_0 + h_1 x + \cdots + h_k x^k$ such that $g(x)h(x) = x^n - 1$ (in $\mathbb{F}_q[x]$). In the ring $\mathbb{F}_q[x]/(x^n - 1)$ we have $g(x)h(x) = 0$, i.e. $g_0 h_i + g_1 h_{i-1} + \cdots + g_{n-k} h_{i-n+k} = 0$ for $i = 0, 1, \ldots, n - 1$. It follows that

$$H := \begin{bmatrix} 0 & 0 & \cdots & 0 & h_k & \cdots & h_1 & h_0 \\ 0 & 0 & \cdots & h_k & \cdots & h_1 & h_0 & 0 \\ \vdots & & & & & & & \vdots \\ h_k & \cdots & h_1 & h_0 & 0 & \cdots & & 0 \end{bmatrix}$$

is a parity check matrix for the code C. We call $h(x)$ the *check polynomial* of C. The code C consists of all $c(x)$ such that $c(x)h(x) = 0$. By comparing G and H we see that the code with generator polynomial $h(x)$ is equivalent to the dual of C (obtained by reversing the order of the symbols). Very often this code is simply called the dual of C (which causes a lot of confusion since it is not equal to $C^\perp$). Notice that in this sense, the "dual" of a maximal cyclic code M_i^+ is the minimal cyclic code M_i^-.

Consider the minimal cyclic code M_i^- with generator $g(x) = (x^n - 1)/f_i(x)$ where $f_i(x)$ has degree k. If $a(x)$ and $b(x)$ are two codewords in M_i^- such that $a(x)b(x) = 0$, then one of them must be divisible by $f_i(x)$ and it is therefore 0. Since M_i^- has no zero divisors, it is a field, i.e. it is isomorphic to $\mathbb{F}_q^k$. A particularly interesting example is obtained if we take $n = 2^k - 1$ and $f_i(x)$ a primitive polynomial of degree k. In that case the n cyclic shifts of the generator polynomial $g(x)$ are apparently all the nonzero codewords of M_i^-. This means that the code is equidistant (cf. Section 5.2) and therefore this distance is 2^{k-1} (by (3.8.5)). As a consequence we see that for every primitive divisor $f(x)$ of $x^n - 1$ (where $n = 2^k - 1$) the polynomial $(x^n - 1)/f(x)$ has exactly 2^{k-1} coefficients equal to 1. An example with $k = 3$ was given in (6.1.5).

§6.3. Zeros of a Cyclic Code

Let $x^n - 1 = f_1(x) \ldots f_t(x)$ and let β_i be a zero of $f_i(x)$ in some extension field of $\mathbb{F}_q$. Then $f_i(x)$ is the minimal polynomial of β_i and therefore the maximal code M_i^+ is nothing but the set of polynomials $c(x)$ for which $c(\beta_i) = 0$. So in general a cyclic code can be specified by requiring that all codewords have certain prescribed zeros. In fact, it is sufficient to take one zero β_i of each irreducible factor f_i of the generator polynomial $g(x)$ and require that all codewords have these points as zeros (all in a suitable extension field of $\mathbb{F}_q$). If we start with any set $\alpha_1, \alpha_2, \ldots, \alpha_s$ and define a code C by $c(x) \in C$ iff $c(\alpha_i) = 0$ for $i = 1, 2, \ldots, s$, then C is cyclic and the generator polynomial of C is the least common multiple of the minimal polynomials of $\alpha_1, \alpha_2, \ldots, \alpha_s$. Suppose that all these zeros lie in $\mathbb{F}_{q^m}$ (which we can represent as a vector space $\mathbb{F}_q^m$). For every i we can consider the m by n matrix with the vector representations of $1, \alpha_i, \alpha_i^2, \ldots, (\alpha_i)^{n-1}$ as columns and put all these together to form the sm by n matrix H which has its entries in $\mathbb{F}_q$. Clearly $\mathbf{c}H^T = \mathbf{0}$, where $\mathbf{c} = (c_0, c_1, \ldots, c_{n-1})$, means the same thing as $c(\alpha_i) = 0$ for $i = 1, 2, \ldots, s$. The rows of H are not necessarily independent. We may obtain a parity check matrix from H by deleting some of the rows. As an illustration of this way of describing cyclic codes, we shall prove that binary (and many other) Hamming codes (cf. (3.3.1)) are (equivalent to) cyclic codes.

(6.3.1) Theorem. *Let $n := (q^m - 1)/(q - 1)$ and let β be a primitive nth root of unity in $\mathbb{F}_{q^m}$. Furthermore, let $(m, q - 1) = 1$. The cyclic code*

$$C := \{c(x) | c(\beta) = 0\}$$

is equivalent to the $[n, n - m]$ Hamming code over $\mathbb{F}_q$.

Proof. Since

$$n = (q - 1)(q^{m-2} + 2q^{m-3} + \cdots + m - 1) + m,$$

we have $(n, q - 1) = (m, q - 1) = 1$. Therefore $\beta^{i(q-1)} \neq 1$ for $i = 1, 2, \ldots, n - 1$, i.e. $\beta^i \notin \mathbb{F}_q$ for $i \doteq 1, 2, \ldots, n - 1$. It follows that the columns of the matrix H, which are the representations of $1, \beta, \beta^2, \ldots, \beta^{n-1}$ as vectors in $\mathbb{F}_q^m$, are pairwise linearly independent over $\mathbb{F}_q$. So H is the parity check matrix of an $[n, n - m]$ Hamming code. □

We illustrate what we have learned so far by constructing binary cyclic codes of length 9.

(6.3.2) Example. The smallest extension field of $\mathbb{F}_2$ which contains a primitive 9th root of unity is $\mathbb{F}_{2^6}$. If α is a primitive element of the field, then $\alpha^{63} = 1$ and $\beta := \alpha^7$ is a primitive 9th root of unity. By Theorem 1.1.22, the minimal polynomial of β has the zeros $\beta, \beta^2, \beta^4, \beta^8, \beta^{16} = \beta^7, \beta^{14} = \beta^5$. This polynomial must be $(x^9 - 1)/(x^3 - 1) = x^6 + x^3 + 1$ (cf. (1.1.28)). So

$$(x^9 - 1) = (x - 1)(x^2 + x + 1)(x^6 + x^3 + 1) = f_1(x)f_2(x)f_3(x).$$

The code M_3^+ has pairwise independent columns in H, i.e. minimum distance ≥ 3. Since M_3^+ clearly consists of the codewords

$$(c_0\, c_1\, c_2 \quad c_0\, c_1\, c_2 \quad c_0\, c_1\, c_2)$$

we immediately see that $d = 3$. The code M_3^- has check polynomial $x^6 + x^3 + 1$, so it is a $[9, 6]$ code. Since $x^3 - 1$ is a codeword, the distance is 2. If we construct $\mathbb{F}_{2^6}$ with $x^6 + x + 1$ and then form the 12 by 9 matrix H for M_3^- in the way described before (6.3.1) it will have six rows of 0s only, the all one row, the row (110 110 110) and four rows (011 011 011). So from this we find a 3 by 9 parity check matrix. Of course from $x^6 + x^3 + 1$ we find a parity check matrix equivalent to (I I I). The reader can work out less trivial examples in a similar way.

(6.3.3) Example. We consider $x^8 - 1$ over $\mathbb{F}_3$. If β is an 8th root of unity in $\mathbb{F}_{3^2}$, then $\beta^9 = \beta$. Therefore $x^8 - 1$ must be the product of $(x - 1)$, $(x + 1)$, and three irreducible polynomials of degree 2. By substituting $x = 0, 1$, or 2 in $x^2 + ax + b$, we see that the only irreducible polynomials of degree 2 in $\mathbb{F}_3[x]$ are $x^2 + 1$, $x^2 + x + 2$, and $x^2 + 2x + 2$. So we know the factorization of $x^8 - 1$. The cyclic code with generator $g(x) := (x^2 + 1)(x^2 + x + 2)$ has minimum distance ≤ 4 since $g(x)$ has weight 4. In §6.6 we demonstrate an easy way to prove that 4 is the minimum distance of this code.

§6.4. The Idempotent of a Cyclic Code

In many applications it turns out to be advantageous to replace the generator polynomial of a cyclic code by a polynomial $c(x)$ called the *idempotent*. The definition is included in the following theorem.

(6.4.1) Theorem. *Let C be a cyclic code. Then there is a unique codeword $c(x)$ which is an identity element for C.*

PROOF. Let $g(x)$ be the generator polynomial of C and $h(x)$ the check polynomial, i.e. $g(x)h(x) = x^n - 1$ in $\mathbb{F}_q[x]$. Since $x^n - 1$ has no multiple zeros we have $(g(x), h(x)) = 1$ and hence there exist polynomials $a(x)$ and $b(x)$ such that $a(x)g(x) + b(x)h(x) = 1$. Now define

$$c(x) := a(x)g(x) = 1 - b(x)h(x).$$

Clearly $c(x)$ is a codeword in C. Furthermore if $p(x)g(x)$ is any codeword in C, then

$$\begin{aligned} c(x)p(x)g(x) &= p(x)g(x) - b(x)h(x)p(x)g(x) \\ &\equiv p(x)g(x) \qquad (\operatorname{mod}(x^n - 1)). \end{aligned}$$

So $c(x)$ is an identity element for C, and hence it is unique. □

Since $c^2(x) = c(x)$, this codeword is called the *idempotent*. Of course, there can be other elements in C that are equal to their squares, but only one of these is an identity for the code. Since every codeword $v(x)$ can be written as $v(x)c(x)$, i.e. as a multiple of $c(x)$, we see that $c(x)$ generates the ideal C.

Let us consider the factorization $x^n - 1 = f_1(x) \ldots f_t(x)$ once more. We now take $q = 2$. From Theorem 1.1.22 we know that these factors correspond to the decomposition of $\{0, 1, \ldots, n - 1\}$ into so-called *cyclotomic cosets*: $\{0\}$, $\{1, 2, 4, \ldots, 2^r\}$, $\ldots$, $\{a, 2a, \ldots, 2^s a\}$, where s is the minimal exponent such that $a(2^{s+1} - 1) \equiv 0 \pmod{n}$. In Example 6.3.2 this decomposition was $\{0\}$, $\{1, 2, 4, 8, 7, 5\}$, $\{3, 6\}$, with $n = 9$. On the other hand, it is obvious that if an idempotent $c(x)$ contains the term x^i, it also contains the term x^{2i}. Therefore an idempotent must be a sum of idempotents of the form $x^a + x^{2a} + \cdots + x^{2^s a}$, where $\{a, 2a, \ldots, 2^s a\}$ is one of the cyclotomic cosets. Since there are exactly 2^t such sums, we see that it is very easy to find all possible idempotents and hence generate all binary cyclic codes of a given length without factoring $x^n - 1$ at all!

We extend the theory a little further. Again we make the restriction $q = 2$. First observe that from the proof of Theorem 6.4.1 it follows that if $c(x)$ is the idempotent of the code C, with generator $g(x)$ and check polynomial $h(x)$, then $1 + c(x)$ is the idempotent of the code with generator $h(x)$. Therefore $1 + x^n c(x^{-1})$ is the idempotent of the dual code.

(6.4.2) Definition. The idempotent of an irreducible cyclic code M_i^- is called a *primitive idempotent* and denoted by $\theta_i(x)$. For example in (6.1.5) the polynomial $(x^2 + 1)g(x) = x^6 + x^5 + x^3 + 1$ is a primitive idempotent.

Let α be a primitive nth root of unity in an extension field of $\mathbb{F}_2$. If the polynomial $c(x)$ is idempotent, then $c(\alpha^i) = 0$ or 1 for all values of i and the converse is clearly also true. If $c(x)$ is a primitive idempotent, then there is an irreducible factor $f(x)$ of $x^n - 1$ such that $c(\alpha^i) = 1$ iff $f(\alpha^i) = 0$, i.e. $c(\alpha^i) = 1$ iff i belongs to one of the cyclotomic cosets $\{a, 2a, \dots\}$. Such a primitive idempotent is often denoted by θ_a, i.e. in (6.4.2) the index i is chosen from the representatives of different cyclotomic cosets. For example, consider $n = 15$ and let α be a zero of $x^4 + x + 1$. Then the primitive idempotent belonging to the minimal cyclic code with check polynomial $x^4 + x + 1$ is denoted by θ_1 and in this case θ_{-1} corresponds to nonzeros $\alpha^{-1}, \alpha^{-2}, \alpha^{-4}, \alpha^{-8}$, i.e. to the check polynomial $x^4 + x^3 + 1$. In the following, if no such α has been fixed, we simply number the irreducible cyclic codes $M_1^-, M_2^-, \dots, M_t^-$.

(6.4.3) Theorem. *If C_1 and C_2 are cyclic codes with idempotents $c_1(x)$ and $c_2(x)$, then:*

(i) *$C_1 \cap C_2$ has idempotent $c_1(x)c_2(x)$;*
(ii) *$C_1 + C_2$, i.e. the set of all words* $\mathbf{a} + \mathbf{b}$ with $\mathbf{a} \in C_1$ *and* $\mathbf{b} \in C_2$, *has idempotent $c_1(x) + c_2(x) + c_1(x)c_2(x)$.*

PROOF.

(i) is a trivial consequence of Theorem 6.4.1;
(ii) follows in the same way since $c_1(x) + c_2(x) + c_1(x)c_2(x)$ is clearly in $C_1 + C_2$ and is again easily seen to be an identity element for this code because all codewords have the form $a(x)c_1(x) + b(x)c_2(x)$. □

(6.4.4) Theorem. *For the primitive idempotents we have:*

(i) *$\theta_i(x)\theta_j(x) = 0$ if $i \neq j$;*
(ii) $\sum_{i=1}^{t} \theta_i(x) = 1$;
(iii) *$1 + \theta_{i_1}(x) + \theta_{i_2}(x) + \cdots + \theta_{i_r}(x)$ is the idempotent of the code with generator $f_{i_1}(x)f_{i_2}(x)\dots f_{i_r}(x)$.*

Proof.

(i) follows from Theorem 6.4.3(i) since $M_i^- \cap M_j^- = \{0\}$;
(ii) follows from Theorem 6.4.3(ii) and Theorem 6.4.4(i) because $M_1^- + M_2^- + \cdots + M_t^-$ is the set of all words of length n; and finally
(iii) is proved by observing that the check polynomial of $M_{i_1}^- + \cdots + M_{i_r}^-$ is $f_{i_1}(x) \ldots f_{i_r}(x)$. □

It is not too difficult to find the primitive idempotents, using these theorems. One then has an easy way of finding the idempotent of a code if the generator is given in the form $f_{i_1}(x) \ldots f_{i_r}(x)$.

In several more advanced topics in coding theory, one finds proof techniques involving idempotents. In this book we shall not reach that stage but nevertheless we wish to show a little more about idempotents. The reader who wishes to study the literature will find the following remarks useful.

Consider a cyclic code C of length n with generator $g(x)$. Let $x^n - 1 = g(x)h(x)$. We consider the formal derivatives of both sides (cf. Section 1.1). We find

$$x^{n-1} = g'(x)h(x) + g(x)h'(x).$$

Here the degree of $g(x)h'(x)$ is $n - 1$ iff the degree of $h(x)$ is odd. Multiply both sides by x and reduce mod $x^n - 1$. We find

$$1 = xg'(x)h(x) + xg(x)h'(x) + (x^n - 1),$$

where the final term cancels the term x^n which occurs in one of the other two polynomials. We see that the idempotent of C is $xg(x)h'(x) + \delta(x^n - 1)$, where $\delta = 1$ if the degree of $h(x)$ is odd, 0 otherwise. As an example consider the minimal code of length 15 with check polynomial $x^4 + x + 1$. The idempotent θ_1 is $xg(x) = x(x^{15} - 1)/(x^4 + x + 1)$.

The following correspondence between idempotents is another useful exercise. The example above can serve as an illustration. Let $f(x)$ be a primitive divisor of $x^n - 1$, where $n = 2^k - 1$. Let α be a primitive element of $\mathbb{F}_{2^k}$ for which $f(\alpha) = 0$. The primitive idempotents θ_1, resp. θ_{-1} correspond to the cyclotomic cosets $\{1, 2, \ldots, 2^{k-1}\}$, resp. $\{-1, -2, \ldots, -2^{k-1}\}$. We claim that

$$\theta_{-1}(x) = \varphi(x) = \sum_{i=0}^{n-1} \operatorname{Tr}(\alpha^i)x^i,$$

where Tr is the trace function (cf. (1.1.29)). In order to show this, we must calculate $\varphi(\alpha^l)$ for $l = 0, 1, \ldots, n - 1$. We have

$$\varphi(\alpha^l) = \sum_{i=0}^{n-1} (\alpha^l)^i \sum_{j=0}^{k-1} (\alpha^i)^{2^j} = \sum_{j=0}^{k-1} \sum_{i=0}^{n-1} (\alpha^{l+2^j})^i.$$

The inner sum is 0 unless $\alpha^{l+2^j} = 1$. Hence $\varphi(\alpha^l) = 1$ if $l = -2^j$ for some value of j and $\varphi(\alpha^l) = 0$ otherwise. This proves the assertion.

Idempotents are used in many places, e.g. to calculate weight enumerators. We do not go into this subject but refer the reader to [42] and [46]. The theory treated in this section, especially Theorem 6.4.4, is a special case of the general theory of idempotents for semi-simple algebras. We refer the reader to [16].

§6.5. Other Representations of Cyclic Codes

There are several other ways of representing cyclic codes than the standard way which was treated in Section 6.1. Sometimes a proof is easier when one of these other representations is used. The first one that we discuss makes use of the trace function (cf. (1.1.29)).

(6.5.1) Theorem. *Let k be the multiplicative order of* p mod n, $q = p^k$, *and let β be a primitive* nth *root of unity in* $\mathbb{F}_q$. *Then the set*

$$V := \{\mathbf{c}(\xi) := (\mathrm{Tr}(\xi), \mathrm{Tr}(\xi\beta), \ldots, \mathrm{Tr}(\xi\beta^{n-1})) | \xi \in \mathbb{F}_q\}$$

is an $[n, k]$ irreducible cyclic code over $\mathbb{F}_p$.

Proof. By Theorem 1.1.30, V is a linear code. Next, observe that $\mathbf{c}(\xi\beta^{-1})$ is a cyclic shift of $\mathbf{c}(\xi)$. Hence V is a cyclic code. Since β is in no subfield of $\mathbb{F}_q$ we know that β is a zero of an irreducible polynomial $h(x) = h_0 + h_1 x + \cdots + h_k x^k$ of degree k. If $\mathbf{c}(\xi) = (c_0, c_1, \ldots, c_{n-1})$ then

$$\sum_{i=0}^{k} c_i h_i = \mathrm{Tr}(\xi h(\beta)) = \mathrm{Tr}(0) = 0,$$

i.e. we have a parity check equation for the code V.

Since $h(x)$ is irreducible, we see that $x^k h(x^{-1})$ is the check polynomial for V and V is therefore an irreducible cyclic $[n, k]$ code. □

We shall now introduce a discrete analog of the Fourier transform, which in coding theory is always referred to as the *Mattson-Solomon polynomial. Let* β be a primitive nth root of unity in the extension field $\mathscr{F}$ of $\mathbb{F}_q$. Let T be the set of polynomials over $\mathscr{F}$ of degree at most $n - 1$. We define $\Phi: T \to T$ as follows. Let $a(x) \in T$. Then $A(X) = (\Phi a)(X)$ is defined by

$$A(X) := \sum_{j=1}^{n} a(\beta^j) X^{n-j}. \tag{6.5.2}$$

If $\mathbf{a} = (a_0, a_1, \ldots, a_{n-1})$, then the polynomial $A(X)$ obtained from $a_0 + a_1 x + \cdots + a_{n-1} x^{n-1}$ is called the Mattson-Solomon polynomial of the vector $\mathbf{a}$.

(6.5.3) Lemma. *The inverse of Φ is given by*

$$a(x) = n^{-1}(\Phi A)(x^{-1}) \qquad (\mathrm{mod}\ x^n - 1).$$

Proof.

$$A(\beta^k) = \sum_{j=1}^{n} \sum_{i=0}^{n-1} a_i \beta^{ij} \beta^{-kj} = \sum_{i=0}^{n-1} a_i \sum_{j=1}^{n} \beta^{(i-k)j} = n a_k.$$ □

Let $\circ$ denote multiplication of polynomials $\bmod(x^n - 1)$ and let $*$ be defined by

$$\left(\sum a_i x^i\right) * \left(\sum b_i x^i\right) := \sum a_i b_i x^i.$$

Then it is easily seen that Φ is an isomorphism of the ring $(T, +, \circ)$ onto the ring $(T, +, *)$.

Now let us use these polynomials to study cyclic codes.

(6.5.4) Lemma. *Let V be a cyclic code over $\mathbb{F}_q$ generated by*

$$g(x) = \prod_{k \in K} (x - \beta^k).$$

Suppose $\{1, 2, \ldots, d-1\} \subset K$ and $\mathbf{a} \in V$. Then the degree of the Mattson-Solomon polynomial A of $\mathbf{a}$ is at most $n - d$.

PROOF. $a(\beta^j) = 0$ for $1 \le j \le d-1$ since $a(x)$ is divisible by $g(x)$. The result follows from (6.5.2). □

(6.5.5) Theorem. *If there are r n-th roots of unity which are zeros of the Mattson-Solomon polynomial A of a word $\mathbf{a}$, then $w(\mathbf{a}) = n - r$.*

PROOF. This is an immediate consequence of Lemma 6.5.3. □

We can also make a link between cyclic codes and the theory of *linear recurring sequences* for which there exists extensive literature (cf. e.g. [61]). A linear recurring sequence with elements in $\mathbb{F}_q$ is defined by an initial sequence $a_0, a_1, \ldots, a_{k-1}$ and a recursion

$$a_l + \sum_{i=1}^{k} b_i a_{l-i} = 0, \qquad (l \ge k). \tag{6.5.6}$$

The standard technique for finding a solution is to try $a_l = \beta^l$. This is a solution of (6.5.6) if β is a zero of $h(x)$, where $h(x) := x^k + \sum_{i=1}^{k} b_i x^{k-i}$. Let us assume that the equation $h(x) = 0$ has k distinct roots $\beta_1, \beta_2, \ldots, \beta_k$ in some extension field of $\mathbb{F}_q$. Then, if $c_1, c_2, \ldots, c_k$ are arbitrary, the sequence $a_l = \sum_{i=1}^{k} c_i \beta_i^l$ is a solution of (6.5.6). We must choose the c_i in such a way that $a_0, a_1, \ldots, a_{k-1}$ have the prescribed values. This amounts to solving a system of k linear equations for which the determinant of coefficients is the Vandermonde determinant

$$\begin{vmatrix} 1 & 1 & \ldots & 1 \\ \beta_1 & \beta_2 & \ldots & \beta_k \\ \beta_1^2 & \beta_2^2 & \ldots & \beta_k^2 \\ \cdots & \cdots & \cdots & \cdots \\ \beta_1^{k-1} & \beta_2^{k-1} & \ldots & \beta_k^{k-1} \end{vmatrix} = \prod_{i>j} (\beta_i - \beta_j) \neq 0. \tag{6.5.7}$$

So we can indeed find the required sequence.

Suppose $h(x)$ is a divisor of $x^n - 1$ (again $(n, q) = 1$). Then the linear recurring sequence is periodic with period a divisor of n. Now consider all partial sequences $(a_0, a_1, \ldots, a_{n-1})$ where $(a_0, a_1, \ldots, a_{k-1})$ runs through $\mathbb{F}_q^k$. We then have an $[n, k]$ cyclic code with $x^k h(x^{-1})$ as check polynomial. So

$$C = \left\{(a_0, \ldots, a_{n-1}) \,\middle|\, a_l = \sum_{i=1}^{k} c_i \beta_i^l (0 \le l < n), (c_1, c_2, \ldots, c_k) \in \mathbb{F}_q^k \right\}$$

is another representation of a cyclic code.

§6.6. BCH Codes

An important class of cyclic codes, still used a lot in practice, was discovered by R. C. Bose and D. K. Ray-Chaudhuri (1960) and independently by A. Hocquenghem (1959). The codes are known as BCH codes.

(6.6.1) Definition. A cyclic code of length n over $\mathbb{F}_q$ is called a *BCH code of designed distance* δ if its generator $g(x)$ is the least common multiple of the minimal polynomials of $\beta^l, \beta^{l+1}, \ldots, \beta^{l+\delta-2}$ for some l, where β is a primitive nth root of unity. Usually we shall take $l = 1$ (sometimes called a *narrow-sense* BCH code). If $n = q^m - 1$, i.e. β is a primitive element of $\mathbb{F}_{q^m}$, then the BCH code is called *primitive*.

The terminology "designed distance" is explained by the following theorem.

(6.6.2) Theorem. *The minimum distance of a* BCH *code with designed distance* d *is at least* d.

First Proof. In the same way as in Section 6.3 we form the $m(d-1)$ by n matrix H:

$$H := \begin{bmatrix} 1 & \beta^l & \beta^{2l} & \cdots & \beta^{(n-1)l} \\ 1 & \beta^{l+1} & \beta^{2(l+1)} & \cdots & \beta^{(n-1)(l+1)} \\ \cdots & \cdots & \cdots & \cdots & \cdots \\ 1 & \beta^{l+d-2} & \beta^{2(l+d-2)} & \cdots & \beta^{(n-1)(l+d-2)} \end{bmatrix}$$

where each entry is interpreted as a column vector of length m over $\mathbb{F}_q$. A word $\mathbf{c}$ is in the BCH code iff $\mathbf{c}H^T = \mathbf{0}$. The $m(d-1)$ rows of H are not necessarily independent. Consider any $d-1$ columns of H and let $\beta^{i_1 l}, \ldots, \beta^{i_{d-1} l}$ be the top elements in these columns. The determinant of the submatrix of H obtained in this way is again a Vandermonde determinant (cf. (6.5.7)) with value $\beta^{(i_1 + \cdots + i_{d-1})l} \prod_{r>s} (\beta^{i_r} - \beta^{i_s}) \neq 0$, since β is a primitive nth root of unity. Therefore any $d-1$ columns of H are linearly independent and hence a codeword $\mathbf{c} \neq \mathbf{0}$ has weight $\geq d$.

SECOND PROOF. W.l.o.g. we take $l = 1$. By Lemma 6.5.4 the degree of the Mattson-Solomon polynomial of a codeword **c** is at most $n - d$. Therefore in Theorem 6.5.5 we have $r \le n - d$, i.e. $w(\mathbf{c}) \ge d$. □

REMARK. Theorem 6.6.2 is usually called the *BCH bound*. From now on we usually consider narrow sense BCH codes. If we start with $l = 0$ instead of $l = 1$ we find the even weight subcode of the narrow sense code.

EXAMPLE. Let $n = 31$, $m = 5$, $q = 2$ and $d = 8$. Let α be a primitive element of $\mathbb{F}_{32}$. The minimal polynomial of α is

$$(x - \alpha)(x - \alpha^2)(x - \alpha^4)(x - \alpha^8)(x - \alpha^{16}).$$

In the same way we find the polynomial $m_3(x)$. But

$$m_5(x) = (x - \alpha^5)(x - \alpha^{10})(x - \alpha^{20})(x - \alpha^9)(x - {}^{18}) = m_9(x).$$

It turns out that $g(x)$ is the least common multiple of $m_1(x)$, $m_3(x)$, $m_5(x)$, $m_7(x)$ and $m_9(x)$. Therefore the minimum distance of the primitive BCH code with designed distance 8 (which was obviously at least 9) is in fact at least 11.

Several generalizations of the BCH bound have been proved. We now describe a method of estimating the minimum distance of a cyclic code. The method is due to J. H. van Lint and R. M. Wilson [76]. Earlier improvements of Theorem 6.6.2 are consequences of the method.

If $A = \{\alpha^{i_1}, \ldots, \alpha^{i_l}\}$ is a set of n-th roots of unity such that for a cyclic code C of length n

$$c(x) \in C \Leftrightarrow \forall_{\xi \in A}[c(\xi) = 0],$$

then we shall say that A is a *defining set* for C. If A is the maximal defining set for C, then we shall call A *complete*.

(6.6.3) Definition. We denote by $M(A)$ or $M(\alpha^{i_1}, \ldots, \alpha^{i_l})$ the matrix of size l by n that has $1, \alpha^{i_k}, \alpha^{2i_k}, \ldots, \alpha^{(n-1)i_k}$ as its kth row; that is

$$M(\alpha^{i_1}, \ldots, \alpha^{i_k}) = \begin{bmatrix} 1 & \alpha^{i_1} & \alpha^{2i_1} & \ldots & \alpha^{(n-1)i_1} \\ 1 & \alpha^{i_2} & \alpha^{2i_2} & \ldots & \alpha^{(n-1)i_2} \\ \cdot & \cdot & \cdot & \ldots & \cdot \\ 1 & \alpha^{i_l} & \alpha^{2i_l} & \ldots & \alpha^{(n-1)i_l} \end{bmatrix}.$$

We refer to $M(A)$ as the parity check matrix corresponding to A. This is the same notation as in Theorem 6.6.2. (Note that over $\mathbb{F}_q$ the matrix $M(A)$ has rows that are not necessarily independent.)

(6.6.4) Definition. A set $A = \{\alpha^{i_1}, \ldots, \alpha^{i_l}\}$ will be called a *consecutive set* of length l if there exists a primitive nth root of unity β and an exponent i such that $A = \{\beta^i, \beta^{i+1}, \ldots, \beta^{i+l-1}\}$.

So, Theorem 6.6.2 states that if a defining set A for a cyclic code contains a consecutive set of length $d-1$, then the minimum distance is at least d. A consequence of our proof of Theorem 6.6.2 is the following lemma.

(6.6.5) Lemma. *If A is a consecutive set of length l, then the submatrix of $M(A)$ obtained by taking any l columns has rank l.*

We shall frequently use the following corollary of this lemma.

(6.6.6) Corollary. *If β is a primitive n-th root of unity and*

$$i_1 < i_2 < \cdots < i_k = i_1 + t - 1,$$

then if we take any t columns of $M(\beta^{i_1}, \beta^{i_2}, \ldots, \beta^{i_k})$, the resulting matrix has rank k.

We now introduce the notation

$$AB := \{\xi\eta \mid \xi \in A, \eta \in B\}.$$

Subsequently, we consider a product operation for matrices that will be applied in the special situation where the matrices are parity check matrices of the form $M(A)$ defined in (6.6.3).

(6.6.7) Definition. The matrix $A * B$ is the matrix that has as its rows all products **ab**, where **a** runs through the rows of A and **b** runs through the rows of B.

The following (nearly trivial) lemma is the basis of the method to be described. We consider matrices A and B with n columns.

(6.6.8) Lemma. *If a linear combination of all the columns of $A * B$ with* non-zero *coefficients is* **0**, *then*

$$rank(A) + rank(B) \leq n.$$

Proof. If the coefficients in the linear combination are λ_j $(j = 1, \ldots, n)$ then multiply column j of B by λ_j $(j = 1, \ldots, n)$. This yields a matrix B' with the same rank as B. The condition of the lemma states that every row of A has inner product 0 with every row of B'. Since this implies that $\text{rank}(A) + \text{rank}(B') \leq n$, we are done. □

Now we are in a position to state the theorem that will enable us to find the minimum distance of a large number of cyclic codes. If **c** is a codeword in a cyclic code, then the *support* I of **c** is the set of coordinate positions i such that $c_i \neq 0$. If A is a matrix, then A_I denotes the submatrix obtained by deleting the positions not in I.

(6.6.9) Theorem. *Let A and B be matrices with entries from a field $\mathbb{F}$. Let $A * B$ be a parity check matrix for the code C over $\mathbb{F}$. If I is the support of a codeword in C, then*

$$rank(A_I) + rank(B_I) \le |I|.$$

PROOF. This is an immediate corollary of Lemma 6.6.8. □

We shall apply this theorem in the following way. Lemma 6.6.5 allows us to say something about the rank of suitable matrices of type A_I, respectively B_I. If the sum of these ranks is $> |I|$ for every subset I of $\{1, 2, \ldots, n\}$ of size $< \delta$, then the code has minimum distance at least δ.

(6.6.10) EXAMPLE. We illustrate the method by proving the so-called *Roos bound.* This bound states that if A is a defining set for a cyclic code with minimum distance d_A and if B is a set of nth roots of unity such that the shortest consecutive set that contains B has length $\le |B| + d_A - 2$, then the code with defining set AB has minimum distance $d \ge |B| + d_A - 1$.

To prove this, we first observe that we have

$$\operatorname{rank}(M(A)_I) = \begin{cases} |I|, & \text{for } |I| < d_A \\ \ge d_A - 1, & \text{for } |I| \ge d_A. \end{cases}$$

Then Corollary 6.6.6 provides us with information on the rank of submatrices of $M(B)$, namely

$$\operatorname{rank}(M(B)_I) \ge \begin{cases} 1, & \text{for } |I| < d_A \\ |I| - d_A + 2, & \text{for } d_A \le |I| \le |B| + d_A - 2. \end{cases}$$

We now apply Theorem 6.6.9. We find from the above that

$$\operatorname{rank}(M(A)_I) + \operatorname{rank}(M(B)_I) > |I| \text{ for } |I| \le |B| + d_A - 2,$$

and hence for these values of $|I|$, the set I cannot be the support of a codeword of a code with defining set AB. (Here we use the fact that the rows of $M(A) * M(B)$ are the same as the rows of $M(AB)$.)

Remark. This bound is due to C. Roos [80]. The special case where B is a consecutive set was proved by C. R. P. Hartmann and K. K. Tzeng in 1972; cf. [33].

EXAMPLE. Consider the cyclic code C of length 35 with generator

$$g(x) = m_1(x)m_5(x)m_7(x).$$

If α is a primitive 35th root of unity, then the defining set of C contains the set $\{\alpha^i | i = 7, 8, 9, 10, 11, 20, 21, 22, 23\}$. This set can be written as AB, where $A = \{\alpha^i | i = 7, 8, 9, 10\}$ and $B = \{\beta^j | j = 0, 3, 4\}$ with $\beta = \alpha^{12}$ (also a primitive 35th root of unity). The set A is the defining set for a cyclic code with minimum distance $d_A = 5$. The set B is contained in a consecutive set of length

5. The condition on $|B|$ of Example 6.6.10 is satisfied. It follows that C has minimum distance at least $3 + 5 - 1 = 7$. This is in fact the minimum distance of this code. Note that the BCH bound only shows that the minimum distance is at least 6.

Before giving one of the nicest examples of our method, we prove a special case of a theorem due to R. J. McEliece.

(6.6.11) Lemma. *Let C be a binary cyclic code of length n with complete defining set R. Suppose that no two nth roots of unity that are not in R have product 1. Then the weight of every codeword in C is divisible by 4.*

PROOF. Clearly $1 \in R$ and for every nth root of unity γ, we have $\gamma \in R$ or $\gamma^{-1} \in R$. Let $c(x) = x^{i_1} + x^{i_2} + \cdots + x^{i_k}$ be a codeword. Since $1 \in R$, k must be even. Since $c(x)c(x^{-1})$ is zero for every nth root of unity, it is the zero polynomial. If $x^{i-j} = x^{l-m}$, then $x^{j-i} = x^{m-l}$, i.e. in the product $c(x)c(x^{-1})$ the terms cancel four at a time. There are k terms equal to 1 and hence $k(k-1) \equiv 0 \pmod 4$, so $4|k$. □

A consequence of Lemma 6.6.11 is that the dual C of the primitive BCH code of length 127 and designed distance 11, has minimum distance divisible by 4. By the BCH bound, C has minimum distance at least 16. By the Roos bound the distance is at least 22, so in fact at least 24 from Lemma 6.6.11. Since the code contains the shortened second order Reed-Muller code $\mathscr{R}(2, 7)$, its minimum distance is at most 32. We shall now show that the method treated above shows that $d \geq 30$, thus showing that in fact $d = 32$.

EXAMPLE. Let R be the defining set of C. Note that R contains the sets $\{\alpha^i | 81 \leq i \leq 95\}, \{\alpha^i | 98 \leq i \leq 111\}, \{\alpha^i | 113 \leq i \leq 127\}$, where α is a primitive 127th root of unity. Let

$$A = \{\alpha^i | 83 \leq i \leq 95\} \cup \{\alpha^i | 98 \leq i \leq 111\}$$

$$B = \{\beta^j | j = -7, 0, 1\}, \qquad \beta = \alpha^{16}.$$

Then $R \supseteqq AB$. The set A contains 14 consecutive powers of α, and furthermore, it is a subset of a set of 29 consecutive powers of α, with the powers α^{96} and α^{97} missing. So, from Lemma 6.6.5 and Corollary 6.6.6 we have

$$\operatorname{rank}(M(A)_I) \geq \begin{cases} |I|, & \text{for } 1 \leq |I| \leq 14 \\ 14, & \text{for } 14 \leq |I| \leq 16 \\ |I| - 2, & \text{for } 17 \leq |I| \leq 29. \end{cases}$$

In the same way we find

$$\operatorname{rank}(M(B)_I) \geq \begin{cases} |I|, & \text{for } 1 \leq |I| \leq 2 \\ 2, & \text{for } 2 \leq |I| \leq 8 \\ 3, & \text{for } |I| \geq 9. \end{cases}$$

By Theorem 6.6.9 a set I with $|I| < 30$ cannot be the support of a codeword in C.

It was shown by Van Lint and Wilson [76] that the method described above gives the exact minimum distance for all binary cyclic codes of length < 63 with only two exceptions.

Finding the actual minimum distance of a BCH code is in general a hard problem. However, something can be said. To illustrate this we shall restrict ourselves to binary primitive BCH codes. We first must prove a lemma. Consider $\mathbb{F}_{2^k}$ as the space $\mathbb{F}_2^k$ and let U be a subspace of dimension l. We define $\sum_i(U) := \sum_{x \in U} x^i$.

(6.6.12) Lemma. *If i has less than l ones in its binary expansion then* $\sum_i(U) = 0$.

PROOF. We use induction. The case $l = 1$ is trivial. Let the assertion be true for some l and let V have dimension $l + 1$ and $V = U \cup (U + b)$, where U has dimension l. Then

$$\sum_i(V) = \sum_i(U) + \sum_{x \in U} (x + b)^i = \sum_{v=0}^{i-1} \binom{i}{v} b^{i-v} \sum_v(U).$$

If the binary expansion of i has at most l ones, then by Theorem 4.5.1 the binomial coefficient $\binom{i}{v}$, where $v < i$, is 0 unless the binary expansion of v has less than l ones, in which case $\sum_v(U)$ is 0 by the induction hypothesis. □

(6.6.13) Theorem. *The primitive binary* BCH *code C of length $n = 2^m - 1$ and designed distance $\delta = 2^l - 1$ has minimum distance δ.*

PROOF. Let U be an l-dimensional subspace of $\mathbb{F}_{2^m}$. Consider a vector $\mathbf{c}$ which has its ones exactly in the positions corresponding to nonzero elements of U, i.e.

$$c(x) = \sum_{j : \alpha^j \in U \setminus \{0\}} x^j.$$

Let $1 \le i < 2^l - 1$. Then the binary expansion of i has less than l ones. Furthermore $c(\alpha^i) = \sum_i(U)$ and hence by Lemma 6.6.12 we have $c(\alpha^i) = 0$ for $1 \le i < 2^l - 1$, i.e. $c(x)$ is a codeword in C. □

(6.6.14) Corollary. *A primitive* BCH *code of designed distance δ has distance $d \le 2\delta - 1$.*

PROOF. In Theorem 6.6.13 take l such that $2^{l-1} \le \delta \le 2^l - 1$. The code of Theorem 6.6.13 is a subcode of the code with designed distance δ. □

Although it is not extremely difficult, it would take us too long to also give reasonable estimates for the actual dimension of a BCH code. In the binary case we have the estimate $2^m - 1 - mt$ if $\delta = 2t + 1$, which is clearly poor for large t although it is accurate for small t (compared to m). We refer the interested reader to [46]. Combining the estimates one can easily show that long primitive BCH codes are bad in the sense of Chapter 5, i.e. if C_ν is a primitive $[n_\nu, k_\nu, d_\nu]$ BCH code for $\nu = 1, 2, \ldots,$ and $n_\nu \to \infty$, then either $k_\nu/n_\nu \to 0$ or $d_\nu/n_\nu \to 0$.

In Section 6.1 we have already pointed out that the automorphism group of a cyclic code of length n over $\mathbb{F}_q$ not only contains the cyclic permutations but also π_q. For BCH codes we can prove much more. Consider a primitive BCH code C of length $n = q^m - 1$ over $\mathbb{F}_q$ with designed distance d (i.e. $\alpha, \alpha^2, \ldots, \alpha^{d-1}$ are the prescribed zeros of the codewords, where α is a primitive element of $\mathbb{F}_{q^m}$).

We denote the *positions* of the symbols in the codewords by X_i $(i = 0, 1, \ldots, n-1)$, where $X_i = \alpha^i$. We extend the code to $\bar{C}$ by adding an overall parity check. We denote the additional position by ∞ and we make the obvious conventions concerning arithmetic with the symbol ∞. We represent the codeword $(c_0, c_1, \ldots, c_\infty)$ by $c_0 + c_1 x + \cdots + c_{n-1}x^{n-1} + c_\infty x^\infty$ and make the further conventions $1^\infty := 1$, $(\alpha^i)^\infty := 0$ for $i \not\equiv 0 \pmod{n}$.

We shall now show that $\bar{C}$ is invariant under the permutations of the *affine permutation group* AGL$(1, q^m)$ acting on the positions (cf. Section 1.1). This group consists of the permutations

$$P_{u,v}(X) := uX + v, \qquad (u \in \mathbb{F}_{q^m},\ v \in \mathbb{F}_{q^m},\ u \neq 0).$$

The group is 2-transitive. First observe that $P_{\alpha,0}$ is the cyclic shift on the positions of C and that it leaves ∞ invariant. Let $(c_0, c_1, \ldots, c_{n-1}, c_\infty) \in \bar{C}$ and let $P_{u,v}$ yield the permuted word $(c_0', c_1', \ldots, c_\infty')$. Then for $0 \le k \le d - 1$ we have

$$\sum_i c_i'\alpha^{ik} = \sum_i c_i(u\alpha^i + v)^k = \sum_i c_i \sum_{l=0}^{k} \binom{k}{l} u^l \alpha^{il} v^{k-l}$$
$$= \sum_{l=0}^{k} \binom{k}{l} u^l v^{k-l} \sum_i c_i (\alpha^l)^i = 0$$

because the inner sum is 0 for $0 \le l \le d - 1$ since $\mathbf{c} \in \bar{C}$. So we have the following theorem.

(6.6.15) Theorem. *Every extended primitive* BCH *code of length* $n + 1 = q^m$ *over* $\mathbb{F}_q$ *has* AGL$(1, q^m)$ *as a group of automorphisms.*

(6.6.16) Corollary. *The minimum weight of a primitive binary* BCH *code is odd.*

PROOF. Let C be such a code. We have shown that Aut$(\bar{C})$ is transitive on the positions. The same is true if we consider only the words of minimum weight in $\bar{C}$. So $\bar{C}$ has words of minimum weight with a 1 in the final check position. □

§6.7. Decoding BCH Codes

Once again consider a BCH code of length n over $\mathbb{F}_q$ with designed distance $\delta = 2t + 1$ and let β be a primitive nth root of unity in $\mathbb{F}_{q^m}$. We consider a codeword $C(x)$ and assume that the received word is

$$R(x) = R_0 + R_1 x + \cdots + R_{n-1}x^{n-1}.$$

Let $E(x) := R(x) - C(x) = E_0 + E_1 x + \cdots + E_{n-1}x^{n-1}$ be the error vector. We define:

$M := \{i | E_i \neq 0\}$, the positions where an error occurs,

$e := |M|$, the number of errors,

$\sigma(z) := \prod_{i \in M} (1 - \beta^i z)$, which we call the *error-locator polynomial*,

$\omega(z) := \sum_{i \in M} E_i \beta^i z \prod_{j \in M\setminus\{i\}} (1 - \beta^j z)$.

It is clear that if we can find $\sigma(z)$ and $\omega(z)$, then the errors can be corrected. In fact an error occurs in position i iff $\sigma(\beta^{-i}) = 0$ and in that case the error is $E_i = -\omega(\beta^{-i})\beta^i/\sigma'(\beta^{-i})$. From now on we assume that $e \leq t$ (if $e > t$ we do not expect to be able to correct the errors). Observe that

$$\frac{\omega(z)}{\sigma(z)} = \sum_{i \in M} \frac{E_i \beta^i z}{1 - \beta^i z} = \sum_{i \in M} E_i \sum_{l=1}^{\infty} (\beta^i z)^l$$

$$= \sum_{l=1}^{\infty} z^l \sum_{i \in M} E_i \beta^{li} = \sum_{l=1}^{\infty} z^l E(\beta^l),$$

where all calculations are with formal power series over $\mathbb{F}_{q^m}$. For $1 \leq l \leq 2t$ we have $E(\beta^l) = R(\beta^l)$, i.e. the receiver knows the first $2t$ coefficients on the right-hand side. Therefore $\omega(z)/\sigma(z)$ is known mod z^{2t+1}. We claim that the receiver must determine polynomials $\sigma(z)$ and $\omega(z)$ such that degree $\omega(z) \leq$ degree $\sigma(z)$ and degree $\sigma(z)$ is as small as possible under the condition

$$\frac{\omega(z)}{\sigma(z)} \equiv \sum_{l=1}^{2t} z^l R(\beta^l) \pmod{z^{2t+1}}. \tag{6.7.1}$$

Let $S_l := R(\beta^l)$ for $l = 1, \ldots, 2l$ and let $\sigma(z) = \sum_{i=0}^{e} \sigma_i z^i$. Then

$$\omega(z) \equiv \left(\sum_{l=1}^{2t} S_l z^l\right)\left(\sum_{i=0}^{e} \sigma_i z^i\right) = \sum_k z^k \left(\sum_{i+l=k} S_l \sigma_i\right) \pmod{z^{2t+1}}.$$

Because $\omega(z)$ has degree $\leq e$ we have

$$\sum_{i+l=k} S_l \sigma_i = 0, \qquad \text{for } e + 1 \leq k \leq 2t.$$

This is a system of $2t - e$ linear equations for the unknowns $\sigma_1, \ldots, \sigma_e$ (we know that $\sigma_0 = 1$). Let $\tilde{\sigma}(z) = \sum_{i=0}^{e} \tilde{\sigma}_i z^i$ (where $\tilde{\sigma}_0 = 1$) be the polynomial of lowest degree found by solving these equations (we know there is at least the solution $\sigma(z)$). For $e + 1 \le k \le 2t$ we have

$$0 = \sum_l S_{k-l}\tilde{\sigma}_l = \sum_{i \in M} \sum_l E_i \beta^{(k-l)i} \tilde{\sigma}_l = \sum_{i \in M} E_i \beta^{ik} \tilde{\sigma}(\beta^{-i}).$$

We can interpret the right-hand side as a system of linear equations for $E_i\tilde{\sigma}(\beta^{-i})$ with coefficients β^{ik}. So the determinant of coefficients is again Vandermonde, hence $\neq 0$. So $E_i\tilde{\sigma}(\beta^{-i}) = 0$ for $i \in M$. Since $E_i \neq 0$ for $i \in M$ we see that $\sigma(z)$ divides $\tilde{\sigma}(z)$, i.e. $\tilde{\sigma}(z) = \sigma(z)$. So indeed, the solution $\tilde{\sigma}(z)$ of lowest degree solves our problem and we have seen that finding it amounts to solving a system of linear equations. The advantage of this approach is that the decoder has an algorithm that does not depend on e. Of course, in practice it is even more important to find a fast algorithm that actually does what we have only considered from a theoretical point of view. Such an algorithm (with implementation) was designed by E. R. Berlekamp (cf. [2], [24]) and is often referred to as the *Berlekamp-decoder*.

If we call the (known) polynomial on the right hand side of (6.7.1) $S(z)$ and define $G(z) := z^{2t+1}$, then (6.7.1) reads

(6.7.1) $$S(z)\sigma(z) \equiv \omega(z) \pmod{G(z)}.$$

We need to find a solution of this congruence with σ of degree $\le t$ and ω of degree smaller than the degree of σ. The (unique) solution makes the error correction possible. In §9.5 we encounter the same congruence.

§6.8. Reed–Solomon Codes

One of the simplest examples of BCH codes, namely the case $n = q - 1$, turns out to have many important applications.

(6.8.1) Definition. A *Reed-Solomon code* (RS code) is a primitive BCH code of length $n = q - 1$ over $\mathbb{F}_q$. The generator of such a code has the form $g(x) = \prod_{i=1}^{d-1}(x - \alpha^i)$ where α is primitive in $\mathbb{F}_q$.

By the BCH bound (6.6.2) the minimum distance of an RS code with this generator $g(x)$ is at least d. By Section 6.2 this code has dimension $k = n - d + 1$. Therefore Corollary 5.2.2 implies that the minimum distance is d and the RS code is a maximum distance separable code.

Suppose we need a code for a channel that does not have random errors (like the B.S.C.) but instead has errors occurring in *bursts* (i.e. several errors close together). This happens quite often in practice (telecommunication, magnetic tapes, compact disc). For such a channel, RS codes are often used. We illustrate this briefly. Suppose binary information is taken in strings of m symbols which are interpreted as elements of $\mathbb{F}_{2^m}$. If these are encoded using RS code, then a burst of several errors (in the 0s and 1s) will influence only a few consecutive symbols in a codeword of the RS code. Of course this idea can be used for any code but since the RS codes are MDS they are particularly useful. A more important application will occur in Section 11.2. In connection with this application we mention the original approach of Reed and Solomon. Let $n = q-1$, α a primitive element of $\mathbb{F}_q$. As usual, identify $\mathbf{a} = (a_0, a_1, \ldots, a_{k-1}) \in \mathbb{F}_q^k$ with $a_0 + a_1x + \ldots + a_{k-1}x^{k-1} = a(x)$. Then

$$C = \{(c_0, c_1, \ldots, c_{n-1}) | c_i = a(\alpha^i), 0 \le i < n, \mathbf{a} \in \mathbb{F}_q^k\}$$

is the RS code with $d = n - k + 1$. To see this first observe that C is obviously cyclic. The definition of C and Lemma 6.5.3 imply that a codeword $\mathbf{c}$ has $na(x)$ as its Mattson-Solomon polynomial. Since the degree of $a(x)$ is $\le k-1$ this means that $c(\alpha^i) = 0$ for $i = 1, 2, \ldots, n-k$. Hence C is an RS code. This representation gives a very efficient encoding procedure for RS codes even though it is not systematic.

If we extend the words of C by adjoining a symbol $c_n = a(0)$, then $\sum_{i=0}^{n} c_i = a_0 q = 0$. So, we indeed obtain $\overline{C}$. If $c_n = 0$, i.e. $c(1) = 0$, then the word has weight $\ge n - k + 2$ and clearly this is also true if $c_n \ne 0$. So, the code $\overline{C}$ is also an MDS code.

The second representation of Reed-Solomon codes allows us to generalize the idea. We now consider $\mathbb{F}_{q^m}$ as alphabet and choose n distinct elements from the field, say $\alpha_1, \alpha_2, \ldots, \alpha_n$. Let $\mathbf{v} = (v_1, v_2, \ldots, v_n)$ be a vector from $\mathbb{F}_{q^m}^n$ with no zero coordinates and write $\mathbf{a} := (\alpha_1, \alpha_2, \ldots, \alpha_n)$.

(6.8.2) Definition. The *generalized Reed-Solomon code* $\mathrm{GRS}_k(\mathbf{a}, \mathbf{v})$ has as codewords all $(v_1 f(\alpha_1), v_2 f(\alpha_2), \ldots, v_n f(\alpha_n))$, where f runs through the set of polynomials of degree less than k in $\mathbb{F}_{q^m}[x]$.

In the same way as above, we see that a generalized Reed-Solomon code and its dual are MDS codes.

Our second description of RS codes also allows us to give a rough idea of the codes that are defined using algebraic geometry. In (1.3.4) we saw that the projective line of order q can be described by giving the points coordinates (x, y), where (x, y) and (cx, cy) are the same point ($c \in \mathbb{F}_q$). If $a(x, y)$ and $b(x, y)$ are homogeneous polynomials of the same degree, then it makes sense to study the rational function $a(x, y)/b(x, y)$ on the projective line (since a change of coordinates does not change the value of the fraction). We pick the point $Q := (1, 0)$ as a special point on the line. The remaining points have as coordinates $(0, 1)$ and $(\alpha^i, 1)$, $(0 \le i < q-1)$, where α again denotes a primitive element of $\mathbb{F}_q$. We now consider those rational functions $a(x, y)/y^l$ for which $l < k$ (and of course $a(x, y)$ is homogeneous of degree l). This is a vector space (say K) of dimension k and one immediately sees that the description of RS codes given above amounts to numbering the points of the line in some fixed order (say $P_0, P_1, \ldots, P_{q-1}$), and taking as codewords $(f(P_0), \ldots, f(P_{q-1}))$, where f runs through the space K. The functions have been chosen in such a way that we can indeed calculate their values in all the points P_i; this is not so for Q. In terms of analysis, the point Q is a *pole* of order at most $k-1$ for the functions in K.

The simplest examples of algebraic geometry codes generalize this construction by replacing the projective line by a projective curve in some projective space. We treat algebraic geometry codes in Chapter 10.

We now look at MDS codes in general. If C is an $[n, k]$ code with minimum distance $d = n-k+1$, then C is systematic on any k positions (cf. Problem 3.8.2).

(6.8.3) Theorem. *The dual of an MDS code is also an MDS code.*

PROOF. Let $G = (I_k \quad P)$ be the generator matrix of C. Since C has minimum weight d, every set of $d-1 = n-k$ columns of the parity check matrix $H := (-P^\top \quad I_{n-k})$ is linearly independent. Hence every square submatrix of H is nonsingular, i. e. no codeword of $C^\perp$ has $n-k$ zeros. So $C^\perp$ is an $[n, n-k, k+1]$ code, i. e. MDS. □

Let C be a $[n, k, d]$ code with $d = n-k+1$. If we consider a set of d positions and then look at the subcode of C with zeros in all other positions, this subcode has dimension $\geq k-(n-d) = 1$. Since this subcode has minimum distance d, it must have dimension exactly 1. It follows that for $n \geq d' > d$, specifying a set of d' positions and requiring the codewords to be zero in all other positions, will define a subcode of C with dimension $d' - d + 1$. We formulate this result as a lemma.

(6.8.4) Lemma. *For $n \geq d' \geq d = n-k+1$, the subcode of an MDS code with parameters n, k, d, consisting of those codewords that have zeros outside a set of d' positions has dimension $d' - d + 1$.*

We shall use this lemma and an analog of the Möbius inversion formula (1.1.4) to find the weight enumerator of an MDS code. We first prove an analog of the Möbius inversion formula.

(6.8.5) Definition. If N is a finite set and $S \subset T \subset N$, then we define

$$\mu(S, T) := (-1)^{|T|-|S|}.$$

(6.8.6) Theorem. *Let N be a finite set and let f be a function defined on the subsets of N. If*

$$g(S) := \sum_{R \subset S} f(R),$$

then

$$f(T) = \sum_{S \subset T} \mu(S, T) g(S).$$

PROOF.

$$\begin{aligned}\sum_{S\subset T}\mu(S,T)g(S) &= \sum_{S\subset T}\mu(S,T)\sum_{R\subset S}f(R)\\ &= \sum_{R\subset T}f(R)\sum_{R\subset S\subset T}\mu(S,T),\end{aligned}$$

and the result then follows from the equality

$$\sum_{R\subset S\subset T}\mu(S,T)=\sum_{j=0}^{|T|-|R|}\binom{|T|-|R|}{j}(-1)^j=(1-1)^{|T|-|R|}=\begin{cases}0, & \text{if } R\neq T\\ 1, & \text{if } R=T.\end{cases}$$

□

We now show that the weight enumerator of an MDS code is determined by its parameters.

(6.8.7) Theorem. *Let C be an $[n,k]$ code with distance $d=n-k+1$. If the weight enumerator of C is $1+\sum_{i=d}^{n}A_iz^i$, then*

$$A_i=\binom{n}{i}(q-1)\sum_{j=0}^{i-d}(-1)^j\binom{i-1}{j}q^{i-j-d}\qquad (i=d,d+1,\dots,n).$$

PROOF. If R is a subset of $N:=\{0,1,\dots,n-1\}$, define $f(R)$ to be the number of codewords $(c_0,c_1,\dots,c_{n-1})$ for which $c_i\neq 0\Leftrightarrow i\in R$. If we define g as in Theorem 6.8.6, then we have by Lemma 6.8.4

$$g(S)=\begin{cases}1, & \text{if } |S|\le d-1\\ q^{|S|-d+1}, & \text{if } n\ge |S|\ge d.\end{cases}$$

By our definition of f, we have $A_i=\sum_{R\subset N,|R|=i}f(R)$ and therefore application of Theorem 6.8.6 yields

$$\begin{aligned}A_i &= \sum_{R\subset N,|R|=i}\sum_{S\subset R}\mu(S,R)g(S)\\ &= \binom{n}{i}\left\{\sum_{j=0}^{d-1}\binom{i}{j}(-1)^{i-j}+\sum_{j=d}^{i}\binom{i}{j}(-1)^{i-j}q^{j-d+1}\right\}\\ &= \binom{n}{i}\sum_{j=d}^{i}\binom{i}{j}(-1)^{i-j}(q^{j-d+1}-1).\end{aligned}$$

The result now follows if we replace j by $i-j$ and then use $\binom{i}{j}=\binom{i-1}{j-1}+\binom{i-1}{j}$.

□

Theorem 6.8.7 gives the following restriction on the size of the alphabet of an MDS code.

(6.8.8) Theorem. *If there exists an MDS code over $\mathbb{F}_q$ with length n and dimension k, then $q\ge n-k+1$ or $k\le 1$.*

Proof. Let $d = n - k + 1$. By Theorem 6.8.7 we have for $d < n$, $0 \le A_{d+1} = \binom{n}{d+1}(q-1)(q-d)$. □

Since the dual of an MDS code is also MDS (Theorem 6.8.3), we find the following corollary.

(6.8.9) Corollary. *If there exists an MDS code over $\mathbb{F}_q$ with length n and dimension k, then $q \ge k+1$ or $d = n-k+1 \le 2$.*

§6.9. Quadratic Residue Codes

In this section we shall consider codes for which the word length n is an odd prime. The alphabet $\mathbb{F}_q$ must satisfy the condition: q is a quadratic residue (mod n), i.e. $q^{(n-1)/2} \equiv 1 \pmod{n}$. As usual α will denote a primitive nth root of unity in an extension field of $\mathbb{F}_q$. Later it will turn out that we shall require that α satisfies one more condition. We define

$$R_0 := \{i^2 \pmod{n} \mid i \in \mathbb{F}_n, i \neq 0\}, \qquad \text{the quadratic residues in } \mathbb{F}_n,$$

$$R_1 := \mathbb{F}_n^* \backslash R_0, \qquad \text{i.e. the set of nonsquares in } \mathbb{F}_n,$$

$$g_0(x) := \prod_{r \in R_0} (x - \alpha^r), \qquad g_1(x) := \prod_{r \in R_1} (x - \alpha^r).$$

Since we have required that $q \pmod{n}$ is in R_0, the polynomials $g_0(x)$ and $g_1(x)$ both have coefficients in $\mathbb{F}_q$ (cf. Theorem 1.1.22). Furthermore

$$x^n - 1 = (x-1)g_0(x)g_1(x).$$

(6.9.1) Definition. The cyclic codes of length n over $\mathbb{F}_q$ with generators $g_0(x)$ resp. $(x-1)g_0(x)$ are both called *quadratic residue codes* (QR codes).

We shall only consider extended QR codes in the binary case, where the definition is as in (3.2.7). Such a code is obtained by adding an overall parity check to the code with generator $g_0(x)$.

For other fields the definition of extended code is usually modified in such a way that the extended code is self-dual if $n \equiv -1 \pmod 4$, resp. dual to the extension of the code with generator $g_1(x)$ if $n \equiv 1 \pmod 4$ (cf. [46]). In the binary case the code with generator $(x-1)g_0(x)$ is the even-weight subcode of the other QR code. If G is a generator matrix for the first of these codes then we obtain a generator matrix for the latter code by adding a row of 1s to G. If we do the same thing after adding a column of 0s to G we obtain a generator matrix for the extended code.

In the binary case the condition that q is a quadratic residue mod n simply means that $n \equiv \pm 1 \pmod 8$ (cf. Section 1.1). The permutation π_j: $i \mapsto ij$ (mod n) acting on the positions of the codewords maps the code with generator $g_0(x)$ into itself if $j \in R_0$ resp. into the code with generator $g_1(x)$ if $j \in R_1$. So the codes with generators $g_0(x)$ resp. $g_1(x)$ are equivalent. If $n \equiv -1 \pmod 4$ then $-1 \in R_1$ and in that case the transformation $x \to x^{-1}$ maps a codeword of the code with generator $g_0(x)$ into a codeword of the code with generator $g_1(x)$.

(6.9.2) Theorem. *If* $\mathbf{c} = c(x)$ *is a codeword in the* QR *code with generator* $g_0(x)$ *and if* $c(1) \neq 0$ *and* $w(\mathbf{c}) = d$, *then*

(i) $d^2 \geq n$,
(ii) *if* $n \equiv -1 \pmod 4$ *then* $d^2 - d + 1 \geq n$,
(iii) *if* $n \equiv -1 \pmod 8$ *and* $q = 2$ *then* $d \equiv 3 \pmod 4$.

PROOF.

(i) Since $c(1) \neq 0$ the polynomial $c(x)$ is not divisible by $(x - 1)$. By a suitable permutation π_j we can transform $c(x)$ into a polynomial $\hat{c}(x)$ which is divisible by $g_1(x)$ and of course again not divisible by $(x - 1)$. This implies that $c(x)\hat{c}(x)$ is a multiple of $1 + x + x^2 + \cdots + x^{n-1}$. Since the polynomial $c(x)\hat{c}(x)$ has at most d^2 nonzero coefficients we have proved the first assertion.

(ii) In the proof above we may take $j = -1$. In that case it is clear that $c(x)\hat{c}(x)$ has at most $d^2 - d + 1$ nonzero coefficients.

(iii) Let $c(x) = \sum_{i=1}^{d} x^{l_i}$, $\hat{c}(x) = \sum_{i=1}^{d} x^{-l_i}$. If $l_i - l_j = l_k - l_l$ then $l_j - l_i = l_l - l_k$. Hence, if terms in the product $c(x)\hat{c}(x)$ cancel then they cancel four at a time. Therefore $n = d^2 - d + 1 - 4a$ for some $a \geq 0$. □

The idempotent of a cyclic code, introduced in Section 6.4, will prove to be a powerful tool in the analysis of QR codes.

(6.9.3) Theorem. *For a suitable choice of the primitive n-th root of unity* α, *the polynomial*

$$\theta(x) := \sum_{r \in R_0} x^r$$

is the idempotent of the binary QR *code with generator* $(x - 1)g_0(x)$ *if* $n \equiv 1 \pmod 8$ *resp. the* QR *code with generator* $g_0(x)$ *if* $n \equiv -1 \pmod 8$.

PROOF. $\theta(x)$ is obviously an idempotent polynomial. Therefore $\{\theta(\alpha)\}^2 = \theta(\alpha)$, i.e. $\theta(\alpha) = 0$ or 1. By the same argument $\theta(\alpha^i) = \theta(\alpha)$ if $i \in R_0$ and

$$\theta(\alpha^i) + \theta(\alpha) = 1$$

if $i \in R_1$. The "suitable choice" of α is such that $\theta(\alpha) = 0$. (The reader should convince himself that it is impossible that all primitive elements of $\mathbb{F}_q$ satisfy $\theta(\alpha) = 1$.) Our choice implies that $\theta(\alpha^i) = 0$ if $i \in R_0$ and $\theta(\alpha^i) = 1$ if $i \in R_1$. Finally we have $\theta(\alpha^0) = (n - 1)/2$. This proves the assertion. □

With the aid of $\boldsymbol{\theta}$ we now make a (0, 1)-matrix C (called a circulant) by taking the word $\boldsymbol{\theta}$ as the first row and all cyclic shifts as the other rows. Let $\mathbf{c} := (00\ldots0)$ if $n \equiv 1 \pmod 8$ and $\mathbf{c} := (11\ldots1)$ if $n \equiv -1 \pmod 8$. We have

$$G := \begin{bmatrix} 1 & 1 & \dots & 1 \\ \mathbf{c}^T & & C & \end{bmatrix}.$$

It follows from Theorem 6.9.3 that the rows of G (which are clearly not independent) generate the extended binary QR code of length $n + 1$. We now number the coordinate places of codewords in this code with the points of the projective line of order n, i.e. $\infty, 0, 1, \dots, n - 1$. The overall parity check is in front and it has number ∞. We make the usual conventions about arithmetic operations involving ∞. The group PSL(2, n) consists of all transformations $x \to (ax + b)/(cx + d)$ with a, b, c, d in $\mathbb{F}_n$ and $ad - bc = 1$. It is not difficult to check that this group is generated by the transformations $S: x \to x + 1$ and $T: x \to -x^{-1}$. Clearly S is a cyclic shift on the positions different from ∞ and it leaves ∞ invariant. By the definition of a QR code, S leaves the extended code invariant. To check the effect of T on the extended QR code it is sufficient to analyse what T does to the rows of G. It is a simple (maybe somewhat tedious) exercise to show that T maps a row of G into a linear combination of at most three rows of G (the reader who does not succeed is referred to [42]). Therefore both S and T leave the extended QR code invariant, proving the following theorem.

(6.9.4) Theorem. *The automorphism group of the extended binary* QR *code of length $n + 1$ contains* PSL(2, n).

The modified definition of extended code which we mentioned earlier ensures that Theorem 6.9.4 is also true for the nonbinary case (cf. [46]).

(6.9.5) Corollary. *A word of minimum weight in a binary* QR *code satisfies the conditions of Theorem* 6.9.2.

PROOF. The proof is the same as for Corollary 6.6.16. In this case we use the fact that PSL(2, n) is transitive. Therefore the minimum weight is odd. □

EXAMPLES. (a) Let $q = 2$, $n = 7$. We find

$$x^7 - 1 = (x - 1)(x^3 + x + 1)(x^3 + x^2 + 1).$$

We take $g_0(x)$ as generator. The choice of α specified in Theorem 6.9.3 implies that $x + x^2 + x^4$ is also a generator. Hence $g_0(x) = 1 + x + x^3$. Of course this code is the (perfect) [7, 4] Hamming code (see Section 3.3 and Theorem 6.31). The corresponding even weight subcode was treated in (6.1.5).

(b) Let $q = 2$, $n = 23$. We have

$$x^{23} - 1 = (x - 1)(x^{11} + x^9 + x^7 + x^6 + x^5 + x + 1)$$
$$\times (x^{11} + x^{10} + x^6 + x^5 + x^4 + x^2 + 1).$$

Again we take $g_0(x)$ to be the multiple of $\theta(x)$, which is

$$x^{11} + x^9 + x^7 + x^6 + x^5 + x + 1.$$

By Corollary 6.9.5 the corresponding QR code C has minimum distance $d \geq 7$.

Since $\sum_{i=0}^{3} \binom{23}{i} = 2^{11}$ and $|C| = 2^{12}$ it follows that d is equal to 7 and by (3.1.6) C is a perfect code. Since the binary Golay code of Section 4.2 is unique we have now shown that it is in fact a QR code.

We leave several other examples as exercises (Section 6.13).

§6.10. Binary Cyclic Codes of Length $2n$ (n odd)

Let n be odd and $x^n - 1 = f_1(x)f_2(x)\dots f_t(x)$ the factorization of $x^n - 1$ into irreducible factors in $\mathbb{F}_2[x]$.

We define $g_1(x) := f_1(x)\dots f_k(x)$, $g_2(x) := f_{k+1}(x)\dots f_l(x)$, where $k < l < t$. Let $r_1 := \deg g_1$, $r_2 := \deg g_1 g_2$.

Let C_1 be the cyclic code of length n and dimension $n - r_1$ with generator $g_1(x)$, and let C_2 be the cyclic code of length n and dimension $n - r_2$ with generator $g_1(x)g_2(x)$, and let d_i be the minimum distance of C_i ($i = 1, 2$). Clearly $d_2 \geq d_1$.

We shall study the cyclic code C of length $2n$ and dimension $2n - r_1 - r_2$ with generator $g(x) := g_1^2(x)g_2(x)$. We claim that this code has the following structure:

Let $\mathbf{a} = (a_0, a_1, \dots, a_{n-1}) \in C_1$ and $\mathbf{c} = (c_0, c_1, \dots, c_{n-1}) \in C_2$. Define $\mathbf{b} := \mathbf{a} + \mathbf{c}$. Since n is odd, we can define words that belong to C by

$$\mathbf{w} := (a_0, b_1, a_2, \dots, b_{n-2}, a_{n-1}, b_0, a_1, \dots, a_{n-2}, b_{n-1})$$

and in this way we find *all* words of C; (the final assertion follows from dimension arguments). To demonstrate this, we proceed as follows. Write

$$\begin{aligned} a(x) &= a_0 + a_1 x + \cdots + a_{n-1}x^{n-1} \\ &= (a_0 + a_2 x^2 + \cdots + a_{n-1}x^{n-1}) + x(a_1 + \cdots + a_{n-2}x^{n-3}) \\ &= a_e(x^2) + x a_o(x^2), \end{aligned}$$

and analogously for $c(x)$ and $b(x)$. We then have the following two (equal) representations for the polynomial $w(x)$ corresponding to the codeword $\mathbf{w}$:

$$w(x) = \{a_e(x^2) + x^{n+1}a_o(x^2)\} + \{x b_o(x^2) + x^n b_e(x^2)\} \tag{6.10.1}$$

and

$$w(x) = \{a(x) + x(x^n + 1)a_o(x^2)\} + \{b(x) + (x^n + 1)b_e(x^2)\}. \tag{6.10.2}$$

Both terms in (6.10.2) are divisible by $g_1(x)$. From (6.10.1) we see that the first term only contains even powers of x, the second one only odd powers of x. Since $g_1(x)$ has no multiple factors, this implies that both terms are actually divisible by $g_1^2(x)$.

From (6.10.2) we find

$$w(x) = (x^n + 1)a(x) + c(x) + (x^n + 1)c_e(x^2)$$

in which every term is divisible by $g_2(x)$.

Since $\mathbf{b} = \mathbf{a} + \mathbf{c}$, the word $\mathbf{w}$ is a permutation of the word $|\mathbf{a}|\mathbf{a} + \mathbf{c}|$, (cf. 4.4.1). We have proved the following theorem.

(6.10.3) Theorem. *Let C_1 be a binary cyclic code of length n (odd) with generator $g_1(x)$, and let C_2 be a binary cyclic code of length n with generator $g_1(x)g_2(x)$. Then the binary cyclic code C of length $2n$ with generator $g_1^2(x)g_2(x)$ is equivalent to the $|\mathbf{u}|\mathbf{u} + \mathbf{v}|$ sum of C_1 and C_2. Therefore C has minimum distance* $\min\{2d_1, d_2\}$.

There are not many good binary cyclic codes of even length. However, the following theorem shows the existence of a class of optimal examples.

(6.10.4) Theorem. *The even weight subcode of a shortened binary Hamming code is cyclic (for a suitable ordering of the symbols).*

PROOF. It is not difficult to see that it makes no difference on which position the code is shortened (all resulting codes are equivalent). Let $n = 2^s - 1$. Let $m_1(x)$ denote the minimal polynomial of a primitive element α of $\mathbb{F}_{2^s}$. Then $m_1(x)$ is the generator polynomial of the $[n, n - s]$ binary Hamming code and $(x + 1)m_1(x)$ is the generator polynomial of the corresponding even weight subcode. In Theorem 6.10.3 we take $g_1(x) = (x + 1)$ and $g_2(x) = m_1(x)$. We then find a cyclic code C of length $2n$, dimension $2n - s - 2$, with minimum distance 4. It follows from the $|\mathbf{u}|\mathbf{u} + \mathbf{v}|$ construction that all weights in C are even. Therefore C has a parity check matrix with a top row of 1's and all columns distinct. Hence C is equivalent to the even weight subcode of a shortened Hamming code. □

We observe that there is a different way of proving the previous theorem. We shall use the Hasse derivative (see Chapter 1). The generator of C has 1 as a zero with multiplicity 2 and α as a zero with multiplicity 1. This means that if $c(x) = \sum c_i x^i$ is a codeword, then

$$\sum c_i = 0, \qquad \sum ic_i = 0, \qquad \sum c_i\alpha^i = 0,$$

i.e.

$$H_1 := \begin{bmatrix} 1 & 1 & 1 & \cdots & 1 & 1 & 1 & \cdots & 1 & 1 \\ 0 & 1 & 0 & \cdots & 0 & 1 & 0 & \cdots & 0 & 1 \\ 1 & \alpha & \alpha^2 & \cdots & \alpha^{n-1} & \alpha^n & \alpha^{n+1} & \cdots & \alpha^{2n-2} & \alpha^{2n-1} \end{bmatrix}$$

is a parity check matrix for C; here the second row is obtained by using the property of the Hasse derivative and multiple zeros. Note that $\alpha^n = 1$. Hence the matrix H_1 consists of all possible columns with a 1 at the top, except for $(1000\ldots0)^{\mathrm{T}}$ and $(1100\ldots0)^{\mathrm{T}}$, i.e. the code is indeed equivalent to the even weight subcode of a shortened Hamming code.

§6.11. Generalized Reed-Muller Codes

We shall define a class of (extended) cyclic codes over $\mathbb{F}_q$ that are equivalent to Reed-Muller codes in the case $q = 2$. First, we generalize the idea of Hamming weight to integers written in the q-ary number system.

(6.11.1) Definition. If q is an integer ≥ 2 and $j = \sum_{i=0}^{m-1} \xi_i q^i$, with $0 \leq \xi_i < q$ for $i = 0, 1, \ldots, m-1$, then we define $w_q(j) := \sum_{i=0}^{m-1} \xi_i$.

Note that the sum is taken in $\mathbb{Z}$. The new class of codes is defined as follows.

(6.11.2) Definition. The *shortened r th order generalized Reed-Muller code* (GRM code) of length $n = q^m - 1$ over $\mathbb{F}_q$ is the cyclic code with generator

$$g(x) := \prod^{(r)} (x - \alpha^j),$$

where α is a primitive element in $\mathbb{F}_{q^m}$ and the upper index (r) indicates that the product is over integers j with $0 \leq j < q^m - 1$ and $0 \leq w_q(j) < (q-1)m - r$.

The *r-th order GRM code* of length q^m has a generator matrix G^* obtained from the generator matrix G of the shortened GRM code by adjoining a column of 0s and then a row of 1s.

Note that the set of exponents in this definition of shortened GRM codes is indeed closed under multiplication by q. Let $h(x)$ be the check polynomial of the shortened r th order GRM code. Then the dual of this code has the polynomial $h^*(x)$ as generator, where $h^*(x)$ is obtained from $h(x)$ by reversing the order of the powers of x. It is defined in the same way as $g(x)$, now with the condition $0 < w_q(j) \leq r$.

We have the following generalization of Theorem 4.5.8.

(6.11.3) Theorem. *The dual of the r-th order GRM code of length q^m is equivalent to a GRM code of order $(q-1)m - r - 1$.*

Proof. We have seen above that $(x-1)h^*(x)$ is the generator of the shortened GRM code of order $(q-1)m - r - 1$. If we now lengthen the cyclic codes to GRM codes, we must show orthogonality of the rows of the generator matrices. The only ones for which this is not a consequence of the duality of the shortened codes are the all one rows. For these, the factor $(x-1)$ in the generators and the fact that the length is q^m takes care of that. Since the dimensions of the two codes add up to q^m, we are done. □.

To handle the binary case, we need a lemma.

(6.11.4) Lemma. *Let C_1 and C_2 be cyclic codes of length n over $\mathbb{F}_q$ with check polynomials $f_1(x) := \prod_{i=1}^{k_1}(x - \alpha_i)$, resp. $\prod_{j=1}^{k_2}(x - \beta_j)$. Let C be the cyclic code of the same length for which the check polynomial has all the products $\alpha_i\beta_j$ as its zeros. Then C contains all the words* **ab**, *where* $\mathbf{a} \in C_1$, $\mathbf{b} \in C_2$.

PROOF. We use the representation of cyclic codes by linear recurring sequences, given at the end of §6.5. We know that the coordinates of **a** and **b** can be represented as sums $a_l = \sum_{i=1}^{k_1} c_i \alpha_i^{-l}$ and $b_l = \sum_{j=1}^{k_2} c'_j \beta_j^{-l}$. The result follows immediately from this representation and the definition of **ab**. □

The following theorem justifies the terminology of this section.

(6.11.5) Theorem. *The rth order binary GRM code of length 2^m is equivalent to the rth order Reed-Muller code of length 2^m.*

PROOF. The proof is by induction. For $r = 0$, the codes defined by (4.5.6) and (6.11.2) are both repetition codes. We know that the binary Hamming code is cyclic. So, for $r = 1$ we are done by the corollary to Theorem 4.5.8. Assume that the assertion is true for some value of r. The check polynomial $h^*(x)$ of the shortened GRM code has zeros α^j, where $w_2(j) \le r$. The zeros of the check polynomial of the shortened 1-st order RM code are the powers α^j with $w_2(j) = 1$. The theorem now follows from the induction hypothesis, Definition 4.5.6 and Lemma 6.11.4. □

We end this section with a theorem on weights in RM codes. It is another application of Theorem 6.8.5.

(6.11.6) Theorem. *Let $F = F(x_1, x_2, \ldots, x_m)$ be a polynomial of degree r defined on $\mathbb{F}_2^m$. We write $G \subset F$ if the monomials of G form a subset of the set of monomials of F. We define $\nu(G)$ to be the number of variables not involved in G and we denote the number of monomials in G by $|G|$. If $N(F)$ is the number of zeros of F in $\mathbb{F}_2^m$, then*

$$N(F) = 2^{m-1} + \sum_{G \subset F} (-1)^{|G|} 2^{|G| + \nu(G) - 1}.$$

PROOF. For every $G \subset F$, we define $f(G)$ to be the number of points in $\mathbb{F}_2^m$ where all the monomials of G have the value 0 and all the other monomials of F have the value 1. Clearly we have

$$\sum_{H \subset G} f(H) = 2^{\nu(F-G)}$$

(because this is the number of points in the affine subspace of $\mathbb{F}_2^m$ defined by $x_{i_1} = x_{i_2} = \ldots = x_{i_\nu} = 1$, where the x_{i_k} are the variables occurring in $F - G$). It follows from Theorem 6.8.6 that

$$f(G) = \sum_{H \subset G} \mu(H, G) 2^{\nu(F-H)}.$$

Furthermore

$$N(F) = \sum_{G \subset F,\, |F-G| \equiv 0 \pmod 2} f(G).$$

Since $\sum_{G \subset F} f(G) = 2^m$, we find

$$\begin{aligned}
N(F) &= 2^{m-1} + \frac{1}{2}\sum_{G\subset F}(-1)^{|F-G|} f(G) \\
&= 2^{m-1} + \frac{1}{2}\sum_{G\subset F}(-1)^{|F-G|}\sum_{H\subset G}\mu(H,G)2^{\nu(F-H)} \\
&= 2^{m-1} + \frac{1}{2}\sum_{H\subset F}(-1)^{|F-H|}2^{\nu(F-H)}\sum_{H\subset G\subset F}1 \\
&= 2^{m-1} + \frac{1}{2}\sum_{H\subset F}(-1)^{|F-H|}2^{\nu(F-H)}2^{|F-H|} \\
&= 2^{m-1} + \frac{1}{2}\sum_{G\subset F}(-1)^{|G|}2^{\nu(G)+|G|}. \qquad \square
\end{aligned}$$

We now apply this to RM codes.

(6.11.7) Theorem. *The weights of the codewords in* $\mathscr{R}(r,m)$ *are divisible by* $2^{\lceil m/r\rceil-1}$.

Proof. The code $\mathscr{R}(r,m)$ consists of the sequences of values taken by polynomials of degree at most r in m binary variables. The codeword corresponding to a polynomial F has weight $2^m - N(F)$. If $G \subset F$ and G has degree d, then $\nu(G) \geq m - |G|\cdot d$, i.e. $|G| \geq \lceil \frac{m-\nu(G)}{d}\rceil$. Since

$$\nu(G) + \lceil \frac{m-\nu(G)}{d} \rceil \geq \lceil \frac{m}{d} \rceil,$$

the result follows from Theorem 6.11.6 $\square$

§6.12. Comments

The reader who is interested in seeing the trace function and idempotents used heavily in proofs should read [46, Chapter 15].

A generalization of BCH codes will be treated in Chapter 9. There is extensive literature on weights, dimension, covering radius, etc. of BCH codes. We mention the Carlitz-Uchiyama bound which depends on a deep theorem in number theory by A. Weil. For the bound we refer to [42]. For a generalization of QR codes to word length n a prime power, in which case the theory is similar to Section 6.9 we refer to a paper by J. H. van Lint and F. J. MacWilliams (1978; [45]).

§6.13. Problems

6.13.1. Show that the [4, 2] ternary Hamming code is a negacyclic code.

6.13.2. Determine the idempotent of the [15, 11] binary Hamming code.

6.13.3. Show that the rth order binary Reed-Muller code of Definition 4.5.6 is equivalent to an extended cyclic code.

6.13.4. Construct a ternary BCH code with length 26 and designed distance 5.

6.13.5. Let α be a primitive element of $\mathbb{F}_{2^5}$ satisfying $\alpha^5 = \alpha^2 + 1$. A narrow-sense BCH code of length 31 with designed distance 5 is being used. We receive

$$(1001 \quad 0110 \quad 1111 \quad 0000 \quad 1101 \quad 0101 \quad 0111 \quad 111).$$

Decode this message using the method of Section 6.7.

6.13.6. Let m be odd. Let β be a primitive element of $\mathbb{F}_{2^m}$. Consider a binary cyclic code C of length $n = 2^m - 1$ with generator $g(x)$ such that $g(\beta) = g(\beta^{-1}) = 0$. Show that the minimum distance d of C is at least 5.

6.13.7. Let C be a $[q + 1, 2, d]$ code over $\mathbb{F}_q$(q odd). Show that $d < q$ (i.e. C is not an MDS code, cf. (5.2.2)).

6.13.8. Show that the [11, 6] ternary QR code is perfect. (This code is equivalent to the code of Section 4.3.)

6.13.9. Determine the minimum distance of the binary QR code with length 47.

6.13.10. Determine all perfect single error-correcting QR codes.

6.13.11. Generalize the ideas of Section 6.9 in the following sense. Let $e > 2$, n a prime such that $e|(n - 1)$ and q a prime power such that $q^{(n-1)/e} \equiv 1 \pmod{n}$. Instead of using the squares in $\mathbb{F}_n$ use the eth powers. Show that Theorem 6.9.2(i) can be generalized to $d^e > n$. Determine the minimum distance of the binary cubic residue code of length 31.

6.13.12. Let m be odd, $n = 2^m - 1$, α a primitive element of $\mathbb{F}_{2^m}$. Let $g(x)$ be a divisor of $x^n - 1$ such that $g(\alpha) = g(\alpha^5) = 0$. Prove that the binary cyclic code with generator $g(x)$ has minimum distance $d \geq 4$ in two ways:

(a) by applying a theorem of this chapter,
(b) by showing that $1 + \xi + \eta = 0$ and $1 + \xi^5 + \eta^5 = 0$ with ξ and η in $\mathbb{F}_{2^m}$ is impossible.
(c) Using the idea of (b), show that in fact $d \geq 5$.

6.13.13. Show that the ternary Golay code has a negacyclic representation.

6.13.14. By Theorem 6.9.2, the [31, 16] QR code has $d \geq 7$, whereas the BCH bound only yields $d \geq 5$. Show that the AB-method of §6.6 also yields $d \geq 7$.

CHAPTER 7

Perfect Codes and Uniformly Packed Codes

§7.1. Lloyd's Theorem

In this chapter we shall restrict ourselves to *binary* codes. To obtain insight into the methods and theorems of this part of coding theory this suffices. Nearly everything can be done (with a little more work) for arbitrary fields $\mathbb{F}_q$. In the course of time many ways of studying perfect codes and related problems have been developed. The algebraic approach which will be discussed in the next section is perhaps the most elegant one. We start with a completely different method. We shall give an extremely elementary proof of a strong necessary condition for the existence of a binary perfect e-error-correcting code. The theorem was first proved by S. P. Lloyd (1957) (indeed for $q = 2$) using analytic methods. Since then it has been generalized by many authors (cf. [44]) but it is still referred to as Lloyd's theorem. The proof in this section is due to D. M. Cvetković and J. H. van Lint (1977; cf. [17]).

(7.1.1) Definition. The square matrix A_k of size 2^k is defined as follows. Number the rows and columns in binary from 0 to $2^k - 1$. The entry $A_k(i, j)$ is 1 if the representations of i and j have Hamming distance 1, otherwise $A_k(i, j) = 0$.

From (7.1.1) we immediately see

(7.1.2) $$A_{k+1} = \begin{pmatrix} A_k & I \\ I & A_k \end{pmatrix}.$$

(7.1.3) Lemma. *The eigenvalues of* A_k *are* $-k + 2j$ $(0 \le j \le k)$ *with multiplicities* $\binom{k}{j}$.

Proof. The proof is by induction. For $k = 1$ it is easily checked. Let the column vector $\mathbf{x}$ be eigenvector of A_k belonging to the eigenvalue λ. Then by (7.1.2) we have

$$A_{k+1}\begin{pmatrix}\mathbf{x}\\ \mathbf{x}\end{pmatrix} = (\lambda + 1)\begin{pmatrix}\mathbf{x}\\ \mathbf{x}\end{pmatrix},$$

$$A_{k+1}\begin{pmatrix}\mathbf{x}\\ -\mathbf{x}\end{pmatrix} = (\lambda - 1)\begin{pmatrix}\mathbf{x}\\ -\mathbf{x}\end{pmatrix}.$$

The proof now follows from well-known properties of binomial coefficients. □

The technically most difficult part of this section is determining the eigenvalues of certain tridiagonal matrices which occur in the proof of the theorem. To keep the notation compact we use the following definition.

(7.1.4) Definition. The matrix $Q_e = Q_e(a, b)$ is the tridiagonal matrix with

$$(Q_e)_{i,i} := a, \qquad 0 \le i \le e,$$

$$(Q_e)_{i,i+1} := b - i, \qquad 0 \le i \le e - 1,$$

$$(Q_e)_{i,i-1} := i, \qquad 1 \le i \le e.$$

Furthermore, we define

$$P_e := P_e(a, b) := \left[\begin{array}{c|c} Q_{e-1}(a, b) & \begin{matrix}1\\1\\ \vdots\\1\end{matrix} \\ \hline 0\,0 \dots 0\; e & 1\end{array}\right].$$

The determinants of these matrices are denoted by $\bar{Q}_e$ resp. $\bar{P}_e$.

(7.1.5) Lemma. *Let $\Psi_e(x)$ be the Krawtchouk polynomial $K_e(x - 1; n - 1, 2)$ defined in* (1.2.1) *and* (1.2.15). *Then*

$$\bar{P}_e(2y - n, n) = (-1)^e e!\Psi_e(y).$$

Proof. By adding all columns to the last one and then developing by the last row we find

$$\bar{Q}_e = (a + e)\bar{Q}_{e-1} - e(a + b)\bar{P}_{e-1}.$$

Developing $\bar{P}_e$ by the last row yields

$$\bar{P}_e = \bar{Q}_{e-1} - e\bar{P}_{e-1}.$$

Combining these relations yields the following recurrence relation for $\bar{P}_e$:

(7.1.6) $$\bar{P}_{e+1} = (a - 1)\bar{P}_e - e(b - e)\bar{P}_{e-1}.$$

It is easy to check that the assertion of the lemma is true for $e = 1$ and $e = 2$. From (1.2.9) and (7.1.6) it follows that the two polynomials in the assertion satisfy the same recurrence relation. This proves the lemma. □

We need one more easy lemma on eigenvalues.

(7.1.7) Lemma. *Let A be a matrix of size m by m which has the form*

$$A = \begin{bmatrix} A_{11} & A_{12} & \cdots & A_{1k} \\ A_{21} & A_{22} & \cdots & A_{2k} \\ \cdots & \cdots & \cdots & \cdots \\ A_{k1} & A_{k2} & \cdots & A_{kk} \end{bmatrix},$$

where A_{ij} has size m_i by m_j ($i = 1, 2, \ldots, k$; $j = 1, 2, \ldots, k$). Suppose that for each i and j the matrix A_{ij} has constant row sums b_{ij}. Let B be the matrix with entries b_{ij}. Then each eigenvalue of B is also an eigenvalue of A.

Proof. Let $B\mathbf{x} = \lambda\mathbf{x}$, where $\mathbf{x} = (x_1, x_2, \ldots, x_k)^T$. Define $\mathbf{y}$ by

$$\mathbf{y}^T := (x_1, x_1, \ldots, x_1, x_2, x_2, \ldots, x_2, \ldots, x_k, x_k, \ldots, x_k)$$

where each x_i is repeated m_i times. By definition of B it is obvious that $A\mathbf{y} = \lambda\mathbf{y}$. □

We now come to the remarkable theorem which will have important consequences.

(7.1.8) Theorem. *If a binary perfect e-error-correcting code of length n exists, then $\Psi_e(x)$ has e distinct zeros among the integers* $1, 2, \ldots, n$.

Proof. The fact that the zeros are distinct is a well-known property of Krawtchouk polynomials (cf. (1.2.13)). To show that they are integers, we assume that C is a code as in the theorem. Consider the matrix A_n (cf. (7.1.1)). Reorder the rows and columns as follows. First take the rows and columns with a number corresponding to a codeword. Then successively those with numbers corresponding to words in $C_i := \{\mathbf{x} \in \mathbb{F}_2^n \mid d(\mathbf{x}, C) = i\}$, $1 \le i \le e$. Since C is perfect, this yields a partitioning of A_n into blocks such as in Lemma 7.1.7, where now

$$B = \begin{bmatrix} 0 & n & 0 & 0 & 0 & \cdots\cdots & & & \\ 1 & 0 & n-1 & 0 & 0 & \cdots\cdots & & & \\ 0 & 2 & 0 & n-2 & 0 & \cdots\cdots & & & \\ \cdots & \cdots & \cdots & \cdots & \cdots & \cdots & \cdots & \cdots & \cdots \\ 0 & \cdots\cdots & & & & 0 & e-1 & 0 & n-e+1 \\ 0 & \cdots\cdots & & & & 0 & 0 & e & n-e \end{bmatrix}.$$

The substitution $x = n - 2y$ in $\det(B - xI_{e+1})$ yields

$$\det(B - xI_{e+1}) = 2y\bar{P}_e(2y - n, n).$$

The result now follows from Lemmas 7.1.3, 7.1.5 and 7.1.7. □

The proof given in this section gives no insight into what is going on (e.g. Do the zeros of Ψ_e have a combinatorial meaning?) but it has the advantage of being completely elementary (except for the unavoidable knowledge of properties of Krawtchouk polynomials). In Section 7.5 we shall use Theorem 7.1.8 to find all binary perfect codes.

§7.2. The Characteristic Polynomial of a Code

We consider a binary code C of length n. In (3.5.1) we defined the weight distribution of a code and in (5.3.2) generalized this to distance distribution or inner distribution $(A_i)_{i=0}^n$. Corresponding to this sequence we have the *distance enumerator*

(7.2.1) $$A_C(z) := \sum_{i=0}^{n} A_i z^i = |C|^{-1} \sum_{\substack{\mathbf{u} \in C \\ \mathbf{v} \in C}} z^{d(\mathbf{u},\mathbf{v})}.$$

To get even more information on distances we now define the *outer distribution* to be a matrix B (rows indexed by elements of $\mathscr{R} = \mathbb{F}_2^n$, columns indexed $0, 1, \ldots, n$), where

(7.2.2) $$B(\mathbf{x}, i) := |\{\mathbf{c} \in C \mid d(\mathbf{x}, \mathbf{c}) = i\}|.$$

The row of B indexed by $\mathbf{x}$ is denoted by $B(\mathbf{x})$. Observe that

(7.2.3) $$(A_0, A_1, \ldots, A_n) = |C|^{-1} \sum_{\mathbf{x} \in C} B(\mathbf{x}),$$

and

(7.2.4) $$\mathbf{x} \in C \Leftrightarrow B(\mathbf{x}, 0) = 1.$$

(7.2.5) Definition. A code C is called a *regular* code if all rows of B with a 1 in position 0 are equal. The code is called *completely regular* if

$$\forall_{\mathbf{x} \in \mathscr{R}} \forall_{\mathbf{y} \in \mathscr{R}} [(\rho(\mathbf{x}, C) = \rho(\mathbf{y}, C)) \Rightarrow (B(\mathbf{x}) = B(\mathbf{y}))],$$

where $\rho(\mathbf{x}, C)$ is the distance from $\mathbf{x}$ to the code C.

Observe that if a code C is regular and $\mathbf{0} \in C$, then the weight enumerator of C is equal to $A_C(z)$.

In order to study the matrix B, we first introduce some algebra (cf. (1.1.11)).

(7.2.6) Definition. If G is an additive group and $\mathbb{F}$ a field, then the *group algebra* $\mathbb{F}G$ (or better $(\mathbb{F}G, \oplus, *)$) is the vector space over $\mathbb{F}$ with elements of

G as basis, with addition $\oplus$ and a multiplication $*$ defined by

$$\sum_{g\in G} \alpha(g)g * \sum_{h\in G} \beta(h)h := \sum_{k\in G}\left(\sum_{g+h=k} \alpha(g)\beta(h)\right)k.$$

Some authors prefer introducing an extra symbol z and using formal multiplication as follows

$$\sum_{g\in G} \alpha(g)z^g * \sum_{h\in G} \beta(h)z^h = \sum_{k\in G}\left(\sum_{g+h=k} \alpha(g)\beta(h)\right)z^k.$$

We shall take G to be $\mathscr{R} = \mathbb{F}_2^n$ and $\mathbb{F} = \mathbb{C}$. We denote this algebra by $\mathscr{A}$. In order not to confuse addition in G with the addition of elements of $\mathscr{A}$ we write the elements of the group algebra as $\boldsymbol{\Sigma}_{\mathbf{x}\in\mathscr{R}}\alpha(\mathbf{x})\mathbf{x}$. If S is a subset of $\mathscr{R}$ we shall identify this subset with the element $\boldsymbol{\Sigma}_{\mathbf{x}\in S}\mathbf{x}$ in $\mathscr{A}$ (i.e. we also denote this element by S). We introduce a notation for the sets of words of fixed weight resp. the spheres around $\mathbf{0}$:

(7.2.7) $$Y_i := \{\mathbf{x}\in\mathscr{R} \mid w(\mathbf{x}) = i\},$$

(7.2.8) $$S_j := \{\mathbf{x}\in\mathscr{R} \mid w(\mathbf{x}) \le j\}.$$

If C is a code with outer distribution B, then the conventions made above imply that

(7.2.9) $$Y_i * C = \boldsymbol{\Sigma}_{\mathbf{x}\in\mathscr{R}} B(\mathbf{x}, i)\mathbf{x}.$$

If $D(\mathbf{x}, j)$ denotes the number of codewords with distance at most j to $\mathbf{x}$ (i.e. $D(\mathbf{x}, j) = \sum_{i\le j} B(\mathbf{x}, i)$), then we have

(7.2.10) $$S_j * C = \boldsymbol{\Sigma}_{\mathbf{x}\in\mathscr{R}} D(\mathbf{x}, j)\mathbf{x}.$$

Let χ be the character of $\mathbb{F}_2$ with $\chi(1) = -1$. For every $\mathbf{u}\in\mathscr{R}$ we define a mapping $\chi_{\mathbf{u}}\colon \mathscr{R}\to\mathbb{C}$ by

(7.2.11) $$\forall_{\mathbf{v}\in\mathscr{R}}[\chi_{\mathbf{u}}(\mathbf{v}) := \chi(\langle\mathbf{u}, \mathbf{v}\rangle) = (-1)^{\langle\mathbf{u},\mathbf{v}\rangle},$$

i.e. $\chi_{\mathbf{u}}(\mathbf{v}) = 1$ if $\mathbf{u}\perp\mathbf{v}$ and -1 otherwise.

We extend this mapping to a linear functional on the group algebra $\mathscr{A}$ by

(7.2.12) $$\chi_{\mathbf{u}}(\boldsymbol{\Sigma}\alpha(\mathbf{x})\mathbf{x}) := \sum \alpha(\mathbf{x})\chi_{\mathbf{u}}(\mathbf{x}).$$

The following two assertions follow immediately from our definition. We leave the proofs as easy exercises for the reader.

(7.2.13) $$\forall_{\mathbf{u}\in\mathscr{R}}\forall_{A\in\mathscr{A}}\forall_{B\in\mathscr{A}}[\chi_{\mathbf{u}}(A*B) = \chi_{\mathbf{u}}(A)\chi_{\mathbf{u}}(B)],$$

(7.2.14) $$(\chi_{\mathbf{0}}(S) = 2^n \text{ and } \forall_{\mathbf{u}\ne\mathbf{0}}[\chi_{\mathbf{u}}(S) = 0]) \Leftrightarrow S = S_n.$$

The result of Lemma 5.3.1 (where we now have $q = 2$) can be written as

(7.2.15) $$\chi_{\mathbf{u}}(Y_k) = K_k(w(\mathbf{u})).$$

From this it follows that if $w(\mathbf{u}) = x$

(7.2.16) $$\chi_{\mathbf{u}}(S_j) = \sum_{k=0}^{j} K_k(x) = \Psi_j(x).$$

(Cf. (1.2.15).)

Let C be a code. We consider the numbers

$$C_j := |C|^{-1} \sum_{\mathbf{u} \in Y_j} \chi_{\mathbf{u}}(C).$$

We have seen these numbers before. If C is a linear code, then the proof of Theorem 3.5.3 shows us that C_j is the number of words of weight j in $C^\perp$. If C is not linear, we can still consider the numbers C_j and continue the proof of Theorem 3.5.3 to find that $2^{-n}|C|\sum_{j=0}^{n} C_j(1-z)^j(1+z)^{n-j}$ is the weight enumerator of C. This relation between the weight enumerator and the numbers C_j, defined with the aid of χ, is a nonlinear form of the MacWilliams relation.

We now define a second sequence of numbers, again using the character χ.

(7.2.17) Definition. The *characteristic numbers* B_j $(0 \le j \le n)$ of the code C are defined by

$$B_j := |C|^{-2} \sum_{\mathbf{u} \in Y_j} |\chi_{\mathbf{u}}(C)|^2.$$

As before we see that B_j is the number of words of weight j in the code $C^\perp$ if C is a linear code. Let $N(C) := \{j | 1 \le j \le n, B_j \ne 0\}$. We define the *characteristic polynomial* F_C of the code C by

(7.2.18) $$F_C(x) := 2^n|C|^{-1} \prod_{j \in N(C)} \left(1 - \frac{x}{j}\right).$$

(7.2.19) Theorem. *Let $\alpha_0, \alpha_1, \ldots, \alpha_n$ be the coefficients in the Krawtchouk expansion of F_C. Then in $\mathscr{A}$ we have*

$$\sum \alpha_i Y_i * C = S_n.$$

Proof. Let $\mathbf{u} \in \mathscr{R}$, $w(\mathbf{u}) = j$. By (7.2.13) and (7.2.15) we have

$$\chi_{\mathbf{u}}(\sum \alpha_i Y_i * C) = \chi_{\mathbf{u}}(\sum \alpha_i Y_i)\chi_{\mathbf{u}}(C) = \chi_{\mathbf{u}}(C) \sum \alpha_i K_i(j) = \chi_{\mathbf{u}}(C) F_C(j).$$

If $\mathbf{u} \ne \mathbf{0}$ the right-hand side is 0 by definition of F_C. If $\mathbf{u} = \mathbf{0}$ then the right-hand side is 2^n. The assertion now follows from (7.2.14). □

(7.2.20) Corollary. *If $\alpha_0, \alpha_1, \ldots, \alpha_n$ are the coefficients in the Krawtchouk expansion of F_C and $\mathbf{u} \in \mathscr{R}$ then*

$$\sum_{i=0}^{n} \alpha_i B(\mathbf{u}, i) = 1.$$

Proof. Apply (7.2.9). □

(7.2.21) Definition. The number $s := |N(C)|$ is called the *external distance* of C.

Note that if C is linear, then s is the number of nonzero weights occurring in $C^\perp$. The somewhat strange name is slightly justified by Corollary 7.2.20 which shows that the covering radius $\rho(C)$ of a code (see (3.14)) is at most equal to s.

§7.3. Uniformly Packed Codes

In this section we consider codes which are generalizations of perfect codes (cf. (5.1.5)). Note that if C is a perfect e-error-correcting code, then in $\mathscr{A}$ we have $S_e * C = S_n$. We now consider codes C with $d \geq 2e + 1$ and $\rho(C) = e + 1$. (If $d = 2e + 3$ then this means that C is perfect.) The spheres with radius $e - 1$ around codewords are disjoint and each word not in one of these spheres has distance e or $e + 1$ to at least one codeword.

(7.3.1) Definition. A code C with $\rho(C) = e + 1$ and $d \geq 2e + 1$ is called *uniformly packed* with parameter r if each word $\mathbf{u}$ with $\rho(\mathbf{u}, C) \geq e$ has distance e or $e + 1$ to exactly r codewords.

Note that if $r = 1$, then C is a perfect $(e + 1)$-error-correcting code. Of course a word $\mathbf{u}$ with $\rho(\mathbf{u}, C) = e$ has distance e to exactly one codeword. Let $\rho(\mathbf{u}, C) = e + 1$ and w.l.o.g. take $\mathbf{u} = \mathbf{0}$. Then the codewords with distance $e + 1$ to $\mathbf{u}$ have weight $e + 1$. Since they must have mutual distances $\geq 2e + 1$, it follows that

$$r \leq \frac{n}{e+1}. \tag{7.3.2}$$

We now assume that $e + 1$ does not divide $n + 1$. A code for which $r = \lfloor n/(e + 1) \rfloor$ is called *nearly perfect*. It is easy to check that this means that C satisfies the Johnson bound (5.2.16) with equality. In a paper by J.-M. Goethals and H. C. A. van Tilborg [25], (7.3.1) is generalized by replacing r by two numbers depending on whether $\rho(\mathbf{u}, C) = e$ or $e + 1$.

(7.3.3) Theorem. *A code C with $\rho(C) = e + 1$ and $d \geq 2e + 1$ is uniformly packed with parameter r iff in $\mathscr{A}$ we have*

$$\left\{ Y_0 \oplus Y_1 \oplus \cdots \oplus Y_{e-1} \oplus \frac{1}{r}(Y_e \oplus Y_{e+1}) \right\} * C = S_n.$$

PROOF. This follows from (7.2.2), (7.2.9) and (7.3.1). □

(7.3.4) Theorem. *A code C with $\rho(C) = e + 1$ and $d \geq 2e + 1$ is uniformly packed with parameter r iff the characteristic polynomial has degree $s = e + 1$*

and Krawtchouk coefficients

$$\alpha_0 = \alpha_1 = \cdots = \alpha_{e-1} = 1, \qquad \alpha_e = \alpha_{e+1} = \frac{1}{r}.$$

PROOF.

(i) The if part follows from Theorem 7.2.19 and Theorem 7.3.3.
(ii) Let C be uniformly packed. We know that F_C has degree $s \geq e + 1$. Let $F(x) := \sum_{i=0}^{e+1} \alpha_i K_i(x)$ with $\alpha_0 = \alpha_1 = \cdots = \alpha_{e-1} = 1$, $\alpha_e = \alpha_{e+1} = 1/r$. If $\mathbf{u} \in \mathscr{R}$, $w(\mathbf{u}) = j \neq 0$ and $\chi_{\mathbf{u}}(C) \neq 0$, then $F(j) = 0$ by (7.2.13), (7.2.14), (7.2.15) and Theorem 7.3.3. Then by (7.2.18) it follows that $F_C(x)$ divides $F(x)$. Hence $s = e + 1$ and $F(x) = aF_C(x)$ for some a. Substituting $x = 0$ we find $a = 1$ (again using Theorem 7.3.3). □

The following is a different formulation of Theorem 7.3.4.

(7.3.5) Theorem. *If a uniformly packed code C with $\rho(C) = e + 1$ and $d \geq 2e + 1$ exists, then the polynomial*

$$F(x) := \sum_{i=0}^{e-1} K_i(x) + \frac{1}{r}[K_e(x) + K_{e+1}(x)]$$

has $e + 1$ distinct integral zeros in $[1, n]$ and $F(0) = 2^n|C|^{-1}$.

First observe that if C is a perfect code, i.e. $d = 2e + 3$, then $r = 1$ and $F(x) = \Psi_{e+1}(x)$ by (1.2.15) and Theorem 7.3.5 is Lloyd's theorem (7.1.8).

Next we remark that the requirement about $F(0)$ can be written as

$$|C|\left\{\sum_{i=0}^{e-1}\binom{n}{i} + \frac{1}{r}\binom{n+1}{e+1}\right\} = 2^n, \tag{7.3.6}$$

which is (3.1.6) if $r = 1$, resp. (5.2.16) if $r = \lfloor n/(e + 1) \rfloor$.

In fact (7.3.6) is true in general if we interpret r as the average number of codewords with distance e or $e + 1$ to a word $\mathbf{u}$ for which $\rho(\mathbf{u}, C) \geq e$.

In general it is not easy to check if a given code is uniformly packed using the definition.

We shall now consider a special case, namely a linear code C with $e = 1$. In order for C to be uniformly packed, the characteristic polynomial must have degree 2 (by Theorem 7.3.4). We have already remarked that this means that in $C^\perp$ only two nonzero weights w_1 and w_2 occur. Now suppose that $C^\perp$ is such a *two-weight code* with weight enumerator

$$A_{C^\perp}(z) = 1 + N_1 z^{w_1} + N_2 z^{w_2}.$$

Consider the MacWilliams relation (cf. Section 7.2) and substitute

$$\sum_{k=0}^{\infty} K_k(x) z^k$$

for $(1+z)^{n-x}(1-z)^x$ (cf. (1.2.3)). Since we have assumed that C has minimum distance $d \geq 3$, we find three equations from the coefficients of z^0, z^1, z^2, namely

$$1 + N_1 + N_2 = 2^n |C|^{-1},$$

$$K_k(0) + N_1 K_k(w_1) + N_2 K_k(w_2) = 0, \qquad (k = 1, 2).$$

By definition we have $F_C(w_1) = F_C(w_2) = 0$ and $F_C(0) = 2^n|C|^{-1}$. For the coefficients $\alpha_0, \alpha_1, \alpha_2$ in the Krawtchouk expansion of $F_C(x)$ we then find, using (1.2.7)

$$\alpha_0 + \alpha_1 n + \alpha_2 \binom{n}{2} = 2^n |C|^{-1},$$

$$\alpha_0 + \alpha_1(n - 2w_i) + \alpha_2 \left\{ 2w_i^2 - 2nw_i + \binom{n}{2} \right\} = 0, \qquad (i = 1, 2).$$

We compare these equations with the equations for N_1 and N_2. This shows that $\alpha_0 = 1$. We define

$$r := 2(n+1)w_1 - 2w_1^2 - \tfrac{1}{2}n(n+1).$$

It then follows that $\alpha_1 = \alpha_2 = 1/r$ if $w_1 + w_2 = n + 1$. We have thus proved the following characterization of 1-error-correcting uniformly packed codes.

(7.3.7) Theorem. *A linear code C with $\rho(C) = 2$ and $d \geq 3$ is uniformly packed iff $C^\perp$ is a two-weight code with weights w_1, w_2 satisfying $w_1 + w_2 = n + 1$.*

In [25] it is shown that if we adopt the more general definition of uniformly packed codes, we can drop the restriction $w_1 + w_2 = n + 1$. The theorem is also true for $e > 1$ with $e + 1$ weights in $C^\perp$ instead of two.

§7.4. Examples of Uniformly Packed Codes

(7.4.1) *A Hadamard code* (cf. Section 4.1)

Consider the (12, 24, 6) Hadamard code. Puncture the code to obtain the (11, 24, 5) code C. It is obvious that any word $\mathbf{z}$ can have distance 2 or 3 to at most four codewords and if this happens we have the following situation (after changing back to $\pm$ notation and suitable multiplications of columns by -1):

$$\mathbf{z} = - - \quad + + + \quad + + + \quad + + +,$$

$$\mathbf{x}_1 = + + \quad + + + \quad + + + \quad + + +,$$

$$\mathbf{x}_2 = - - \quad - - - \quad + + + \quad + + +,$$

$$\mathbf{x}_3 = - - \quad + + + \quad - - - \quad + + +,$$

$$\mathbf{x}_4 = - - \quad + + + \quad + + + \quad - - -,$$

This would mean that the original Hadamard matrix of order 12 had the four rows $(+, \mathbf{x}_1)$, $(-, \mathbf{x}_2)$, $(-, \mathbf{x}_3)$, $(-, \mathbf{x}_4)$. The row vector $(-4, -4, -4, 0, 0, \ldots, 0)$ is a linear combination of these four rows and must therefore be orthogonal to the remaining rows of the Hadamard matrix, which is clearly impossible. It follows that a word $\mathbf{z}$ has distance 2 or 3 to at most three codewords. From (7.3.6) it follows that the average number of codewords with distance 2 or 3 to a word $\mathbf{z}$ with $\rho(\mathbf{z}, C) > 1$ is three. Hence this number is always three. So C is uniformly packed with $r = 3$. In this example C is nonlinear.

Note that in this example, $e+1$ divides $n+1$. So (7.3.6) holds but is not equal to (5.2.16). This is *not* a nearly perfect code.

(7.4.2) *A Punctured* RM *Code*

Let V be the six-dimensional vector space over $\mathbb{F}_2$. Let W be the set of 35 points $\mathbf{x}$ in $V \backslash \{\mathbf{0}\}$ on the quadric with equation $x_1x_2 + x_3x_4 + x_5x_6 = 0$. We take these vectors as columns of a 6 by 35 matrix G. As in Section 4.5 we see that the ith row of G is the characteristic function of the intersection of W and the hyperplane with equation $x_i = 1$ $(1 \le i \le 6)$. Hence the weight of a linear combination $\mathbf{a}^T G$ $(\mathbf{a} \in V)$ is the number of solutions of

$$x_1x_2 + x_3x_4 + x_5x_6 = 0 \quad \text{and} \quad \sum_{i=1}^{6} a_i x_i = 1.$$

W.l.o.g. we may take $a_1 = 1$ (unless $\mathbf{a} = \mathbf{0}$). By substitution and the affine transformation

$$y_2 = x_2, \qquad y_3 = x_3 + a_4x_2, \qquad y_4 = x_4 + a_3x_2,$$
$$y_5 = x_5 + a_6x_2, \qquad y_6 = x_6 + a_5x_2$$

(which is invertible) we see that we must count the number of solutions of the equation

$$(1 + a_2 + a_3a_4 + a_5a_6)y_2 + y_3y_4 + y_5y_6 = 0.$$

If the coefficient of y_2 is 1 this number is 16, if it is 0 the number of solutions is 20. Therefore the code C which has G as parity check matrix has a dual $C^\perp$ which is a two-weight code with weights 16 and 20. The code C has $d \ge 3$ since it is projective. By the remark following (7.2.21) we have $\rho(C) = 2$. So by Theorem 7.3.7, C is uniformly packed with $r = 10$ (by (7.3.6)). The same method works in higher dimensions.

(7.4.3) *Preparata-codes*

In 1968, F. P. Preparata [57] introduced a class of nonlinear double-error-correcting codes which turned out to have many interesting properties. His definition was based on a combination of Hamming codes and 2-error-correcting BCH codes. The analysis of the codes involves tedious calculation (cf. [11]). The following description of the Preparata codes is due to R. D. Baker, R. M. Wilson and the author (cf. [72]).

In the following m is odd ($m \geq 3$), $n = 2^m - 1$. We shall define a code $\bar{\mathscr{P}}$ of length $2n + 2 = 2^{m+1}$. The words will be described by pairs (X, Y), where $X \subset \mathbb{F}_{2^m}$ and $Y \subset \mathbb{F}_{2^m}$. As usual, we interpret the pair (X, Y) as the corresponding characteristic function, which is a (0, 1)-vector of length 2^{m+1}.

(7.4.4) Definition. The extended *Preparata code* $\bar{\mathscr{P}}$ of length 2^{m+1} consists of the codewords described by all pairs (X, Y) satisfying

(i) $|X|$ is even, $|Y|$ is even,

(ii) $\sum_{x \in X} x = \sum_{y \in Y} y$,

(iii) $\sum_{x \in X} x^3 + (\sum_{x \in X} x)^3 = \sum_{y \in Y} y^3$.

The code $\mathscr{P}$ is obtained by leaving out the coordinate corresponding to the zero position in the first half.

We first show that $\bar{\mathscr{P}}$ has 2^{2n-2m} words. We can choose X, satisfying (i) in 2^n ways. Next, we observe that since m is odd the minimal polynomial $m_3(x)$ for $\mathbb{F}_{2^m}$ has degree m. Therefore the BCH code of length n and designed distance 5 has dimension $n - 2m$. This in turn implies that for a given X the equations (ii) and (iii) have 2^{n-2m} solutions $Y \subset \mathbb{F}_{2^m}^*$. We can add the zero element to this Y if necessary to satisfy (i). This proves our assertion.

The next claim is that $\bar{\mathscr{P}}$ has minimum distance 6. From (7.4.4)(i) we see that the minimum distance is even and also that if (X, Y) satisfies the conditions, then so does (Y, X). Suppose we have two words (X, Y_1) and (X, Y_2) and let $Y := Y_1 \triangle Y_2$. Then from (7.4.4)(ii) and (iii) we find that

$$\sum_{y \in Y} y = \sum_{y \in Y} y^3 = 0,$$

i.e. $|Y| \geq 5$ by the BCH bound. So in this case the two words have distance ≥ 6. It remains to consider the possibility (X_1, Y_1), (X_2, Y_2) with

$$|X_1 \triangle X_2| = |Y_1 \triangle Y_2| = 2.$$

Let $X_1 \triangle X_2 = \{\alpha, \beta\}$, $Y_1 \triangle Y_2 = \{\gamma, \delta\}$ and let $s + \alpha$ be the sum of the elements in X_1. Then (7.4.4)(ii) and (iii) imply

$$\alpha + \beta = \gamma + \delta,$$

$$s^2(\alpha + \beta) + s(\alpha + \beta)^2 = \gamma^3 + \delta^3.$$

From these we find $(s + \gamma)^3 + (s + \delta)^3 = 0$, i.e. $\gamma = \delta$, a contradiction. This proves our claim. We have proved the following theorem.

(7.4.5) Theorem. *The Preparata code* $\mathscr{P}$ *of length* $2^{m+1} - 1$ *(*m *odd,* $m \geq 3$*) has* $|\mathscr{P}| = 2^k$, *where* $k = 2^{m+1} - 2m - 2$, *and minimum distance* 5.

From (7.3.6) we find that the average value of r for the code $\mathscr{P}$ is $(2^{m+1} - 1)/3$ and then (7.3.2) implies that r is constant and equal to $(2^{m+1} - 1)/3$, i.e. $\mathscr{P}$ is nearly perfect.

If we take $m = 3$ in (7.4.4) then we find the Nordstrom-Robinson code which was introduced in Section 4.4.

Remark. The exponents 3 in (7.4.4)(iii) are not essential. We can replace 3 by $s := 2^t + 1$, where we require that $x \to x^s$ and $x \to x^{s-2}$ are 1-1 mappings of $\mathbb{F}_m$ to itself. The first part of the argument concerning minimum distance is then replaced by one involving Theorem 6.6.3. We leave this as an easy exercise for the reader.

Observe that the code satisfying only (i) and (ii) in (7.4.4) is the extended Hamming code of length 2^{m+1}. From this and a counting argument it follows that if we take $\mathscr{P}$ and adjoin to it all words at distance 3 to $\mathscr{P}$, we obtain the Hamming code of length $2^{m+1} - 1$. In Theorem 7.4.6 we give a direct construction.

(7.4.6) Theorem. *The union of the Preparata code $\mathscr{P}$ of length $n = 2^{m+1} - 1$ (m odd) and the set of words with distance 3 to $\mathscr{P}$ is the Hamming code of length n.*

Proof. Since $\mathscr{P}$ satisfies (5.2.16) with equality, we know that there are $|\mathscr{P}| \cdot \frac{n-1}{2}$ words with distance 3 to $\mathscr{P}$. Hence the union C has 2^{n-m-1} words, which is the cardinality of the Hamming code of length n.

We now define $C_0 := \mathscr{P}$ and for $\alpha \in \mathbb{F}^*_{2^m}$, we define C_α to be the code obtained by adding the word corresponding to $(\{0, \alpha\}, \{0, \alpha\})$ to the words of C_0. Clearly, each C_α has only even weight vectors. If weight 2 would occur, then C_0 would have a word corresponding to $X = \{0, \alpha\}$, $Y = \{0, \alpha, \beta, \gamma\}$, contradicting (ii) in (7.4.4). So, each C_α has minimum weight 4. From the proof of Theorem 7.4.5, it follows that the C_α are pairwise disjoint ($\alpha \in \mathbb{F} := \mathbb{F}_{2^m}$). We claim that $\overline{H} := \bigcup_{\alpha \in \mathbb{F}} C_\alpha$ is linear. To show this, using (7.4.4) (iii), comes down to solving an equation of type $x^3 = a$, which is possible since m is odd. From the parameters and the linearity we conclude that $\overline{H}$ is the extended Hamming code of length $n + 1$. This proves the assertion about C. □

Note that it follows from Theorem 7.4.6 that the linear span of the Preparata code is contained in the Hamming code.

§7.5. Nonexistence Theorems

It was shown by A. Tietäväinen [68] and J. H. van Lint [41] that the Golay codes are the only nontrivial e-error-correcting perfect codes with $e > 1$ over any alphabet Q for which $|Q|$ is a prime power. For $e > 2$ the restriction on Q can be dropped as was shown by M. R. Best [7] and Y. Hong [74] but that is much more difficult to prove. For $e = 1$ we have seen the Hamming codes. There are also examples of nonlinear perfect codes with $e = 1$ (cf. (7.7.4)).

In 1975 Van Tilborg [69] showed that e-error-correcting uniformly packed codes with $e > 3$ do not exist and those with $e \leq 3$ are all known. In this section we wish to give some idea of the methods which were used to establish these results. It suffices to consider the binary case.

(7.5.1) Theorem. *If C is a perfect e-error-correcting binary code with $e > 1$, then C is a repetition code or the binary Golay code.*

Proof. By Lloyd's theorem (7.1.8) the polynomial Ψ_e has zeros $x_1 < x_2 < \cdots < x_e$ which are integers in $[1, n]$. By the definition of Ψ_e and (1.2.8), the elementary symmetric functions of degree 1 and 2 of the zeros are known:

$$\sum_{i=1}^{e} x_i = \frac{1}{2}e(n+1), \tag{7.5.2}$$

$$\sum_{i<j} x_i x_j = \frac{1}{24}e(e-1)\{3n^2 + 3n + 2e + 2\}. \tag{7.5.3}$$

Observe that (7.5.2) also follows from (1.2.2) which also shows that

$$x_{e-i+1} = n + 1 - x_i. \tag{7.5.4}$$

From (7.5.2) and (7.5.3) we find

$$\sum_{i=1}^{e}\sum_{j=1}^{e} (x_i - x_j)^2 = \frac{1}{2}e^2(e-1)\left\{n - \frac{2e-1}{3}\right\}. \tag{7.5.5}$$

To find the product of the zeros, we calculate $\Psi_e(0)$. From (1.2.1) we find $\Psi_e(0) = \sum_{j=0}^{e}\binom{n}{j}$. Combining this with (3.1.6) and (1.2.8), we find

$$\prod_{i=1}^{e} x_i = e!\,2^l, \quad \text{(for some integer } l). \tag{7.5.6}$$

In a similar way we calculate $\Psi_e(1)$ and $\Psi_e(2)$, which leads to

$$\prod_{i=1}^{e} (x_i - 1) = 2^{-e}(n-1)(n-2)\ldots(n-e), \tag{7.5.7}$$

$$\prod_{i=1}^{e} (x_i - 2) = 2^{-e}(n-1-2e)(n-2)(n-3)\ldots(n-e). \tag{7.5.8}$$

We now draw conclusions about $x_1, x_2, \ldots, x_e$ from these relations. Let $A(x)$ denote the largest odd divisor of x. Then (7.5.6) shows that

$$\prod_{i=1}^{e} A(x_i) = A(e!) < e!.$$

This implies that there must be two zeros x_i and x_j such that $A(x_i) = A(x_j)$ $(i < j)$. Hence $2x_i \le x_j$ and therefore $2x_1 \le x_e$ and then (7.5.4) implies

$$x_e - x_1 \ge \tfrac{1}{3}(n+1). \tag{7.5.9}$$

If we fix x_1 and x_e, then the left-hand side of (7.5.5) is minimal if $x_2 = x_3 = \cdots = x_{e-1} = \frac{1}{2}(x_1 + x_e)$.

Substitution of these values in (7.5.5) leads to

$$(x_e - x_1)^2 \le \frac{1}{2}e(e-1)\left(n - \frac{2e-1}{3}\right), \tag{7.5.10}$$

which we combine with (7.5.9). The result is

(7.5.11) $$n + 1 \le \tfrac{9}{2}e(e-1).$$

Now consider (7.5.7) and (7.5.8). Since $(x-1)(x-2)$ is always even if $x \in \mathbb{N}$, we find

(7.5.12) $$(n-1-2e)(n-1)(n-2)^2(n-3)^2\cdots(n-e)^2 \equiv 0 \qquad (\mathrm{mod}\ 2^{3e}).$$

This is a triviality if $e = \frac{1}{2}(n-1)$, i.e. C is the repetition code. Suppose $e < \frac{1}{2}(n-1)$. Let 2^α be the highest power of 2 in any factor $n-j$ on the left-hand side of (7.5.12), including $n-1-2e$. Then the highest power of 2 that divides the left-hand side of (7.5.12) is at most $2^{3\alpha+2e-3}$, which implies that $\alpha \ge \frac{1}{3}e + 1$.

Hence

(7.5.13) $$n > 2^{1+(1/3)e}.$$

If e is large then (7,5.13) contradicts (7.5.11). Small values of e which by (7.5.11) imply small values of n are easily checked. In fact, if we are a little more accurate in estimating n, there are only very few cases left to analyze. It turns out that $e = 3$ is the only possibility. Actually $e = 3$ can be treated completely without even using Lloyd's theorem. This was shown in Problem 3.7.1. □

The reasoning used to show that all uniformly packed codes are known (i.e. those satisfying (7.3.1)) follows the same lines but a few extra tricks are necessary because of the occurrence of the parameter r.

(7.5.14) Theorem. *Table* 7.5.18 *lists all uniformly packed codes.*

Proof. We start from the generalization of Lloyd's theorem, i.e. Theorem 7.3.5. In exactly the same way as we proved (7.5.10) resp. (7.5.13), we find

(7.5.15) $$x_{e+1} - x_1 \le (e+1)\left(\frac{n+1}{2}\right)^{1/2},$$

(7.5.16) $$n > 2^{e/7}.$$

The argument which led to (7.5.9) has to be modified. We number the zeros in a different way as $y_1, \ldots, y_{e+1}$, where $y_j = A(y_j)2^{\alpha_j}$ and $\alpha_1 \le \alpha_2 \le \cdots \le \alpha_{e+1}$. On the one hand we have (writing (a, b) for the g.c.d. of a and b):

$$\prod_{i=1}^{e} \frac{|y_i - y_{i+1}|}{y_i} \ge \prod_{i=1}^{e} \frac{(y_i, y_{i+1})}{y_i} = \prod_i \frac{(A(y_i), A(y_{i+1}))2^{\alpha_i}}{y_i}$$
$$\ge \prod_{i=1}^{e} \frac{1}{A(y_i)} \ge \frac{1}{A(y_1 \cdots y_{e+1})} = \frac{A(|C|)}{A(r)A((e+1)!)}$$
$$\ge \frac{1}{rA((e+1)!)}.$$

Here, the last equality follows from (7.3.6). On the other hand, we have

$$\prod_{i=1}^{e} \frac{|y_i - y_{i+1}|}{y_i} \le \frac{(x_{e+1} - x_1)^e}{y_1 \cdots y_e} \le \frac{n(x_{e+1} - x_1)^e}{x_1 x_2 \cdots x_{e+1}}$$

$$= \frac{n(x_{e+1} - x_1)^e}{r(e+1)!} \cdot \frac{2^{e+1}|C|}{2^n}.$$

Combining these two inequalities we find

$$(x_{e+1} - x_1)^e \ge \frac{(e+1)!}{A((e+1)!)} \cdot \frac{1}{2^{e+1}} \cdot \frac{2^n}{n|C|}$$

$$\ge \frac{(e+1)!}{A((e+1)!)} \cdot \frac{1}{2^{e+1}} \cdot \frac{1}{n} \sum_{i=0}^{e} \binom{n}{i} \ge \frac{(e+1)!}{A((e+1)!)} \frac{1}{2^{e+1}} \frac{1}{n} \binom{n}{e},$$

and hence

$$\text{(7.5.17)} \quad (x_{e+1} - x_1)^e \ge \frac{(e+1)}{A((e+1)!)} \frac{1}{2^{e+1}} (n-1)(n-2)\ldots(n-e+1).$$

We now compare (7.5.15), (7.5.16) and (7.5.17). If $e \ge 3$ only a finite number of pairs (e, n) satisfy all three inequalities. The cases $e = 1$ and $e = 2$ can easily be treated directly with Theorem 7.3.5. As a result finitely many cases have to be analyzed separately. We omit the details (cf. [69]). As a result we find the codes listed in the table below. □

(7.5.18) Table of all Perfect, Nearly Perfect, and Uniformly Packed Binary codes

e	n	$\|C\|$	Type	Description
0	n	2^n	perfect	$\{0, 1\}^n$
1	$2^m - 1$	2^{n-m}	perfect	Hamming code (and others)
1	$2^m - 2$	2^{n-m}	nearly perfect	shortened Hamming code (cf. (7.7.1))
1	$2^{2m-1} \pm 2^{m-1} - 1$	2^{n-2m}	uniformly packed	cf. (7.4.2)
2	$2^{2m} - 1$	2^{n+1-4m}	nearly perfect	Preparata code
2	$2^{2m+1} - 1$	2^{n-4m-2}	uniformly packed	BCH code (cf. (7.7.2))
2	11	24	uniformly packed	cf. (7.4.1)
3	23	2^{12}	perfect	Golay code
e	$2e + 1$	2	perfect	repetition code
e	e	1	perfect	$\{0\}$

(Here the entries "Hamming code" and "Preparata code" are to be interpreted as all codes with the same parameters as these codes.)

§7.6. Comments

Perfect codes have been generalized in several directions (e.g. other metrics than Hamming distance, mixed alphabets). For a survey (including many references) the reader should consult [44].

A modification of Tietäväinen's nonexistence proof can be found in [46], also many references.

The most complete information on uniformly packed codes can be found in [69]. For connections to design theory we refer to [11] and [46].

The problem of the possible existence of unknown 2-error-correcting codes over alphabets Q with $|Q|$ not a prime power seems to be very hard but probably not impossible. The case $e = 1$ looks hopeless.

Many of the ideas and methods which were used in this chapter (e.g. Section 7.2) were introduced by P. Delsarte (cf. [18]).

§7.7. Problems

7.7.1. Show that the $[2^m - 2, 2^m - m - 2]$ shortened binary Hamming code is nearly perfect.

7.7.2. Let C be the binary BCH code of length $n = 2^{2m+1} - 1$ with designed distance 5. Show that C is uniformly packed with parameter $r = \frac{1}{6}(n - 1)$ by explicitly calculating the number of codewords with distance 3 to a word $\mathbf{u}$ with $\rho(\mathbf{u}, C) \geq 2$.

7.7.3. Show that there is a nearly perfect code C for which $\overline{C}$ is the Nordstrom-Robinson code.

7.7.4. Let H be the [7, 4] binary Hamming code. Define f on H by $f(\mathbf{0}) = 0$, $f(\mathbf{c}) = 1$ if $\mathbf{c} \neq \mathbf{0}$. Let C be the code of length 15 with codewords

$$\left(\mathbf{x}, \mathbf{x} + \mathbf{c}, \sum_{i=1}^{7} x_i + f(\mathbf{c})\right), \quad \text{where } \mathbf{c} \in H, \mathbf{x} \in \mathbb{F}_2^7.$$

Show that C is perfect and C is not equivalent to a linear code.

7.7.5. Show that a perfect binary 2-error-correcting code is trivial.

7.7.6. Let C be a uniformly packed code of length n with $e = 1$ and $r = 6$. Show that $n = 27$ and give a construction of C.

7.7.7. In (3.3.3) we saw that the lines of PG(2,2) generate the [7,4] Hamming code. The extended code is an [8,4,4] code. One might hope that by using PG(2, q), one could find a code C of length $q^2 + q + 2$ over $\mathbb{F}_q$ with $d = 4$ and $|C| = q^{n-4}$. This would in fact be an example of a uniformly packed code in the more general sense (as treated in [25]). Consider the case $q = 3$, $|C| = 3^{10}$. Show that such a code does not exist. Hint: Count in two ways the pairs $(\mathbf{x}, \mathbf{c})$ with $\mathbf{x} \in \mathbb{F}_3^{14}$, $\mathbf{c} \in C$, $d(\mathbf{x}, \mathbf{c}) = 2$. Calculate A_4. Then calculate A_5.

7.7.8. Consider the alphabet $\mathbb{Z}_m$ (m odd) with Lee distance. Construct a perfect single-error-correcting code of length $n = \frac{1}{2}(m^2 - 1)$.

CHAPTER 8

Codes over $\mathbb{Z}_4$

§8.1. Quaternary Codes

In 1994 it was shown (see [88], which we use as guideline for this chapter) that several well known good binary codes can be constructed by first constructing a code over the alphabet $\mathbb{Z}_4$ and then mapping the coordinates to $\mathbb{Z}_2^2$. We first study codes over $\mathbb{Z}_4$ in general.

(8.1.1) Definition. If C is an additive subgroup of $\mathbb{Z}_4^n$, then we shall call C a *linear* block code of length n over $\mathbb{Z}_4$ or a *quaternary* code.

Although C is a $\mathbb{Z}_4$-module and not a vector space, we follow the terminology of coding theory and misuse the word "linear".

The inner product $\langle \mathbf{a}, \mathbf{b} \rangle$ of two words in $\mathbb{Z}_4^n$ is defined in the usual way. We can then define the *dual* code $C^\perp$ in the same way as in (3.2.4).

As usual, we call two codes *equivalent* if one can be obtained from the other by a permutation of coordinate positions. Sometimes this definition is extended by also allowing a change of signs in some positions; (this interchanges the symbols 1 and 3).

In generalizing the concept of generator matrix, we have to be careful. We first give an example.

(8.1.2) EXAMPLE. Consider the additive subgroup of $\mathbb{Z}_4^3$ consisting of all the words (x, x, x) and $(y, y+2, y+2)$, $(x, y \in \mathbb{Z}_4)$. We can consider this quaternary code as the set of linear combinations $a(1, 1, 1) + b(0, 2, 2)$, where $a \in \mathbb{Z}_4$ and $b \in \mathbb{Z}_2$, (addition mod 4).

In general, a quaternary code is a direct product of subcodes of order 4 or 2 (additive cyclic groups of order 4 or 2). This means that there is an equivalent code with *generator matrix* of the form

(8.1.3)
$$G := \begin{pmatrix} I_{k_1} & A & B \\ O & 2I_{k_2} & 2C \end{pmatrix},$$

where the entries in A and C are 0 or 1, and those in B are in $\mathbb{Z}_4$. A codeword has the form $\mathbf{a}G$, where a_1 to a_{k_1} are in $\mathbb{Z}_4$ and a_{k_1+1} to $a_{k_1+k_2}$ are in $\mathbb{Z}_2$. The dual code $C^\perp$ has a generator matrix of the form

(8.1.4)
$$H := \begin{pmatrix} -B^\top - C^\top A^\top & C^\top & I_{n-k_1-k_2} \\ 2A^\top & 2I_{k_2} & O \end{pmatrix}.$$

Following the treatment of linear codes in Chapter 3, we now should look at weight enumerators. As a preparation for this chapter, these were treated in §3.6. We introduced the symmetric weight enumerator of a code in $\mathbb{Z}_4^n$ and the Lee weight enumerator. The important result was the fact that the Lee weight enumerators of a code and its dual satisfy the MacWilliams relations.

§8.2. Binary Codes Derived from Codes over $\mathbb{Z}_4$

There is a natural mapping ϕ from $\mathbb{Z}_4$ to $\mathbb{Z}_2^2$, mapping Lee distance in $\mathbb{Z}_4$ to Hamming distance. It is

$$\phi(0) = (0,0), \quad \phi(1) = (0,1), \quad \phi(2) = (1,1), \quad \phi(3) = (1,0).$$

We now extend this to codes in $\mathbb{Z}_4^n$. To make future notation easy, we introduce three functions from $\mathbb{Z}_4$ to $\mathbb{Z}_2$ as follows.

$i \in \mathbb{Z}_4$	$\alpha(i)$	$\beta(i)$	$\gamma(i)$
0	0	0	0
1	1	0	1
2	0	1	1
3	1	1	0

Note that i written in the binary system with most significant digit first is $(\beta(i), \alpha(i))$. Furthermore $\gamma(i) = \alpha(i) + \beta(i)$. The map ϕ defined above is generalized to $\mathbb{Z}_4^n$ in the obvious way.

(8.2.1) Definition

$$\phi(\mathbf{c}) := (\beta(\mathbf{c}), \gamma(\mathbf{c})), \quad (\mathbf{c} \in \mathbb{Z}_4^n).$$

This map is called the *Gray map* and for a quaternary code C, the code $C' := \phi(C)$ is called the *binary image* of C. Such a binary code is called $\mathbb{Z}_4$-linear. In the following, C' will always denote the binary image of a quaternary code C.

(8.2.2) EXAMPLE. Consider the quaternary code C of length 3 generated by (1,1,3) and (0,2,2). (These are then the rows of G in (8.1.3) with $k_1 = k_2 = 1$.) The contribution of the first row to the binary image are the linear combinations of

(001110) and (110001). The second row only contributes (011011). Finally, we have linear combinations of these. So in this case, the binary image C' is a linear code with generator

$$\begin{pmatrix} 1 & 1 & 1 & 1 & 1 & 1 \\ 0 & 1 & 1 & 0 & 1 & 1 \\ 0 & 0 & 1 & 1 & 1 & 0 \end{pmatrix}.$$

By comparing with this example, the reader can check that if C is a quaternary code with generator as in (8.1.3), and *if* its binary image C' is a binary linear code, then C' has generator matrix

$$\text{(8.2.3)} \qquad G' := \begin{pmatrix} I_{k_1} & A & \alpha(B) & I_{k_1} & A & \alpha(B) \\ O & I_{k_2} & C & O & I_{k_2} & C \\ O & O & \beta(B) & I_{k_1} & A & \gamma(B) \end{pmatrix}.$$

In general, C' will not be linear because ϕ is *not* a linear map. The most important fact for us is that the map ϕ preserves distance, i. e. the Lee distance of two codewords in C equals the Hamming distance of their images.

One of the reasons for the interest in codes over $\mathbb{Z}_4$ was the fact that they gave the explanation for a remarkable "coincidence". A Kerdock code and a Preparata code of the same length are both nonlinear. Their distance enumerators satisfy the MacWilliams relations. In other words : they are trying to be duals although this concept does not make sense for nonlinear codes. We are on our way to the explanation. Both these codes are binary images of quaternary codes that are indeed each others dual . We therefore define $\mathbb{Z}_4$-duality as follows.

(8.2.4) Definition. If C is a quaternary code and $C^\perp$ is its dual code, then $C' = \phi(C)$ and $(C^\perp)' := \phi(C^\perp)$ are called $\mathbb{Z}_4$-duals.

From the fact that C is linear, it follows that its binary image C' is distance invariant.

(8.2.5) Theorem. *If C and $C^\perp$ are dual quaternary codes, then the weight distributions of their binary images C' and $(C^\perp)'$ satisfy the MacWilliams relations of Theorem 3.5.3.*

PROOF. We have observed that the Hamming weight of $\phi(\mathbf{c})$ equals $w_L(\mathbf{c})$. The code C' has length $2n$ if C has length n. So, from (3.6.3) and Theorem 3.6.8 we indeed find

$$Ham_{(C^\perp)'}(x, y) = \frac{1}{|C'|} Ham_{C'}(x + y, x - y). \qquad \square$$

We shall now establish a necessary and sufficient condition for a binary code to be the binary image of a quaternary code. First, observe that $\phi(-\mathbf{c}) = (\gamma(\mathbf{c}), \beta(\mathbf{c}))$. This implies that a $\mathbb{Z}_4$-linear code is fixed by the permutation σ given by $(1, n+1)$ $(2, n+2) \dots (n, 2n)$. This permutation interchanges the left and right halves of each codeword.

(8.2.6) Lemma. *For all* $\mathbf{a}, \mathbf{b} \in \mathbb{Z}_4^n$ *we have*

$$\phi(\mathbf{a}+\mathbf{b}) = \phi(\mathbf{a}) + \phi(\mathbf{b}) + (\phi(\mathbf{a}) + \sigma(\phi(\mathbf{a})))(\phi(\mathbf{b}) + \sigma(\phi(\mathbf{b}))).$$

PROOF. This follows from the following facts :

(1) $\alpha(a) + \beta(a) + \gamma(a) = 0$;

(2) $\alpha(a)\alpha(b) = 1$ iff a and b are odd;

(3) $\beta(a+b) = \beta(a) + \beta(b) + \epsilon$, where $\epsilon = 1$ iff a and b are both odd (0 otherwise), and γ satisfies the same relation.

□

(8.2.7) Theorem. A binary, not necessarily linear code of even length is $\mathbb{Z}_4$-linear iff it is equivalent to a code C for which

$$\mathbf{a}, \mathbf{b} \in C \Rightarrow \mathbf{a} + \mathbf{b} + (\mathbf{a} + \sigma(\mathbf{a}))(\mathbf{b} + \sigma(\mathbf{b})) \in C.$$

PROOF. This follows immediately from Lemma 8.2.6. □

(8.2.8) EXAMPLE. Consider the first order Reed-Muller code $\mathcal{R}(1, m)$ of length 2^m. Every codeword has the form $\mathbf{a} = (\mathbf{x}, \mathbf{x} + \epsilon)$, where $\epsilon = \mathbf{0}$ or $\mathbf{1}$ (of length 2^{m-1}). If $\mathbf{b} = (\mathbf{y}, \mathbf{y} + \epsilon')$ is a second codeword, then $(\mathbf{a} + \sigma(\mathbf{a}))(\mathbf{b} + \sigma(\mathbf{b})) = \mathbf{0}$ or $\mathbf{1}$ and hence by Theorem 8.2.7, $\mathcal{R}(1, m)$ is $\mathbb{Z}_4$-linear. If we take the example $m = 3$, then the corresponding quaternary code has generator

$$G := \begin{pmatrix} 1 & 1 & 1 & 1 \\ 0 & 2 & 0 & 2 \\ 0 & 0 & 2 & 2 \end{pmatrix}.$$

From G and (8.1.3), we find the standard basis vectors of $\mathcal{R}(1, 3)$.

We now consider binary linear codes that are images of quaternary codes. The following statement is a direct consequence of Theorem 8.2.7.

(8.2.9) Corollary. *The binary image* $\phi(C)$ *of a quaternary code is linear iff*

$$\mathbf{a}, \mathbf{b} \in C \Rightarrow 2(\alpha(\mathbf{a})\alpha(\mathbf{b})) \in C.$$

PROOF. By two applications of Lemma 8.2.6, we find

$$\phi(\mathbf{a} + \mathbf{b} + 2(\alpha(\mathbf{a})\alpha(\mathbf{b}))) = \phi(\mathbf{a}) + \phi(\mathbf{b}).$$ □

This means that the word with 2s in the positions where both $\mathbf{a}$ and $\mathbf{b}$ have odd entries, and 0s elsewhere, is in the code.

We shall now show that the extended Hamming code of length $n = 2^m$ is not $\mathbb{Z}_4$-linear for $m \geq 5$.

(8.2.10) Theorem. *The code $\mathcal{R}(m-2, m)$ is not $\mathbb{Z}_4$-linear for $m \geq 5$.*

Proof. Suppose C is a quaternary code of length 2^{m-1} that has the extended binary Hamming code H_m of length $n = 2^m$ as its binary image. Let G of (8.1.3) be the generator matrix of C. Call the upper part G_{k_1} and the lower part $2G_{k_2}$. Since H_m is an even weight code, G_{k_1} has rows with an even number of odd entries. We define

$$G'_{k_1} := (I_{k_1} \quad A \quad \alpha(B)).$$

We shall study the binary code H' generated by G'_{k_1}. Note that $2G'_{k_1}$ and $2G_{k_2}$ generate the subcode of C consisting of the words with even coordinates only. Clearly $k_1 + k_2 \leq 2^{m-1} - 1$. Since $2k_1 + k_2 = 2^m - m - 1$, we see that H' has dimension at least $2^{m-1} - m$.

From Corollary 8.2.9 it follows that if H' has words of weight 2, then the supports of these words are disjoint. We can take these words as basis vectors of the code and we may assume that the other basis vectors have supports disjoint from those of vectors of weight 2 (again because the intersection of the supports cannot be one point). If there are a words of weight 2, we delete these from G'_{k_1} and puncture by the positions of these words. We find the generator of a binary code of length $2^{m-1} - 2a$ with dimension at least $2^{m-1} - m - a$ and minimum distance 4. This contradicts the Hamming bound unless $a = 0$. We have thus shown that H' is itself an extended Hamming code. Since Hamming codes are perfect, one easily sees that the words of weight 4 in an extended Hamming code form a $3-(2^m, 4, 1)$ design. For $m \geq 4$, this design has blocks that meet in one point (an easy calculation). This was excluded by Corollary 8.2.9, so we have a contradiction. This completes the proof. □

§8.3. Galois Rings over $\mathbb{Z}_4$

We wish to generalize the concept of cyclic codes to codes over $\mathbb{Z}_4$. To do this, we need some algebra. For cyclic codes of length n over $\mathbb{F}_2$, we needed an extension field that contained an nth root of unity. The present situation is similar. We must consider an extension of $\mathbb{Z}_4$ by an nth root of unity. We call such an extension a *Galois Ring*.

We need some preparation on irreducible polynomials in $\mathbb{Z}_4[x]$. Consider a polynomial $f(x)$ in $\mathbb{Z}_2[x]$ and write this as $f(x) = a(x^2) - xb(x^2)$ (where we use the minus sign because we will shortly do calculations over $\mathbb{Z}_4$). Define the map ϕ by

$$\phi(f)(x) = F(x) := \pm(a(x)^2 - xb(x)^2) \in \mathbb{Z}_4[x],$$

where the sign in $\pm$ is chosen in such a way that the coefficient of the highest power of x is 1. Note that $\phi(x^n - 1) = x^n - 1$ if n is odd. Clearly the inverse mapping is $f(x) \equiv F(x) \pmod 2$.

It is a trivial calculation to check that $\phi(fg) = \phi(f)\phi(g)$. Since both ϕ and its inverse do not change the degree of a polynomial nor the coefficient of the highest power of x, irreducible polynomials correspond to irreducible polynomials. It follows that if $x^n - 1$ (where $n = 2^m - 1$) can be written as a product $f_1(x)f_2(x)\dots f_l(x)$ of irreducible polynomials in $\mathbb{Z}_2[x]$, then $F_1(x)F_2(x)\dots F_l(x)$, where $F_i := \phi(f_i)$, is the (unique) factorization of $x^n - 1$ in $\mathbb{Z}_4[x]$. An irreducible factor $h(x)$ of degree m is called a *basic primitive polynomial* in $\mathbb{Z}_4[x]$.

This method of *lifting* an irreducible polynomial in $\mathbb{Z}_2[x]$ to an irreducible polynomial in $\mathbb{Z}_4[x]$ (with the map ϕ) is known as *Graeffe's method* (see [100]). It is a special case of a result known as Hensel's lemma (see [94]).

(8.3.1) EXAMPLE. Consider the primitive polynomial $f(x) = x^3 + x + 1$, a factor of $x^7 - 1$. With the notation used above, we have $a(x) = 1$ and $b(x) = -1 - x$, so $F(x) = x^3 + 2x^2 + x - 1 \equiv x^3 + x + 1 \pmod 2$. In this way we find the factorization

$$x^7 - 1 = (x^3 + 2x^2 + x - 1)(x^3 - x^2 + 2x - 1)(x - 1),$$

where the first two factors are basic primitive polynomials in $\mathbb{Z}_4[x]$.

(8.3.2) Definition. The Galois Ring $GR(4, m)$ is defined to be $\mathbb{Z}_4[\xi]$, where ξ is a zero of $h(x)$ (so ξ is an nth root of unity; $\mathbb{Z}_4[\xi] = \mathbb{Z}_4[x]/(h(x))$).

Note that we again have

$$h(x) = x^3 + 2x^2 + x - 1 = (x - \xi)(x - \xi^2)(x - \xi^4).$$

Just as in Example 1.1.23, one can represent elements of $GR(4, m)$ as polynomials of degree $< m$ in ξ, with coefficients in $\mathbb{Z}_4$.

(8.3.3) EXAMPLE. For $GR(4, 3)$, generated by $x^3 + 2x^2 + x - 1$, we have the following table for the set $\{0, 1, \xi, \xi^2, \dots, \xi^6\}$. Here

$$c = \sum_{i=0}^{2} a_i \xi^i, \qquad (\xi^3 = 1 + 3\xi + 2\xi^2).$$

c	a_0	a_1	a_2
0	0	0	0
1	1	0	0
ξ	0	1	0
ξ^2	0	0	1
ξ^3	1	3	2
ξ^4	2	3	3
ξ^5	3	3	1
ξ^6	1	2	1

Such a table does not tell us how to express elements of $GR(4, m)$ that are not in the set

$$\mathcal{T} := \{0, 1, \xi, \xi^2, \ldots, \xi^{n-1}\}.$$

Observe that from the representation of ξ^i as polynomial of degree $< m$ in ξ, we see that $2\xi^i = 2$ implies that $\xi^i = 1$, i. e. $\xi = 1$. Hence if t runs through $\mathcal{T}$, then the elements $2t$ are all different.

Consider the 4^m sums of the form $a + 2b$, where a and b are in the set $\mathcal{T}$. If $a + 2b = a' + 2b'$, then $2a = 2a'$, so $a = a'$, and then it follows that $b = b'$. Therefore the sums $a + 2b$ are all different and hence represent all the elements of $GR(4, m)$. The set $\mathcal{T}$ is the set of squares of elements of $GR(4, m)$.

To find the representation of a given element, we use the following lemma.

(8.3.4) Lemma. *We have*

$$(x + y)^{2^k} \equiv x^{2^k} + 2x^{2^{k-1}} y^{2^{k-1}} + y^{2^k} \pmod 4.$$

PROOF. For $k = 1$, this is trivial. Squaring both sides, the result follows by induction. □

So, if $c = a + 2b, a \in \mathcal{T}, b \in \mathcal{T}$, then

$$c^{2^m} = a^{2^m} = a.$$

Therefore, the map $\tau : c \mapsto a$ is given by

$$\tau(c) = c^{2^m} \qquad (c \in GR(4, m), n = 2^m - 1). \tag{8.3.5}$$

Once a is known, b follows from $c = a + 2b$.

Clearly $\tau(cd) = \tau(c)\tau(d)$ and by Lemma 8.3.4 we have

$$\tau(c + d) = \tau(c) + \tau(d) + 2(cd)^{2^{m-1}}.$$

We can now describe the structure of the ring $R := GR(4, m)$. The unique maximal ideal in R is the set $2\mathcal{T}$. The product of any two elements in this set is 0 (so in R there are zero divisors). The remaining set, $R^* := R\backslash(2\mathcal{T})$, consists of the *invertible* elements. They form a multiplicative group of order $(2^m - 1)2^m$. This group is a direct product of the cyclic group H of order n generated by ξ and the group $\mathcal{E}$ consisting of the elements of the form $1 + 2t, t \in \mathcal{T}$; (these are the *principal units* of R).

To understand the structure of $\mathcal{E}$, observe that if t_i and t_j are in $\mathcal{T}$, then $t_i + t_j = a_{ij} + 2b_{ij}$ with $a_{ij} \in \mathcal{T}, b_{ij} \in \mathcal{T}$. Therefore $(1+2t_i)(1+2t_j) = 1+2a_{ij}$. Furthermore, the additive representation of a_{ij} (as in the table of Example 8.3.3) is congruent mod 2 to the sum of the representations of t_i and t_j. Indeed, mod 2 the elements in the table are the additive group of $\mathbb{F}_{2^m}$. So, $\mathcal{E}$ is isomorphic to this group. Every element of R^* has a unique representation of the form $\xi^r(1 + 2t), 0 \le r < n, t \in \mathcal{T}$. The elements with fixed r form a residue class of the ring $R/(2R) = R/(2\mathcal{T})$ and $2\mathcal{T}$ is the class corresponding to 0. Let θ be a zero of $h_2(x)$, i. e. a primitive element of $\mathbb{F}_{2^m}$. We have already observed that the table for the values of ξ^r, taken mod 2, is the addition table for $\mathbb{F}_{2^m}$ with generator θ. It follows that the map μ that sends ξ^r to θ^r and 0 to 0 is an isomorphism of the residue class ring $R/(2\mathcal{T})$ onto the field $\mathbb{F}_{2^m}$.

Below, we shall need this isomorphism and the fact that $2\xi^r = 2\xi^s$ implies that $r = s$. We shall need a number of results on dependencies among the powers ξ^r. We formulate these as a lemma.

(8.3.6) Lemma. *Let $m \geq 2$. Consider $GR(4, m) = \mathbb{Z}_4[\xi]$ with $\xi^n = 1$, ($n = 2^m - 1$). We have*

(1) *$\pm\xi^j \pm \xi^k$ is invertible for $0 \leq j < k < n$;*

(2) *$\xi^j - \xi^k \neq \pm\xi^l$ for distinct j, k, l in $[0, n-1]$;*

(3) *if $m \geq 3$ and $i \neq j, k \neq l$ in $[0, n-1]$, then*

$$\xi^i - \xi^j = \xi^k - \xi^l \Rightarrow i = k \text{ and } j = l;$$

(4) *if $m \geq 3$, m odd, then*

$$\xi^i + \xi^j + \xi^k + \xi^l = 0 \Rightarrow i = j = k = l.$$

Proof. (1) If $\pm\xi^j \pm \xi^k = 2\lambda$ for some $\lambda \in R$, then squaring both sides of the equation $\pm\xi^j = \mp\xi^k + 2\lambda$ yields $j = k$.

(2) An equation of this kind can be reduced to $1 + \xi^a = \xi^b$. Raise both sides to the power 2^m and apply Lemma 8.3.4. We find $2\xi^{a \cdot 2^{m-1}} = 0$, a contradiction.

(3) An equation of this kind can be reduced to $1 + \xi^a = \xi^b + \xi^c$. Applying Lemma 8.3.4, we find $2(\xi^a)^{2^{m-1}} = 2(\xi^{b+c})^{2^{m-1}}$, so $\xi^a = \xi^{b+c}$. The equation becomes $(\xi^b - 1)(\xi^c - 1) = 0$ and we are done by (1).

(4) Here it is necessary to work in $\mathbb{F}_{2^m}$. We use the isomorphism μ. If four powers of ξ add up to 0, there is a similar relation in which one of the powers is 1. Furthermore, since $\xi^n = 1$, we may assume that the other exponents are even. So we have

$$\xi^{2a} + \xi^{2b} = -\xi^{2c} - 1.$$

Raising both sides to the power 2^m and using Lemma 8.3.4, we find

$$2(\xi^{2a} + \xi^{a+b} + \xi^{2b}) = 2\xi^c.$$

From this equation we cannot draw a conclusion in R but it does have an implication for $\mathbb{F}_{2^m}$. Apply the map μ and write $x := \theta^a, y := \theta^b, z := \theta^c$. From the last equation, we conclude that $x^2 + xy + y^2 = z$. The original equation implies $x^2 + y^2 + z^2 + 1 = 0$, and hence $x + y = z + 1$. From these two we find $x^2 + y^2 = (x+1)(y+1)$. We must show that this implies $x = y = z$, because that shows that $a = b = c = 0$.

Assume that $x \neq 1$. Write $x = u + 1$, $y = ut + 1$. We find the equation $u^2(t^2 + t + 1) = 0$. Since $u \neq 0$, we must have $t^2 + t + 1 = 0$, which contradicts the fact that m is odd. □

§8.4. Cyclic Codes over $\mathbb{Z}_4$

In the same way as we did for $\mathbb{F}_q$, we now study cyclic codes over $\mathbb{Z}_4$. Again, we identify a word $\mathbf{c} = (c_0, c_1, \ldots, c_{n-1})$ with the polynomial $c(x) := c_0 + c_1 x + \ldots + c_{n-1}x^{n-1}$ in the ring $\mathcal{R} := \mathbb{Z}_4[x]/(x^n - 1)$. Not everything generalizes! $\mathcal{R}$ is not a unique factorization domain, i.e. some polynomials in $\mathcal{R}$ can be written as a product of irreducible polynomials in more than one way. This will not influence our arguments.

As usual, we define codes either by giving a generator matrix or a parity check matrix. These have a more compact form than those in §8.1 (as always with cyclic codes; now, elements of R are to be replaced by column vectors, corresponding to their representation of the type of Example 8.3.3). If we are interested in the extension of a code, we first add a column of 0's to the parity check matrix and then a row of 1's. This increases the length of the $\mathbb{Z}_4$-code by one, but the binary image becomes two symbols longer.

(8.4.1) EXAMPLE. Consider the quaternary cyclic code C of length $n = 2^m - 1$ with generator $(2 \quad 2\xi \quad \ldots \quad 2\xi^{n-1})$. This is a trivial code, a binary code in disguise. All the codewords of the binary image have the form $(\mathbf{c}, \mathbf{c})$, where $\mathbf{c}$ is in the dual of the $[2^m - 1, 2^m - m - 1, 3]$ Hamming code.

Next, consider the $\mathbb{Z}_4$-code with generator matrix

$$G := \begin{pmatrix} 1 & 1 & 1 & \cdots & 1 \\ 0 & 2 & 2\xi & \cdots & 2\xi^{n-1} \end{pmatrix}.$$

Both C and C' have 2^{m+2} codewords. The binary image has length 2^{m+1} and is clearly the first order Reed-Muller code $\mathcal{R}(1, m+1)$. We had already seen that this code is $\mathbb{Z}_4$-linear in Example 8.2.8.

We now come to the codes that caused the sudden surge of interest in codes over $\mathbb{Z}_4$. Consider an extended cyclic $\mathbb{Z}_4$-code C_m of length $n + 1 = 2^m$ (m odd) with parity check matrix

$$H := \begin{pmatrix} 1 & 1 & 1 & \cdots & 1 \\ 0 & 1 & \xi & \cdots & \xi^{n-1} \end{pmatrix},$$

ξ is a primitive nth root of unity in $GR(4, m)$.

(8.4.2) Lemma. *C_m has Lee distance $d \geq 6$.*

PROOF. The first row of H implies that every codeword in C_m has an even number of odd entries. Therefore it has even Lee weight. So, d is even. In the four cases below, we consider possible codewords with nonzero coordinates not in the initial position. For the corresponding situations, where there is a nonzero coordinate in front, similar equations with one term less have to be considered and we leave that to the reader.

(1) If a codeword in C_m had Lee weight 2, the two nonzero coordinates would be 1 and -1. Then the second row of H implies that there are i and j such that $\xi^i = \xi^j$, which is false.

(2) If a codeword in C_m had Lee weight 4 and the nonzero coordinates were 1,1,2 (or 3,3,2), we would have $\xi^i + \xi^j = \pm 2\xi^k$ for some i, j, k. By Lemma 8.3.6 (1), this is impossible.

(3) If a codeword in C_m had Lee weight 4 with two coordinates 1 and two coordinates 3, there would be indices i, j, k, l such that $\xi^i - \xi^j = \xi^k - \xi^l$. By Lemma 8.3.6 (3), this is impossible.

(4) If a codeword of weight 4 existed with four equal nonzero coordinates, we would contradict Lemma 8.3.6 (4).

The reader can now easily check the four other cases. This completes the proof. □

From H and (8.1.4), we see that C_m has 4^{n-m-1} codewords. We now look at the binary image of C_m. It is a binary code of length 2^{m+1}, cardinality 2^k, where $k = 2^{m+1} - 2m - 2$ and it has minimum distance at least 6. From §7.4 we see that if we shorten this binary code, we obtain a code of length $2^{m+1} - 1$ with *the same parameters as the Preparata code*! So it is also a nearly perfect code and hence has the same weight enumerator as the Preparata code of the same length. The authors of [88] called C'_m a "Preparata" code because it is *not* equivalent to the code of §7.4.

Let us now consider the dual code $C_m^\perp$. It has H as generator matrix. By Theorem 8.2.5, the weight enumerators of the two binary images satisfy the MacWilliams relations. It was known that the weight enumerators of the Kerdock code and the extended Preparata code of the same length satisfy the MacWilliams relations; (this was the "coincidence" that has puzzled coding theorists for years). (For a proof see [46].) We now know that the binary code that is the image of the code over $\mathbb{Z}_4$ with generator H must have the same weight enumerator as a Kerdock code of that length. In [88] it is shown that the codes defined by Kerdock in [75] are in fact binary images of the $\mathbb{Z}_4$-codes we have discussed here.

We give one example of a good binary code that is the binary image of a $\mathbb{Z}_4$-code that is cyclic but not linear.

(8.4.3) EXAMPLE. Look at (11100) and its five cyclic shifts. When we compare two of these, we find three possibilities : (1) three odd-odd pairs and two even-even pairs , (2) four odd-even pairs and one odd-odd pair, (3) two odd-even pairs, two odd-odd, and one even-even. We now take this vector and three others, obtained by replacing 111 by 113 or some permutation, and 00 by 02, 20, or 22. We claim that the choice (11120), (31100), (13102), (11322) has the property that these four vectors, their negatives, and all the cyclic shifts form a $\mathbb{Z}_4$-code with Lee distance 4. When calculating Lee distance, an odd-even pair contributes 1, a pair with the same parity contributes 0 or 2. Our substitution guarantees that if there are two even-even combinations, then one contributes 2 to the distance or the other coordinates yield distance 6. A 1-3 pair also contributes 2 to the Lee distance. How could there be two codewords with Lee distance < 4? This can only happen if we have two words $(o_1, o_2, o_3, e_1, e_2)$ and $(e'_1, o'_2, o'_3, o'_1, e'_2)$ (o is odd, e is even), where $o_2 = o'_2, o_3 = o'_3$, $e_2 = e'_2$. From our choice, it is easy to see that only six pairs of words have to be checked to show that this does not happen.

Now consider the binary image of this code. It has 40 words, length 10, and minimum distance 4. Since it is known that the *Best* code of §4.4 is unique, this code must be equivalent to the Best code. This construction (cf. [83]) has the advantage that very few pairs of codewords have to be checked by hand.

§8.5. Problems

8.5.1. Construct a selfdual quaternary code of length 6. Show that such a code must contain the word (222222). Is the binary image of the code linear? Is it also selfdual?

8.5.2. Prove that the Reed-Muller code $\mathcal{R}(2, m)$ is $\mathbb{Z}_4$-linear.

8.5.3. Let $c = \xi^i + 2\xi^j$ be an element of $GR(4, m)$. Determine its inverse.

8.5.4. Consider the weight enumerator of the Nordstrom-Robinson code. Show that it is equal to its MacWilliams transform.

8.5.5. Consider the Preparata code $\mathcal{P}$ of Theorem 7.4.5. Determine the number of words of weight 5, resp. 6. Use this to find the weight enumerator of the extended Preparata code of length 16.

8.5.6. Show that the first order Reed-Muller code is a subcode of the binary image of the dual of the code C_m of §8.4.

8.5.7. Show that the code C'_m and the code of (7.4.5) are not equivalent for $m \geq 5$. (Hint: show that the linear span of C'_m has words of weight 2.)

CHAPTER 9

Goppa Codes

§9.1. Motivation

Consider once again the parity check matrix of a (narrow sense) BCH code as given in the proof of Theorem 6.6.2, i.e.

$$H = \begin{bmatrix} 1 & \beta & \beta^2 & \dots & \beta^{n-1} \\ 1 & \beta^2 & \beta^4 & \dots & \beta^{2(n-1)} \\ \vdots & \vdots & \vdots & & \vdots \\ 1 & \beta^{d-1} & \beta^{2(d-1)} & \dots & \beta^{(d-1)(n-1)} \end{bmatrix},$$

where β is a primitive nth root of unity in $\mathbb{F}_{q^m}$ and each entry is interpreted as a column vector of length m over $\mathbb{F}_q$; existence of β implies that $n|(q^m - 1)$.

In Theorem 6.6.2 we proved that the minimum distance is at least d by using the fact that any submatrix of H formed by taking $d - 1$ columns is a Vandermonde matrix and therefore has determinant $\neq 0$. Many authors have noticed that the same argument works if we replace H by

$$\hat{H} = \begin{bmatrix} h_0\beta_0 & h_1\beta_1 & \dots & h_{n-1}\beta_{n-1} \\ \vdots & \vdots & & \vdots \\ h_0\beta_0^{d-1} & h_1\beta_1^{d-1} & \dots & h_{n-1}\beta_{n-1}^{d-1} \end{bmatrix},$$

where $h_j \in \mathbb{F}_{q^m}^*$ and the β_i are different elements of $\mathbb{F}_{q^m}^*$. If $h_j \in \mathbb{F}_q$ $(0 \le j \le n-1)$ then the factors h_j have no essential effect; the code is replaced by an equivalent code. However, if the h_j are elements of $\mathbb{F}_{q^m}$, then the terms $h_j\beta_j^i$ considered as column vectors over $\mathbb{F}_q$ can be very different from the original entries.

We shall consider two ways of generalizing BCH codes in this manner. That we really get something more interesting will follow from the fact that the new classes of codes contain sequences meeting the Gilbert bound whereas long BCH codes are known to be bad.

§9.2. Goppa Codes

Let $(c_0, c_1, \ldots, c_{n-1})$ be a codeword in a BCH code with designed distance d (word length n, β a primitive nth root of unity). Then, by definition $\sum_{i=0}^{n-1} c_i(\beta^j)^i = 0$ for $1 \le j < d$. We wish to write this condition in another way. Observe that

$$(9.2.1) \qquad \frac{z^n - 1}{z - \beta^{-i}} = \sum_{k=0}^{n-1} z^k(\beta^{-i})^{n-1-k} = \sum_{k=0}^{n-1} \beta^{i(k+1)} z^k.$$

It follows that

$$(9.2.2) \qquad \sum_{i=0}^{n-1} \frac{c_i}{z - \beta^{-i}} = \frac{z^{d-1}p(z)}{z^n - 1}$$

for some polynomial $p(z)$.

If $g(z)$ is any polynomial and $g(\gamma) \ne 0$, we define $1/(z - \gamma)$ to be the unique polynomial mod $g(z)$ such that $(z - \gamma) \cdot 1/(z - \gamma) \equiv 1 \pmod{g(z)}$, i.e.

$$(9.2.3) \qquad \frac{1}{z - \gamma} = \frac{-1}{g(\gamma)} \cdot \left(\frac{g(z) - g(\gamma)}{z - \gamma}\right).$$

These observations serve as preparation for the following definition.

(9.2.4) Definition. Let $g(z)$ be a (monic) polynomial of degree t over $\mathbb{F}_{q^m}$. Let $L = \{\gamma_0, \gamma_1, \ldots, \gamma_{n-1}\} \subset \mathbb{F}_{q^m}$, such that $|L| = n$ and $g(\gamma_i) \ne 0$ for $0 \le i \le n - 1$. We define the *Goppa code* $\Gamma(L, g)$ with *Goppa polynomial* $g(z)$ to be the set of codewords $\mathbf{c} = (c_0, c_1, \ldots, c_{n-1})$ over the alphabet $\mathbb{F}_q$ for which

$$(9.2.5) \qquad \sum_{i=0}^{n-1} \frac{c_i}{z - \gamma_i} \equiv 0 \pmod{g(z)}.$$

Observe that Goppa codes are linear.

(9.2.6) EXAMPLE. From the introductory remarks we see that if we take the Goppa polynomial $g(z) = z^{d-1}$ and $L := \{\beta^{-i} | 0 \le i \le n - 1\}$, where β is a primitive nth root of unity in $\mathbb{F}_{q^m}$, the resulting Goppa code $\Gamma(L, g)$ is the narrow sense BCH code of designed distance d (cf. Problem 9.8.2).

In order to establish the connection with Section 9.1, we try to find a suitable parity check matrix for $\Gamma(L, g)$. From (9.2.5) and (9.2.3), we see that

$$\left(\frac{1}{g(\gamma_0)} \cdot \frac{g(z) - g(\gamma_0)}{(z - \gamma_0)}, \ldots, \frac{1}{g(\gamma_{n-1})} \cdot \frac{g(z) - g(\gamma_{n-1})}{(z - \gamma_{n-1})}\right),$$

with each entry interpreted as a column vector, is, in a sense, a parity check matrix. Let $h_j := g(\gamma_j)^{-1}$ and hence $h_j \ne 0$. If $g(z) = \sum_{i=0}^{t} g_i z^i$, then in the same way as (9.2.1) we have

$$\frac{g(z) - g(x)}{z - x} = \sum_{i+j \le t-1} g_{i+j+1} x^j z^i.$$

Leaving out the factors z^i, we find as parity check matrix for $\Gamma(L, g)$

$$\begin{bmatrix} h_0 g_t & \cdots & h_{n-1} g_t \\ h_0(g_{t-1} + g_t\gamma_0) & \cdots & h_{n-1}(g_{t-1} + g_t\gamma_{n-1}) \\ \vdots & \cdots & \vdots \\ h_0(g_1 + g_2\gamma_0 + \cdots + g_t\gamma_0^{t-1}) & \cdots & h_{n-1}(g_1 + g_2\gamma_{n-1} + \cdots + g_t\gamma_{n-1}^{t-1}) \end{bmatrix}$$

By a linear transformation we find the matrix we want, namely

$$H = \begin{bmatrix} h_0 & \cdots & h_{n-1} \\ h_0\gamma_0 & \cdots & h_{n-1}\gamma_{n-1} \\ \vdots & & \vdots \\ h_0\gamma_0^{t-1} & \cdots & h_{n-1}\gamma_{n-1}^{t-1} \end{bmatrix}$$

(here we have used the fact that $g_t \neq 0$).

This does not have the full generality of the matrix $\hat{H}$ of Section 9.1, where the h_j were arbitrary, since we now have $h_j = g(\gamma_j)^{-1}$.

(9.2.7) Theorem. *The Goppa code* $\Gamma(L, g)$ *defined in* (9.2.4) *has dimension* $\geq n - mt$ and minimum distance $\geq t + 1.5$.

PROOF. This follows from the parity check matrix H in exactly the same way as for BCH codes. □

The example given in (9.2.6) shows that the BCH bound (for narrow-sense BCH codes) is a special case of Theorem 9.2.7.

As a preparation for the generalization of these codes to codes on algebraic curves (see Chapter 10), we reformulate the definition of Goppa codes. Start with the field $\mathbb{F}_{q^m}$. Consider the vector space of all rational functions $f(z)$ with the following properties:

(i) $f(z)$ has zeros in all the points where $g(z)$ has zeros, each with at least the same multiplicity as the zero of $g(z)$,
(ii) $f(z)$ has no poles, except possibly in some points $\gamma_0, \gamma_1, \ldots, \gamma_{n-1}$ and in that case poles of order 1.

A code over $\mathbb{F}_{q^m}$ is defined by taking as codewords the n-tuples

$$(\mathrm{Res}_{\gamma_0} f, \mathrm{Res}_{\gamma_1} f, \ldots, \mathrm{Res}_{\gamma_{n-1}} f),$$

where the residue of $f(z)$ in a point γ_i is defined in the usual way. The Goppa code $\Gamma(L, g)$ is the subfield subcode (over $\mathbb{F}_q$) of this code.

Consider the parity check matrix H defined above. Compare the situation with Definition 6.8.2, where we take $\mathbf{v} := (h_0, h_1, \ldots, h_{n-1})$ and $\mathbf{a} := (\gamma_0, \gamma_1, \ldots, \gamma_{n-1})$, $k = t$. We see that H is the generator matrix of the code $\mathrm{GRS}_k(\mathbf{a}, \mathbf{v})$. So the Goppa code $\Gamma(L, g)$ is a subfield subcode of the dual of a generalized Reed-Solomon code.

We have thus seen that the codes, defined using polynomials as in Definition 6.8.2, and the codes that we have defined here, using residues in first order poles, are dual codes. We shall encounter the same phenomenon in the chapter on algebraic geometry codes.

§9.3. The Minimum Distance of Goppa Codes

If we change our point of view a little, it is possible to obtain (9.2.7) in another way and also some improvements. As before, let $\mathscr{R}$ denote $(\mathbb{F}_q)^n$ with Hamming distance. Let L be as in (9.2.4). We define

$$\mathscr{R}^* := \left\{ \xi(z) = \sum_{i=0}^{n-1} \frac{b_i}{z - \gamma_i} \middle| (b_0, b_1, \ldots, b_{n-1}) \in \mathscr{R} \right\}$$

and we define the distance of two rational functions $\xi(z)$, $\eta(z)$ by

$$d(\xi(z), \eta(z)) := \|\xi(z) - \eta(z)\|,$$

where $\|\xi(z)\|$ denotes the degree of the denominator when $\xi(z)$ is written as $n(z)/d(z)$ with $(n(z), d(z)) = 1$. This is easily seen to be a distance function and in fact the mapping $(b_0, \ldots, b_{n-1}) \to \sum_{i=0}^{n-1} b_i/(z - \gamma_i)$ is an isometry from $\mathscr{R}$ onto $\mathscr{R}^*$. We shall use this terminology to study Goppa codes by considering the left-hand side of (9.2.5) as an element of $\mathscr{R}^*$ (i.e. we do not apply (9.2.3)). If $\xi(z) = n(z)/d(z)$ corresponds to a nonzero codeword, then degree $d(z) \geq$ deg $n(z) + 1$. The requirement $\xi(z) \equiv 0 \pmod{g(z)}$ implies that $g(z)$ divides $n(z)$, i.e. degree $d(z) \geq t + 1$. So we have $\|\xi(z)\| \geq t + 1$, which is the result of Theorem 9.2.7.

If we write $\xi(z)$ as $n(z)/d(z)$, where now $d(z)$ equals the product of all n factors $(z - \gamma_i)$, then we can improve our estimate for the minimum distance if the degrees of $n(z)$ and $d(z)$ differ by more than 1. The coefficient of z^{n-1} in $n(z)$ is $\sum_{i=0}^{n-1} b_i$. It follows that if we add an extra parity check equation $\sum_{i=0}^{n-1} b_i = 0$, the estimate for the minimum distance will increase by 1 and the dimension of the code will decrease by at most 1. We can use the same idea for other coefficients. The coefficient of z^{n-s-1} in the numerator $n(z)$ is $(-1)^s \sum_{i=0}^{n-1} b_i \sum'_{j_1, j_2, \ldots, j_s} \gamma_{j_1}\gamma_{j_2}\cdots\gamma_{j_s}$ (where $\sum'$ indicates that $j_\nu \neq i$ for $\nu = 1, 2, \ldots, s$). This coefficient is a linear combination of the sums $\sum_{i=0}^{n-1} b_i\gamma_i^r$ $(0 \leq r \leq s)$. It follows that if we add $s + 1$ parity check equations, namely $\sum_{i=0}^{n-1} b_i\gamma_i^r = 0$ $(0 \leq r \leq s)$, we find a code with dimension at least $n - tm - (1 + sm)$ and minimum distance at least $t + s + 2$. How does this compare to simply replacing $g(z)$ by another Goppa polynomial with degree $t + s$? The first method has the advantage that $\sum_{i=0}^{n-1} b_i\gamma_i = 0$ implies that $\sum_{i=0}^{n-1} b_i\gamma_i^q = 0$. Hence, once in q times we are sure that the dimension does not decrease.

(9.3.1) Theorem. *Let $q = 2$ and let $g(z)$ have no multiple zeros. Then $\Gamma(L, g)$ has minimum distance at least $2t + 1$ (where $t =$ degree $g(z)$).*

PROOF. Let $(c_0, c_1, \ldots, c_{n-1})$ be a codeword. Define $f(z) = \prod_{i=0}^{n-1} (z - \gamma_i)^{c_i}$. Then $\xi(z) = \sum_{i=0}^{n-1} c_i/(z - \gamma_i) = f'(z)/f(z)$, where $f'(z)$ is the formal derivative. In $f'(z)$ only even powers of z occur, i.e. $f'(z)$ is a perfect square. Since we require that $g(z)$ divides $f'(z)$ we must actually have $g^2(z)$ dividing $f'(z)$. So our previous argument yields $d \geq 2t + 1$. □

Of course one can combine Theorem 9.3.1 with the idea of intersecting with a BCH code.

§9.4. Asymptotic Behaviour of Goppa Codes

In Section 6.6 we pointed out that long primitive BCH codes are bad. This fact is connected with Theorem 6.6.7. It was shown by T. Kasami (1969; [39]) that a family of cyclic codes for which the extended codes are invariant under the affine group is bad in the same sense: a subsequence of codes C_i with length n_i, dimension k_i and distance d_i must have lim inf$(k_i/n_i) = 0$ or lim inf$(d_i/n_i) = 0$. We shall now show that the class of Goppa is considerably larger.

(9.4.1) Theorem. *There exists a sequence of Goppa codes over $\mathbb{F}_q$ which meets the Gilbert bound.*

PROOF. We first pick parameters $n = q^m$, t and d, choose $L = \mathbb{F}_{q^m}$, and we try to find an irreducible polynomial $g(z)$ of degree t over $\mathbb{F}_{q^m}$ such that $\Gamma(L, g)$ has minimum distance at least d. Let $\mathbf{c} = (c_0, c_1, \ldots, c_{n-1})$ be any word of weight $j < d$, i.e. a word we do not want in $\Gamma(L, g)$. Since $\sum_{i=0}^{n-1} c_i/(z - \gamma_i)$ has numerator of degree at most $j - 1$, there are at most $\lfloor (j-1)/t \rfloor$ polynomials $g(z)$ for which $\Gamma(L, g)$ does contain $\mathbf{c}$. This means that to ensure distance d, we have to exclude at most $\sum_{j=1}^{d-1} \lfloor (j-1)/t \rfloor (q-1)^j \binom{n}{j}$ irreducible polynomials of degree t. This number is less than $(d/t)V_q(n, d-1)$ (cf. (5.1.4)). By (1.1.19) the number of irreducible polynomials of degree t over $\mathbb{F}_{q^m}$ exceeds $(1/t)q^{mt}(1 - q^{-(1/2)mt+m})$. So a sufficient condition for the existence of the code $\Gamma(L, g)$ that we are looking for is

$$\frac{d}{t} V_q(n, d-1) < \frac{1}{t} q^{mt}(1 - q^{-(1/2)mt+m}). \tag{9.4.2}$$

By Theorem 9.2.7, the code has at least q^{n-mt} words. On both sides of (9.4.2) we take logarithms to base q and divide by n. Suppose n is variable, $n \to \infty$, and $d/n \to \delta$. Using Lemma 5.1.6 we find

$$H_q(\delta) + o(1) < \frac{mt}{n} + o(1).$$

The information rate of the code $\Gamma(L, g)$ is $\geq 1 - mt/n$. It follows that we can find a sequence of polynomials $g(z)$ such that the corresponding Goppa codes have information rate tending to $1 - H_q(\delta)$. This is the Gilbert bound (5.1.9). □

§9.5. Decoding Goppa Codes

The Berlekamp decoder for BCH codes which was mentioned at the end of Section 6.7 can also be used to decode Goppa codes. In order to show this, we shall proceed in the same way as in Section 6.7.

Let $(C_0, C_1, \ldots, C_{n-1})$ be a codeword in $\Gamma(L, g)$ as defined in (9.2.4) and suppose we receive $(R_0, R_1, \ldots, R_{n-1})$. We denote the error vector by $(E_0, E_1, \ldots, E_{n-1}) = \mathbf{R} - \mathbf{C}$. Let $M := \{i \mid E_i \neq 0\}$. We denote the degree of $g(x)$ by t and assume that $|M| = e < \frac{1}{2}t$. Again using the convention (9.2.3) we define a polynomial $S(x)$, called the *syndrome*, by

(9.5.1) $$S(x) \equiv \sum_{i=0}^{n-1} \frac{E_i}{x - \gamma_i} \pmod{g(x)}.$$

Observe that $S(x)$ can be calculated by the receiver using $\mathbf{R}$ and (9.2.5). We now define the *error-locator polynomial* $\sigma(z)$ and a companion polynomial $\omega(z)$ in a manner similar to Section 6.7 (but this time using the locations themselves instead of inverses).

(9.5.2) $$\sigma(z) := \prod_{i \in M} (z - \gamma_i),$$

(9.5.3) $$\omega(z) := \sum_{i \in M} E_i \prod_{j \in M \setminus \{i\}} (z - \gamma_j).$$

From the definitions it follows that $\sigma(z)$ and $\omega(z)$ have no common factors, $\sigma(z)$ has degree e, and $\omega(z)$ has degree $< e$. The computation of $\omega(z)/\sigma(z)$ of Section 6.7 is replaced by the following argument.

(9.5.4) $$S(z)\sigma(z) \equiv \sum_{i=0}^{n-1} \frac{E_i}{z - \gamma_i} \prod_{i \in M} (z - \gamma_i)$$
$$\equiv \omega(z) \pmod{g(z)}.$$

Now suppose we have an algorithm which finds the monic polynomial $\sigma_1(z)$ of lowest degree ($\sigma_1(z) \neq 0$) and a polynomial $\omega_1(z)$ of lower degree, such that

(9.5.5) $$S(z)\sigma_1(z) \equiv \omega_1(z) \pmod{g(z)}.$$

It follows that

$$\omega_1(z)\sigma(z) - \omega(z)\sigma_1(z) \equiv 0 \pmod{g(z)}.$$

Since the degree of the left-hand side is less than the degree of $g(z)$, we find that the left-hand side is 0. Then $(\sigma(z), \omega(z)) = 1$ implies that $\sigma(z)$ divides $\sigma_1(z)$ and therefore we must have $\sigma_1(z) = \sigma(z)$. Once we have found $\sigma(z)$ and $\omega(z)$, it is clear that we know **E**. The Berlekamp-algorithm is an efficient way of computing $\sigma_1(z)$. There are other methods, based on Euclid's algorithm for finding the g.c.d. of two polynomials (cf. [51]).

§9.6. Generalized BCH Codes

Let us take another look at Goppa codes. We consider $L = \{1, \beta, \beta^2, \ldots, \beta^{n-1}\}$ where β is a primitive nth root of unity in $\mathbb{F}_{2^m}$, $g(z)$ a suitable polynomial. Let $(a_0, a_1, \ldots, a_{n-1}) \in \Gamma(L, g)$. As in (6.5.2) we denote by $A(X)$ the Mattson-Solomon polynomial of $a_0 + a_1 x + \cdots + a_{n-1} x^{n-1}$.

Consider the polynomial

$$(X^n - 1) \sum_{i=0}^{n-1} \frac{A(\beta^i)}{X - \beta^i} = (X^n - 1) n \sum_{i=0}^{n-1} \frac{a_i}{X - \beta^i}$$

(by Lemma 6.5.3). The left-hand side is a polynomial of degree $\leq n - 1$ which takes on the value $n\beta^{-i}A(\beta^i)$ for $X = \beta^i$ $(0 \leq i \leq n - 1)$. We can replace n by 1 since we are working over $\mathbb{F}_2$. Therefore the left-hand side is the polynomial $X^{n-1} \circ A(X)$ (using the notation of Section 6.5) because this also has degree $\leq n - 1$ and takes on the same values in the nth roots of unity. Hence we have proved the following theorem.

(9.6.1) Theorem. *If $L = \{1, \beta, \ldots, \beta^{n-1}\}$ where β is a primitive nth root of unity in $\mathbb{F}_{2^m}$ and $(g(z), z^n - 1) = 1$, then the binary Goppa code $\Gamma(L, g)$ consists of the words $(a_0, a_1, \ldots, a_{n-1})$ such that the Mattson-Solomon polynomial $A(X)$ of $a(x)$ satisfies*

$$X^{n-1} \circ A(X) \equiv 0 \pmod{g(X)}.$$

In Theorem 6.6.2 we proved the BCH bound by applying Theorem 6.5.5 and by using the fact that the Mattson-Solomon polynomial of a codeword has sufficiently small degree. For Goppa codes a similar argument works. The polynomial $g(X)$ has degree t and $(g(X), X^n - 1) = 1$. It then follows from Theorem 9.6.1 that at most $n - 1 - t$ nth roots of unity are zeros of $A(X)$. This means that **a** has weight at least $t + 1$, yielding a second proof of Theorem 9.2.7.

The argument above shows how to generalize these codes. The trick is to ensure that the Mattson-Solomon polynomial of a codeword has few nth roots of unity as zeros. This idea was used by R. T. Chien and D. M. Choy (1975; [13]) in the following way.

(9.6.2) Definition. Let $(T, +, \circ)$ be as in Section 6.5 with $\mathscr{F} = \mathbb{F}_{q^m}$ and $S = \mathbb{F}_q[x] \bmod (x^n - 1)$. Let $P(X)$ and $G(X)$ be two polynomials in T such that $(P(X), X^n - 1) = (G(X), X^n - 1) = 1$. The *generalized* BCH *code* (= GBCH code) of length n over $\mathbb{F}_q$ with polynomial pair $(P(X), G(X))$ is defined as

$$\{a(x) \in S \mid P(X) \circ (\Phi a)(X) \equiv 0 \pmod{G(X)}\}.$$

A GBCH code is obviously linear.

(9.6.3) Theorem. *The minimum distance of the* GBCH *code of* (8.6.2) *is at least* $1 + \text{degree } G(X)$.

Proof. We apply Theorem 6.5.5. A common factor $f(X)$ of Φa and $X^n - 1$ is also a factor of $P(X) \circ (\Phi a)(X)$. But $(G(X), f(X)) = 1$. So the degree of $f(X)$ must be at most $n - 1 - \text{degree } G(X)$. □

Notice that the special Goppa codes of Theorem 8.6.1 are examples of GBCH codes. If we take $P(X) = X^{n-1}$ and $G(X) = X^{d-1}$ we obtain a BCH code.

The GBCH codes have a parity check matrix like $\hat{H}$ of Section 8.1. In order to show this, we consider the polynomials $p(x) = (\Phi^{-1}P)(x) = \sum_{i=0}^{n-1} p_i x^i$ and $g(x) = (\Phi^{-1}G)(x) = \sum_{i=0}^{n-1} g_i x^i$. By Lemma 6.5.3, all the coefficients of $p(x)$ and $g(x)$ are nonzero since nth roots of unity are not zeros of $P(X)$ or $G(X)$. Let $a(x)$ be a codeword and $A(X) = (\Phi a)(X)$. By (9.6.2) there is a polynomial $B(X)$ of degree at most $n - 1 - \text{degree } G(X)$ such that

$$P(X) \circ A(X) = B(X)G(X) = B(X) \circ G(X).$$

Define $b(x) := (\Phi^{-1}B)(x) = \sum_{i=0}^{n-1} b_i x^i$. Then we have

$$p(x) * a(x) = b(x) * g(x),$$

i.e.

$$\sum_{i=0}^{n-1} p_i a_i x^i = \sum_{i=0}^{n-1} b_i g_i x^i.$$

So we have found that $b_i = p_i g_i^{-1} a_i \ (0 \le i \le n - 1)$. Let $h_i := p_i g_i^{-1}$ and define

$$H := \begin{bmatrix} h_0 & h_1\beta & \dots & h_{n-1}\beta^{n-1} \\ h_0 & h_1\beta^2 & \dots & h_{n-1}\beta^{2(n-1)} \\ \vdots & \vdots & & \vdots \\ h_0 & h_1\beta^t & \dots & h_{n-1}\beta^{t(n-1)} \end{bmatrix}$$

Let $1 \le j \le t = \text{degree } G(X)$. Then $B_{n-j} = 0$. By (6.5.2)

$$B_{n-j} = b(\beta^j) = \sum_{i=0}^{n-1} b_i \beta^{ij} = \sum_{i=0}^{n-1} h_i a_i \beta^{ij}.$$

So $\mathbf{a}H^T = \mathbf{0}$. Conversely, if $\mathbf{a}H^T = \mathbf{0}$ we find that the degree of $B(X)$ is at most $n - 1 - t$. Hence $B(X)G(X) = B(X) \circ G(X) = P(X) \circ A(X)$, i.e. $\mathbf{a}$ is in the GBCH code.

§9.7. Comments

The codes described in Section 9.1 are called *alternant codes*. The codes treated in this chapter are special cases depending on the choice of h_i and β_i. The first interesting subclass seems to be the class of *Srivastava codes* introduced by J. N. Srivastava in 1967 (unpublished). E. R. Berlekamp [2] recognized their possibilities and recommended further study of this area. The alternant codes were introduced by H. J. Helgert (1974; [35]). The most interesting turned out to be the Goppa codes introduced by V. D. Goppa [27] in 1970 (cf. [4]).

BCH codes are the only cyclic Goppa codes (cf. Problem 9.8.2) but E. R. Berlekamp and O. Moreno [5] showed that extended 2-error-correcting binary Goppa codes are cyclic. Later K. K. Tzeng and K. P. Zimmerman [71] proved a similar result for other Goppa codes and the same authors have generalized the idea of Goppa codes.

§9.8. Problems

9.8.1. Let L consist of the primitive 15th roots of unity in $\mathbb{F}_{2^4}$ (take $\alpha^4 + \alpha + 1 = 0$). Let $g(z) := z^2 + 1$. Analyze the binary Goppa code $\Gamma(L, g)$.

9.8.2. Let α be a primitive nth root of unity in $\mathbb{F}_{2^m}$ and let $L := \{1, \alpha, \alpha^2, \ldots, \alpha^{n-1}\}$. Let the binary Goppa code $C = \Gamma(L, g)$ be a cyclic code. Show that $g(z) = z^t$ for some t, i.e. C is a BCH code.

9.8.3. Let $n = q^m - 1$, $L = \mathbb{F}_{q^m}\backslash\{0\}$. Let C_1 be the BCH code of length n over $\mathbb{F}_q$ obtained by taking $l = 0$ and $\delta = d_1$ in (6.6.1). Let C_2 be a Goppa code $\Gamma(L, g)$. Show that $C_1 \cap C_2$ has minimum distance $d \geq d_1 + d_2 - 1$, where $d_2 := 1 + \deg g$.

9.8.4. Consider the GBCH code with polynomial pair $(P(X), G(X))$ where $G(X)$ has degree t. Show that there is a polynomial $\hat{P}(X)$ such that the pair $(\hat{P}(X), X^t)$ defines the same code.

9.8.5. Let C be the binary cyclic code of length 15 with generator $x^2 + x + 1$. Show that C is a BCH code but not a Goppa code.

CHAPTER 10
Algebraic Geometry Codes

§10.1. Introduction

One of the major developments in the theory of error-correcting codes in the 80's was the introduction of methods from algebraic geometry to construct *good* codes. The ideas are based on generalizations of the Goppa codes of the previous chapter. The algebraic geometry codes were also inspired by ideas of Goppa. In fact, the codes of Chapter 9 are now sometimes called "classical" Goppa codes, and those of this chapter "geometric" Goppa codes. We use the terminology " algebraic geometry" codes.

An intriguing development was a paper by Tsfasman, Vlăduţ, and Zink [99]. By using codes from algebraic curves and deep results from algebraic geometry, a sequence of error-correcting codes was constructed that led to a new lower bound on the information rate of good codes. The bound improved the Gilbert-Varshamov bound (5.1.9). This was the first improvement of that bound in thirty years.

This chapter is based on expository lectures given by the author in 1988 (see [92]) and a course given by G. van der Geer and the author in 1987 (see [73]). We shall only treat the necessary algebraic geometry superficially, often omitting proofs. We point out that in a number of places, more algebraic background is necessary than was treated in Chapter 1.

Much of the recent work on these codes concerns decoding methods. In this book, we have not been concerned too much with decoding, so we do not go into these methods. Actually, the only decoding method that is used extensively in practice (for block codes) is the one of §6.7. It is not clear when algebraic geometry codes will become of practical importance.

We point out that many of the results concerning these codes can be derived without using the heavy machinery of algebraic geometry. For an exposition of that approach we refer to [90].

In §6.9, we saw that it was possible to define Reed-Solomon codes by considering points with coordinates in $\mathbb{F}_q$ on the projective line (possibly over the algebraic closure of $\mathbb{F}_q$). Codewords were defined by considering rational functions with a pole of restricted order at a specified point and taking the values of these functions at the given points as coordinates. In §9.2, we defined Goppa codes by calculating residues of certain functions at given points. The set of functions was restricted by requirements on their zeros and poles. These two ideas are what we shall generalize in this chapter. We must study algebraic curves, find a way to describe the restrictions on the set of functions that we use, and generalize the concept of residue. We describe two classes of codes that are duals.

§10.2. Algebraic Curves

In the following, k is an algebraically closed field. In our applications, k will be the algebraic closure of $\mathbb{F}_q$. (In this section, the reader may think of k as $\mathbb{C}$ if that makes things easier; however, one must be careful since some situations are quite different for the fields we consider.) $\mathbb{A}^n$ will denote n-dimensional affine space (with coordinates $x_1, x_2, \dots, x_n$). Similarly, $\mathbb{P}^n$ will be n-dimensional projective space (with homogeneous coordinates $x_0, x_1, \dots, x_n$). First, we discuss the affine case. The situation for projective spaces is slightly more complicated.

In the space $\mathbb{A}^n$, we introduce a topology, the so-called *Zariski topology*. The *closed* sets are the sets of zeros of ideals a of $k[x_1, x_2, \dots, x_n]$, i.e.

$$B = V(a) := \{(x_1, x_2, \dots, x_n) \in \mathbb{A}^n \mid f(x_1, x_2, \dots, x_n) = 0 \text{ for all } f \in a\}.$$

We always assume that a is radical, i.e. a consists of *all* the polynomials that vanish on B. (An ideal a is called radical if, for all $n \in \mathbb{N}$, $f^n \in a \Rightarrow f \in a$.) A closed subset B is called *irreducible* if B cannot be written as the union of two proper closed subsets of B. The set $V(a)$ is irreducible iff a is a prime ideal. An *open* set is the complement of a closed set.

(10.2.1) Example. In the affine plane, consider the principal ideal generated by $x^2 - y^2$. The corresponding closed set is the union of two lines with equations $y = x$, respectively $y = -x$. Each of these lines is an irreducible closed set in the plane $\mathbb{A}^2$.

All the curves in affine or projective space in this paragraph are required to be irreducible.

Consider a prime ideal p in the ring $k[x_1, x_2, \dots, x_n]$. The set $\mathcal{X}$ of zeros of p is called an *affine variety*.

(10.2.2) Example. In 3-dimensional space, we consider the unit sphere, i.e. the set with equation $x^2+y^2+z^2 = 1$. In our terminology, this is the affine variety consisting of the zeros of the ideal p, generated by the polynomial $x^2 + y^2 + z^2 - 1$. We are just using algebraic terminology to describe geometric objects that are defined by equations.

Two polynomials that differ by an element of p will have the same value in each point of $\mathcal{X}$. This is the reason for introducing the following ring.

(10.2.3) Definition. The ring $k[x_1, x_2, \ldots, x_n]/p$ is called the *coordinate ring* $k[\mathcal{X}]$ of the variety $\mathcal{X}$.

The coordinate ring is an integral domain since p is a prime ideal. Therefore, we can make the following definition.

(10.2.4) Definition. The quotient field of the ring $k[\mathcal{X}]$ is denoted by $k(\mathcal{X})$. It is called the *function field* of $\mathcal{X}$. The *dimension* of the variety $\mathcal{X}$ is the transcendence degree of $k(\mathcal{X})$ over k. If this dimension is 1, $\mathcal{X}$ is called an algebraic curve.

(10.2.5) EXAMPLE. In the affine plane over the field k, we consider the parabola $\mathcal{X}$ with equation $y^2 = x$. In this example, the coordinate ring $k[\mathcal{X}]$ consists of all the expressions of the form $A + By$, where A and B are in $k[x]$ and y satisfies $y^2 = x$. So, $k(\mathcal{X})$ is an algebraic extension of $k(x)$ by the element y, satisfying this equation of degree 2.

In projective space $\mathbb{P}^n$, the situation is complicated by the fact that we must use homogeneous coordinates. This means that it only makes sense to study rational functions for which numerator and denominator are homogeneous polynomials of the same degree. A projective variety $\mathcal{X}$ is the zero set in $\mathbb{P}^n$ of a homogeneous prime ideal p in $k[x_0, x_1, \ldots, x_n]$. Consider the subring $R(\mathcal{X})$ of $k(x_0, x_1, \ldots, x_n)$ consisting of the fractions f/g, where f and g are homogeneous polynomials of the same degree and $g \notin p$. Then $R(\mathcal{X})$ has a unique maximal ideal $M(\mathcal{X})$ consisting of all those f/g with $f \in p$. The function field $k(\mathcal{X})$ is by definition $R(\mathcal{X})/M(\mathcal{X})$.

Now, let $\mathcal{X}$ be an affine or a projective variety. Let P be a point on $\mathcal{X}$ and let U be a neighborhood of this point. Let f and g be polynomials, respectively homogeneous polynomials of the same degree, and let $g(P) \neq 0$. Then the quotient $\phi = f/g$, defined on U, is called *regular* in the point P. The functions that are regular in every point of the set U form a ring, denoted by $k[U]$. Since k is algebraically closed, there are no regular functions on $\mathcal{X}$ except constant functions, if $\mathcal{X}$ is projective.

(10.2.6) Definition. The *local ring* O_P (sometimes denoted by $O_P(\mathcal{X})$) of the point P on the variety $\mathcal{X}$ is the set of rational functions on $\mathcal{X}$ that are regular in P.

The reader familiar with algebraic terminology will realize that this is indeed a "local ring" in the algebraic sense, i. e. it has a unique maximal ideal, namely the set m_P of functions in O_P that are zero in P.

An affine variety can be embedded in a projective variety in the following way. If $f \in k[x_1, x_2, \ldots, x_n]$, then we associate with f the homogeneous polynomial

$$f^*(x_0, x_1 \ldots, x_n) := x_0^l f(x_1/x_0, \ldots, x_n/x_0),$$

where l is the degree of f.

Let $\mathcal{X}$ be an affine variety in $\mathbb{A}^n$ defined by the prime ideal p. Let p^* be the homogeneous prime ideal generated by the set $\{f^* | f \in p\}$. Then p^* defines a projective variety $\mathcal{X}^*$ in $\mathbb{P}^n$. We define $\mathcal{X}_0^* := \{(x_0, x_1, \ldots, x_n) \in \mathcal{X}^* | x_0 \neq 0\}$. Then $\mathcal{X}$ is isomorphic with $\mathcal{X}_0^*$ under the map $(x_1, \ldots, x_n) \mapsto (1 : x_1 : \ldots : x_n)$. The points $(x_0 : \ldots : x_n) \in \mathcal{X}^*$ with $x_0 = 0$ are called *points at infinity* of $\mathcal{X}$. Furthermore, the function fields $k(\mathcal{X})$ and $k(\mathcal{X}^*)$ are isomorphic under the map $f/g \mapsto f^* x_0^m / g^*$, where m is $\deg(g) - \deg(f)$.

(10.2.7) EXAMPLE. In $\mathbb{P}^2$ with coordinates $(x : y : z)$, consider the variety $\mathcal{X}$ defined by $xz - y^2 = 0$. (This is the parabola of (10.2.5), now with a point at infinity, namely Q:=(1:0:0).) The function $(2xz + z^2)/(y^2 + z^2)$ is regular in the point $P = (0 : 0 : 1)$. By replacing y^2 by xz, we see that the function is equal to $(2x + z)/(x + z)$ and therefore also regular in Q.

Note that the function $(x^3 + y^3)/z^3$ which is 0 in P, can be written as the product of y^3/z^3 and $(y^3 + z^3)/z^3$, where the second factor is regular and not 0 in P. If $k = \mathbb{C}$ with the usual topology, then for points near P, there is a one to one correspondence with the value of y/z but this is not true for x/z. This is an example of what will be called a local parameter below.

The examples at the end of this paragraph will clarify things, but we must first introduce all the terminology that we need. From now on, we only consider curves.

Consider a curve in $\mathbb{A}^2$, defined by an equation $F(x, y) = 0$. Let $P = (a, b)$ be a point on this curve. If at least one of the derivatives F_x or F_y is not zero in P, then P is called a *simple* or *nonsingular* point of the curve. The curve then has a tangent at P with equation $F_x(P)(x - a) + F_y(P)(y - b) = 0$. We now define

$$d_P F := F_x(a, b)(x - a) + F_y(a, b)(y - b).$$

Then the tangent T_P at P is defined by $d_P F = 0$. This is well known. If $G \in k[\mathcal{X}]$, it would not make sense to define $d_P G$ in the same way because G is only defined modulo multiples of F. However, on T_P the linear function $d_P G := G_x(a, b)(x - a) + G_y(a, b)(y - b)$ is well defined. Given P, the map d_P maps an element of $k[\mathcal{X}]$ to a linear function defined on the tangent T_P, i. e. an element of T_P^*. We can extend this mapping to $\mathcal{O}_P$. Since $d_P f = 0$ if f is constant, we can restrict ourselves to rational functions f in m_P. Then from the product rule for differentiation, we see that m_P^2 is in the kernel of this mapping. Without proof we state that it is in fact the kernel. Therefore m_P/m_P^2 is isomorphic to T_P^* and for a nonsingular point that is a 1-dimensional space. This means that we can define a simple point of a curve by requiring that the k-vector space m_P/m_P^2 has dimension 1. From now on we consider only *nonsingular* curves (also called *smooth* curves), i. e. curves for which all the points are nonsingular. This restriction has the following consequence. Let P be a point of $\mathcal{X}$. We remind the reader that the maximal ideal m_P of the local ring $\mathcal{O}_P$ consists of the "functions" that are 0 in P. The other elements of $\mathcal{O}_P$ are units. Since m_P/m_P^2 has dimension 1, there is a generating element t for this space. We also use the symbol t for a corresponding element in m_P. We can then write every element z of $\mathcal{O}_P$ in a unique way as $z = ut^m$, where u is a unit and $m \in \mathbb{N}_0$. The function t is called a *local parameter* or *uniformizing parameter* in P. A function f is a local parameter at P if $d_P f$ is not zero on T_P.

If $m > 0$, then P is a zero of multiplicity m of z. (We saw an example with $m = 3$ in Example 10.2.7.) We write $m = \text{ord}_P(z) = v_P(z)$. (For readers familiar with the terminology: $\mathcal{O}_P$ is a discrete valuation ring and elements t with $v_P(t) = 1$ are local parameters.) We extend the order function to $k(\mathcal{X})$ by defining $v_P(f/g) := v_P(f) - v_P(g)$. If $v_P(z) = -m < 0$, then we say that z has a *pole* of order m in P. If z is an element of $k(\mathcal{X})$ with $v_P(z) = m$, then we can write $z = at^m + z'$, where $a \in k, a \neq 0$, and $v_P(z') > m$. In this way, one can show that z can be expanded as a Laurent series. Later, we shall use this to define the "residue" of z.

(10.2.8) Example. Let $\mathcal{X}$ be the circle in $\mathbb{A}^2$ with equation $x^2 + y^2 = 1$ and let $P = (1, 0)$, $(\text{char}(k) \neq 2)$. Let $z = z(x, y) = 1 - x$. This function is 0 in P, so it is in m_P. We claim that z has order 2. To see this, observe that y is a local parameter in P. Note that $d_P x = x - 1$ which is 0 on T_P, so x is not a local parameter at P. On $\mathcal{X}$ we have $1 - x = y^2/(1 + x)$ and the funcion $1/(1 + x)$ is a unit in $\mathcal{O}_P$. In Example 10.2.7 we saw a similar situation for $\mathbb{P}^2$.

(10.2.9) Example. Consider once again the parabola of Example 10.2.7. Let Q be the point at infinity, i. e. $Q = (1 : 0 : 0)$. The field $k(\mathcal{X})$ consists of the fractions $(A_l + B_{l-1}y)/(C_l + D_{l-1}y)$, where the coefficients are homogeneous polynomials of degree l (resp. $l - 1$) in x and z, and y satisfies $y^2 = xz$. Such a function is regular in Q if the coefficient of x^l in C_l is not 0. It is easy to see that y/x is a local parameter in Q. What can we say about the behavior of the function $g := (z^3 + xyz)/x^3$ in Q? On $\mathcal{X}$ we have

$$\frac{z^3 + xyz}{x^3} = \left(\frac{y}{x}\right)^3 \left(\frac{x^2 + yz}{x^2}\right).$$

The second factor on the right is a unit in $\mathcal{O}_Q$, so g has a zero of multiplicity 3 in Q.

When we construct codes, we will be interested in points that have their coordinates in our alphabet $\mathbb{F}_q$. We give these a special name.

(10.2.10) Definition. If k is the algebraic closure of $\mathbb{F}_q$ and $\mathcal{X}$ is a curve over k, then points on $\mathcal{X}$ with all their coordinates in $\mathbb{F}_q$ are called *rational points*. (We shall only use this terminology for curves over k with equations that have all coefficients in $\mathbb{F}_q$.)

We give three more examples

(10.2.11) Example. Let $\mathbb{P}$ be the projective line over k. A local parameter in the point $P = (1 : 0)$ is y/x. The rational function $(x^2 - y^2)/y^2$ has a pole of order 2 in P. If k does not have characteristic 2, then $(1 : 1)$ and $(-1 : 1)$ are zeros with multiplicity 1.

(10.2.12) Example. The plane curve with equation $x^3y + y^3z + z^3x = 0$ is called the *Klein quartic*. We consider the curve over the algebraic closure of $\mathbb{F}_2$. Look at a few of the subfields. Over $\mathbb{F}_2$ the rational points are (1:0:0), (0:1:0), and (0:0:1). If we go to $\mathbb{F}_4$, there are two more rational points, namely $(1 : \alpha : 1+\alpha)$ and $(1 : 1+\alpha : \alpha)$ if $\mathbb{F}_4 = \{0, 1, \alpha, \alpha^2\}$, where $\alpha^2 = 1 + \alpha$.

In later examples, this curve will be studied over $\mathbb{F}_8$. As usual, we define this field as $\mathbb{F}_2(\xi)$, where $\xi^3 = \xi + 1$. If a rational point has a coordinate 0, it must be one of the points over $\mathbb{F}_2$. If $xyz \neq 0$, we can take $z = 1$. If $y = \xi^i$ $(0 \leq i \leq 6)$, then write $x = \xi^{3i}\eta$. Substitution in the equation gives $\eta^3 + \eta + 1 = 0$, i. e. η is one of the elements ξ, ξ^2, or ξ^4. So we find a total of 24 rational points over $\mathbb{F}_8$.

(10.2.13) Example. Let $\mathcal{X}$ be the plane curve with equation $x^3 + y^3 + z^3 = 0$ over the closure of $\mathbb{F}_2$ and look at the subfield $\mathbb{F}_4$. Since a third power of an element of $\mathbb{F}_4$ is 0 or 1, all the rational points have one coordinate 0. We can take one of the others to be 1, and the third one any nonzero element of $\mathbb{F}_4$. So we find nine (projective) points. In $Q = (0 : 1 : 1)$, we can take $t = x/z$ as local parameter. We consider a difficulty that will come up again. The expression $f := x/(y + z)$ looks like a perfectly reasonable function and in fact on most of $\mathcal{X}$ it is. However, in Q the fraction does not make sense. We must find an equivalent form for f near Q. On $\mathcal{X}$ we have

$$\frac{x}{y+z} = \frac{x(y^2+yz+z^2)}{y^3+z^3} = t^{-2} \cdot \frac{y^2+yz+z^2}{z^2},$$

where the second factor on the right is regular and not 0 in Q. By our earlier conventions, we say that f has a pole of order 2 in Q. Similarly, $y/(y + z)$ has a pole of order 3 in Q.

As a preparation for §10.6, we now consider the intersections of plane curves. We assume that the reader is familiar with the fact that a polynomial of degree m in one variable, with coefficients in a field has at most m zeros. If the field is algebraically closed and if the zeros are counted with multiplicities, then the number of zeros is equal to m. We shall now state a theorem, known as *Bézout's theorem*, which is a generalization of these facts to polynomials in several variables. We only consider the case of two variables, i. e. we consider plane curves. Again, we assume that the reader knows how multiplicities are attached to points of intersection of two plane curves. (If P is a nonsingular point of a curve with equation $F(x, y) = 0$ and the curve with equation $G(x, y) = 0$ contains P, then the multiplicity of intersection is $v_P(G)$.) In the following, we consider two affine plane curves defined by equations $F(x, y) = 0$ and $G(x, y) = 0$ of degree l respectively m. We assume that F and G do not have a nontrivial common factor, i. e. the curves do not have a component in common. We consider the case where the coefficients are from an algebraically closed field k.

(10.2.14) Theorem. *Two plane curves of degree l and m that do not have a component in common, intersect in exactly lm points (if counted with multiplicity and the points at infinity are also considered).*

If k is not algebraically closed, the curves intersect in at most lm points. We do not prove this theorem.

(10.2.15) EXAMPLE. Clearly the affine plane curve over the closure of $\mathbb{F}_2$ with equation $x^3 + y^3 = 1$ and the line with equation $x = y$ do not meet. However, when considered projectively, we have the curve $\mathcal{X}$ of Example 10.2.13 and the reader can easily check that $\mathcal{X}$ and the line with equation $x + y = 0$ intersect in $P := (1 : 1 : 0)$ (at infinity) with multiplicity 3; (this is done in the same way as in Example 10.2.13).

We now generalize the idea of Reed-Solomon codes as defined in §6.8 (second description). Let V_l be the vector space of polynomials of degree at most l in two variables x, y and coefficients in $\mathbb{F}_q$. Consider an irreducible element G of degree m in $\mathbb{F}_q[x, y]$. Let $P_1, P_2, \ldots, P_n$ be points on the plane curve defined by the equation $G(x, y) = 0$, i.e. $G(P_i) = 0$ for $1 \le i \le n$. We define a code C by

$$C := \{(F(P_1), F(P_2), \ldots, F(P_n)) \mid F \in \mathbb{F}_q[x, y], \deg(F) \le l\}.$$

We shall use d for the minimum distance of this code and (as usual) call the dimension k (not to be confused with the field that was considered earlier in this section).

(10.2.16) Theorem. *Let $lm < n$. For the minimum distance d and the dimension k of C, we have*

$$d \ge n - lm,$$

$$k = \begin{cases} \binom{l+2}{2} & \text{if } l < m, \\ lm + 1 - \binom{m-1}{2} & \text{if } l \ge m. \end{cases}$$

PROOF. The monomials of the form $x^\alpha y^\beta$ with $\alpha + \beta \le l$ form a basis of V_l. Hence V_l has dimension $\binom{l+2}{2}$.

Let $F \in V_l$. If G is a factor of F, then the codeword in C corresponding to F is zero. Conversely, if this codeword is zero, then the curves with equation $F = 0$ and $G = 0$ have degree $l' \le l$ and m respectively, and they have the n points $P_1, P_2, \ldots, P_n$ in their intersection. Bézout's theorem and the assumption $lm < n$ imply that F and G have a common factor. Since G is irreducible, F must be divisible by G. Hence the functions $F \in V_l$ that yield the zero codeword form the subspace GV_{l-m}. This implies that if $l < m$, then $k = \binom{l+2}{2}$, and if $l \ge m$, then

$$k = \binom{l+2}{2} - \binom{l-m+2}{2} = lm + 1 - \binom{m-1}{2}.$$

The same argument with Bézout's theorem shows that a nonzero codeword has at most lm coordinates equal to 0, i.e. it has weight at least $n - lm$. Hence $d \ge n - lm$.
□

§10.3. Divisors

In the following, $\mathcal{X}$ is a smooth projective curve over k.

(10.3.1) Definition.

(1) A *divisor* is a formal sum $D = \sum_{P\in\mathcal{X}} n_P P$, with $n_P \in \mathbb{Z}$ and $n_P = 0$ for all but a finite number of points P;
(2) $\mathrm{Div}(\mathcal{X})$ is the additive group of divisors with formal addition (the free abelian group on $\mathcal{X}$);
(3) A divisor D is called *effective* if all coefficients n_P are non-negative (notation $D \succcurlyeq 0$);
(4) The *degree* $\deg(D)$ of the divisor D is $\sum n_P$.

Let $v_P = \mathrm{ord}_P$ be the discrete valuation defined for functions on $\mathcal{X}$ in §10.2.

(10.3.2) Definition. If f is a rational function on $\mathcal{X}$, not identically 0, we define the divisor of f to be

$$(f) := \sum_{P\in\mathcal{X}} v_P(f)P.$$

So, in a sense, the divisor of f is a bookkeeping device that tells us where the zeros and poles of f are and what their multiplicities and orders are. Since f is a rational function for which the numerator and denominator have the same degree, and since k is algebraically closed, it is intuitively clear that f has the same number of zeros as poles, if counted properly. We do not give a proof but state the consequence as a theorem.

(10.3.3) Theorem. *The degree of a divisor of a rational function is 0.*

The divisor of a rational function is called a *principal divisor.*

(10.3.4) Definition. We shall call two divisors D and D' *linearly equivalent* iff $D - D'$ is a principal divisor ; notation $D \equiv D'$.

This is indeed an equivalence relation.

In §9.2, we gave a definition of Goppa codes, involving a vector space of functions with prescribed zeros and possible poles. We now have a mechanism available to generalize this to curves.

(10.3.5) Definition. Let D be a divisor on a curve $\mathcal{X}$. We define a vector space $\mathcal{L}(D)$ over k by

$$\mathcal{L}(D) := \{f \in k(\mathcal{X})^* : (f) + D \succcurlyeq 0\} \cup \{0\}.$$

Note that if $D = \sum_{i=1}^{r} n_i P_i - \sum_{j=1}^{s} m_j Q_j$ with all $n_i, m_j > 0$, then $\mathcal{L}(D)$ consists of 0 and the functions in the function field that have zeros of multiplicity at least m_j at Q_j $(1 \le j \le s)$ and that have no poles except possibly at the points P_i, with order at most n_i $(1 \le i \le r)$. We shall show that this vector space has finite dimension.

First we note that if $D \equiv D'$ and g is a rational function with $(g) = D - D'$, then the map $f \mapsto fg$ shows that $\mathcal{L}(D)$ and $\mathcal{L}(D')$ are isomorphic.

(10.3.6) Theorem.
(i) $\mathcal{L}(D) = 0$ *if* $deg(D) < 0$;
(ii) $l(D) := dim_k \mathcal{L}(D) \le 1 + deg(D)$.

PROOF. (i) If $\deg(D) < 0$, then for any function $f \in k(\mathcal{X})^*$, we have $\deg((f)+D) < 0$, i.e. $f \notin \mathcal{L}(D)$.

(ii) If f is not 0 and $f \in \mathcal{L}(D)$, then $D' := D + (f)$ is an effective divisor for which $\mathcal{L}(D')$ has the same dimension as $\mathcal{L}(D)$ by our observation above. So w. l. o. g. D is effective, say $D = \sum_{i=1}^{r} n_i P_i$, $(n_i \ge 0$ for $1 \le i \le r)$. Again, assume that f is not 0 and $f \in \mathcal{L}(D)$. In the point P_i, we map f onto the corresponding element of the n_i-dimensional vector space $(t_i^{-n_i} \mathcal{O}_{P_i})/\mathcal{O}_{P_i}$, where t_i is a local parameter at P_i. We thus obtain a mapping of f onto the direct sum of these vector spaces ; (map the 0-function onto 0). This is a linear mapping. Suppose that f is in the kernel. This means that f does not have a pole in any of the points P_i, i. e. f is a constant function. It follows that $\dim_k \mathcal{L}(D) \le 1 + \sum_{i=1}^{r} n_i = 1 + \deg(D)$. □

(10.3.7) EXAMPLE. Look at the curve of Example 10.2.13. We saw that $f = x/(y+z)$ has a pole of order 2 in $Q = (0 : 1 : 1)$. The function has two zeros, each with multiplicity 1, namely $P_1 = (0 : \alpha : 1)$ and $P_2 = (0 : 1 + \alpha : 1)$. From the representation $f = (y^2 + yz + z^2)/x^2$ we see that Q is the only pole. So $(f) = P_1 + P_2 - 2Q$ and $\deg((f)) = 0$ in accordance with Theorem 10.3.3. It is not trivial, but one can show that there cannot be a function in $k(\mathcal{X})$ that has a pole of order 1 in Q and no other poles. So in this example, the space $\mathcal{L}(2Q)$ has dimension 2 and f and the function that is identically 1 form a basis.

§10.4. Differentials on a Curve

Consider a smooth affine curve $\mathcal{X}$ in $\mathbb{A}^2$ defined by the equation $F(x, y) = 0$, and let $P = (a, b)$ be a point on $\mathcal{X}$. The tangent T_P at P is defined by $d_P F = 0$. In Section 10.2 we defined the map d_P that maps an element of $k[\mathcal{X}]$ to a linear function on T_P (i. e. an element of T_P^*). We now consider the set $\Phi[\mathcal{X}]$ of *all* mappings that associate with each point P of $\mathcal{X}$ an element of T_P^*.

(10.4.1) Definition. An element $\phi \in \Phi[\mathcal{X}]$ is called a *regular differential form* (on the curve $\mathcal{X}$) if every point P of $\mathcal{X}$ has a neighborhood U such that in this neighborhood, ϕ can be represented as $\phi = \sum_{i=1}^{n} f_i dg_i$, where all the functions f_i and g_i are regular in U.

The regular differential forms on $\mathcal{X}$ form a $k[\mathcal{X}]$-module, which we denote by $\Omega[\mathcal{X}]$. This module is generated by elements df, where $f \in k[\mathcal{X}]$, with the relations $d(f+g) = df + dg$ and $d(fg) = fdg + gdf$ and $da = 0$ for $a \in k$. For the extension to *rational differential forms* we must add the (well known) relation $d(f/g) = (gdf - fdg)/g^2$. We wish to define a rational differential form on a smooth projective curve $\mathcal{X}$. To do this, consider pairs (U, ω), where U is a nonempty affine set in $\mathcal{X}$ and ω has the form $g\,df$ on U. We call pairs (U, ω) and (V, η) equivalent if $\omega = \eta$ on the set $U \cap V$. An equivalence class for this relation is called a rational differential form. From now on, we call the rational differential forms on $\mathcal{X}$ *differentials* and denote the space of differentials by $\Omega(\mathcal{X})$. We state without proof:

(10.4.2) Theorem. *The space $\Omega(\mathcal{X})$ has dimension 1 over $k(\mathcal{X})$; in a neighborhood of a point P with local parameter t, a differential ω can be represented as $\omega = f\,dt$, where f is a rational function.* The reader might think this is unnecessarily complicated. Why not just use functions? The next example shows that on a projective curve, one can have a nonzero rational differential form that is regular on the whole curve, this in contrast to rational functions.

(10.4.3) Example. We again look at the curve $\mathcal{X}$ in $\mathbb{P}^2$ given by $x^3 + y^3 + z^3 = 0$ (char$(k) \neq 3$). We define the open set U_x by $U_x := \{(x : y : z) \in \mathcal{X} : y \neq 0, z \neq 0\}$ and similarly U_y and U_z. Then U_x, U_y, and U_z cover $\mathcal{X}$ since there is no point on $\mathcal{X}$ where two coordinates are zero. It is easy to check that the three representations

$$\omega := \left(\frac{y}{z}\right)^2 d\left(\frac{x}{y}\right) \text{ on } U_x, \quad \eta := \left(\frac{z}{x}\right)^2 d\left(\frac{y}{z}\right) \text{ on } U_y, \quad \zeta := \left(\frac{x}{y}\right)^2 d\left(\frac{z}{x}\right) \text{ on } U_z$$

define *one* differential on $\mathcal{X}$. For instance, to show that η and ζ agree on $U_y \cap U_z$, one takes the equation $(x/z)^3 + (y/z)^3 + 1 = 0$, differentiates, and applies the formula $d(f^{-1}) = -f^{-2}\,df$ to $f := z/x$. A regular function on $\mathcal{X}$ is constant, so one cannot represent this differential as $g\,df$ with f and g regular functions on $\mathcal{X}$.

(10.4.4) Definition. The divisor (ω) of the differential ω is defined by

$$(\omega) := \sum_{P \in \mathcal{X}} v_P(f_P)P,$$

where $\omega = f_P\,dt_P$ is the local representation of ω and v_P is the valuation on $\mathcal{O}_P$ (extended to $k(\mathcal{X})$).

Of course, one must show that this does not depend on the choice of local parameters and also that only finitely many coefficients are not 0.

Let ω be a differential and $W = (\omega)$. Then W is called a *canonical divisor*. If ω' is another nonzero differential, then $\omega' = f\omega$ for some rational function f. So $(\omega') = W' \equiv W$ and therefore the canonical divisors form one equivalence class. This class is also denoted by W. Now consider the space $\mathcal{L}(W)$. This space of rational functions (cf.(10.3.5)) can be mapped onto an isomorphic space of differential forms by $f \mapsto f\omega$. By the definition of $\mathcal{L}(W)$, the image of f under the mapping is a regular differential form, i. e. $\mathcal{L}(W)$ is isomorphic to $\Omega[\mathcal{X}]$.

(10.4.5) Definition. Let $\mathcal{X}$ be a smooth projective curve over k. We define the *genus* g of $\mathcal{X}$ by $g := l(W)$.

The genus of a curve will play an important role in the following sections. For methods with which one can determine the genus of a curve, we must refer to textbooks on algebraic geometry. We mention one formula without proof, the so-called *Plücker formula.*

(10.4.6) Theorem. *If $\mathcal{X}$ is a nonsingular projective curve of degree d in $\mathbb{P}^2$, then*

$$g = \frac{1}{2}(d-1)(d-2).$$

So the curve of Example 10.4.3 has genus 1 and by the definition of genus, $\mathcal{L}(W) = k$, so regular differentials on $\mathcal{X}$ are scalar multiples of the differential ω of Example 10.4.3.

For the construction of codes over algebraic curves that generalize Goppa codes, we shall need the concept of "*residue*" of a differential at a point P. This is defined in accordance with our treatment of local behavior of a differential ω. Let P be a point on $\mathcal{X}$, t a local parameter at P and $\omega = f\,dt$ the local representation of ω. The function f can be written as $\sum_i a_i t^i$. We define the residue $\text{Res}_P(\omega)$ of ω in the point P to be a_{-1} (as was to be expected). One can show that this algebraic definition of the residue does not depend on the choice of the local parameter t.

One of the basic results in the theory of algebraic curves is known as the "*residue theorem*". We only state the theorem.

(10.4.7) Theorem. *If ω is a differential on a smooth projective curve $\mathcal{X}$, then*

$$\sum_{P \in \mathcal{X}} \text{Res}_P(\omega) = 0.$$

§10.5. The Riemann-Roch Theorem

The following famous theorem, known as the *Riemann-Roch theorem* is not only a central result in algebraic geometry with applications in other areas, but it is also the key to the new results in coding theory.

(10.5.1) Theorem. *Let D be a divisor on a smooth projective curve of genus g. Then, for any canonical divisor W*

$$l(D) - l(W-D) = deg(D) - g + 1.$$

We do not give the (quite complicated) proof. The theorem allows us to determine the degree of canonical divisors.

(10.5.2) Corollary. *For a canonical divisor W, we have* $\deg(W) = 2g - 2$.

PROOF. Everywhere regular functions on a projective curve are constant, i.e. $\mathcal{L}(0) = k$, so $l(0) = 1$. Substitute $D = W$ in Theorem 10.5.1 and the result follows from Definition 10.4.5. □

It is now clear why in Example 10.3.7 the space $\mathcal{L}(2Q)$ only had dimension 2. By Theorem 10.4.6, the curve $\mathcal{X}$ has genus 1, the degree of $W - 2Q$ is negative, so $l(W - 2Q) = 0$. By Theorem 10.5.1, we have $l(2Q) = 2$.

At first, Theorem 10.5.1 does not look too useful. However, Corollary 10.5.2 provides us with a means to use it successfully.

(10.5.3) Corollary. *Let D be a divisor on a smooth projective curve of genus g and let $\deg(D) > 2g - 2$. Then*

$$l(D) = \deg(D) - g + 1.$$

PROOF. By Corollary 10.5.2, $\deg(W - D) < 0$, so by Theorem 10.3.6(i), $l(W - D) = 0$. □

(10.5.4) EXAMPLE. Consider the code of Theorem 10.2.16. We embed the affine plane in a projective plane and consider the rational functions on the curve defined by G. By Bézout's theorem, this curve intersects the line at infinity, i.e. the line defined by $z = 0$, in m points. These are the possible poles of our rational functions, each with order at most l. So, in the terminology of Definition 10.3.5, we have a space of rational functions, defined by a divisor D of degree lm. By the Plücker formula (10.4.6), the curve defined by G has genus equal to $\binom{m-1}{2}$. If $l \geq m - 2$ we may apply Corollary 10.5.3 and we find the same result as from Theorem 10.2.16.

The term $l(W - D)$ in Theorem 10.5.1 can be interpreted in terms of differentials. We introduce a generalization of Definition 10.3.5 for differentials.

(10.5.5) Definition. Let D be a divisor on a curve $\mathcal{X}$. We define

$$\Omega(D) := \{\omega \in \Omega(\mathcal{X}) : (\omega) - D \succcurlyeq 0\}$$

and we denote $\dim_k \Omega(D)$ by $\delta(D)$, called the *index of speciality* of D.

The connection with functions is established by the following theorem.

(10.5.6) Theorem. $\delta(D) = l(W - D)$.

PROOF. If $W = (\omega)$, we define a linear map $\phi : \mathcal{L}(W - D) \to \Omega(D)$ by $\phi(f) := f\omega$. This is clearly an isomorphism. □

(10.5.7) EXAMPLE. If we take $D = 0$, then by Definition 10.4.5 there are exactly g linearly independent regular differentials on a curve $\mathcal{X}$. So the differential of

Example 10.4.2 is the only regular differential on $\mathcal{X}$ (up to a constant factor) as was already observed after Theorem 10.4.6.

§10.6. Codes from Algebraic Curves

We now come to the applications to coding theory. Our alphabet will again be $\mathbb{F}_q$. We shall apply the theorems of the previous sections. A few adaptations are necessary, since e. g. the space $\mathcal{L}(D)$ will not be considered over an algebraically closed field but over $\mathbb{F}_q$. All that we need to know is that Theorem 10.5.1 remains true. In a number of examples this will be obvious from the basis of $\mathcal{L}(D)$ (a basis over the closure k, consisting of polynomials over $\mathbb{F}_q$).

Let $\mathcal{X}$ be a non-singular projective curve over $\mathbb{F}_q$. We shall define two kinds of algebraic geometry codes from $\mathcal{X}$. The first kind generalizes Reed-Solomon codes, the second kind generalizes Goppa codes. In the following, $P_1, P_2, \ldots, P_n$ are rational points on $\mathcal{X}$ and D is the divisor $P_1 + P_2 + \ldots + P_n$. Furthermore G is some other divisor that has support disjoint from D. Although it is not necessary to do so, we shall make more restrictions on G, namely that the support of G also consists of rational points and furthermore

(10.6.1) $$2g - 2 < \deg(G) < n.$$

(10.6.2) Definition. The linear code $C(D, G)$ of length n over $\mathbb{F}_q$ is the image of the linear map $\alpha : \mathcal{L}(G) \to \mathbb{F}_q^n$ defined by $\alpha(f) := (f(P_1), f(P_2), \ldots, f(P_n))$.

Codes of this kind are called "*geometric generalized RS codes*".

(10.6.3) Theorem. *The code $C(D, G)$ has dimension $k = \deg(G) - g + 1$ and minimum distance $d \geq n - \deg(G)$.*

Proof. (i) If f belongs to the kernel of α, then $f \in \mathcal{L}(G - D)$ and by Theorem 10.3.6(i), this implies $f = 0$. The result follows from (10.6.1) and Corollary 10.5.3.

(ii) If $\alpha(f)$ has weight d, then there are $n - d$ points P_i, say $P_{i_1}, P_{i_2}, \ldots, P_{i_{n-d}}$, for which $f(P_i) = 0$. Therefore $f \in \mathcal{L}(G - E)$, where $E = P_{i_1} + \ldots + P_{i_{n-d}}$. Hence $\deg(G) - n + d \geq 0$. □

Note the analogy with the proof of Theorem 10.2.16.

(10.6.4) Example. Let $\mathcal{X}$ be the projective line over $\mathbb{F}_q$. Take $G := mQ$ where Q is the point (1:0), $n = q$, $P_i = (\alpha_i : 1)$, where $\mathbb{F}_q = \{\alpha_1, \alpha_2, \ldots, \alpha_q\}$. Then, if $m = k - 1$, we see that $C(D, G)$ is the extended Reed-Solomon code as described in §6.8.

(10.6.5) Example. Let $\mathcal{X}$ be the curve of Examples 10.2.13 and 10.3.7, $G := 2Q$, where $Q := (0 : 1 : 1)$. We take $n = 8$ (so D is the sum of the remaining rational points). The coordinates are given by

	Q	P_1	P_2	P_3	P_4	P_5	P_6	P_7	P_8
x	0	0	0	1	α	$\overline{\alpha}$	1	α	$\overline{\alpha}$
y	1	α	$\overline{\alpha}$	0	0	0	1	1	1
z	1	1	1	1	1	1	0	0	0

where $\overline{\alpha} = 1 + \alpha$. We saw in Example 10.3.7 that 1 and $x/(y+z)$ are a basis of $\mathcal{L}(2Q)$ over k and hence also over $\mathbb{F}_4$. This leads to the following generator matrix for $C(D, G)$:

$$\begin{pmatrix} 1 & 1 & 1 & 1 & 1 & 1 & 1 & 1 \\ 0 & 0 & 1 & \alpha & \overline{\alpha} & 1 & \alpha & \overline{\alpha} \end{pmatrix}.$$

By Theorem 10.6.2, the minimum distance is at least 6 and of course, one immediately sees from the generator matrix that $d = 6$.

We now come to the second class of algebraic geometry codes. We shall call these codes "*geometric Goppa codes*".

(10.6.6) Definition. The linear code $C^*(D, G)$ of length n over $\mathbb{F}_q$ is the image of the linear map $\alpha^* : \Omega(G - D) \to \mathbb{F}_q^n$ defined by

$$\alpha^*(\eta) := (\mathrm{Res}_{P_1}(\eta), \mathrm{Res}_{P_2}(\eta), \ldots, \mathrm{Res}_{P_n}(\eta)).$$

The parameters are given by the following theorem.

(10.6.7) Theorem. *The code $C^*(D, G)$ has dimension $k^* = n - \deg(G) + g - 1$ and minimum distance $d^* \geq \deg(G) - 2g + 2$.*

Proof. Just as in Theorem 10.6.3, these assertions are direct consequences of Theorem 10.5.1 (Riemann-Roch), using Theorem 10.5.5 (making the connection between the dimension of $\Omega(G)$ and $l(W - G)$) and Corollary 10.5.2 (stating that the degree of a canonical divisor is $2g - 2$). □

(10.6.8) Example. Consider the projective line over $\mathbb{F}_{q^m}$. Let $P_i := (\gamma_i : 1)$, where γ_i $(0 \leq i \leq n-1)$ are as in Definition 9.2.4. We define $D := P_0 + P_1 + \ldots + P_{n-1}$ and $G := (g)$ where $g(x, y)$ is the homogeneous form of the Goppa polynomial $g(z)$ of (9.2.4). Then the Goppa code $\Gamma(L, g)$ of (9.2.4) is the subfield subcode (over $\mathbb{F}_q$) of the geometric Goppa code $C^*(D, G)$. We observed in §9.2 that this code is a subcode of the dual of a generalized Reed-Solomon code. This is a special case of the following theorem.

(10.6.9) Theorem *The codes $C(D, G)$ and $C^*(D, G)$ are dual codes.*

Proof. From Theorem 10.6.3 and Theorem 10.6.7 we know that $k + k^* = n$. So it suffices to take a word from each code and show that the inner product of the two words is 0. Let $f \in \mathcal{L}(G)$, $\eta \in \Omega(G - D)$. By Definitions 10.6.2 and 10.6.6, the differential $f\eta$ has no poles except possibly poles of order 1 in the points

$P_1, P_2, \ldots, P_n$. The residue of $f\eta$ in P_i is equal to $f(P_i)\mathrm{Res}_{P_i}(\eta)$. By Theorem 10.4.7, the sum of the residues of $f\eta$ over all the poles (i.e. over the points P_i) is equal to zero. Hence we have

$$0 = \sum_{i=1}^{n} f(P_i)\mathrm{Res}_{P_i}(\eta) = \langle \alpha(f), \alpha^*(\eta) \rangle. \qquad \square$$

Several authors prefer the codes $C^*(D, G)$ over geometric RS codes but the nonexperts in algebraic geometry probably feel more at home with polynomials than with differentials. In [73] it is shown that the codes $C(D, G)$ suffice to get all the codes. However, it is useful to have both classes when treating decoding methods. These use parity checks, so one needs a generator matrix for the dual code.

In the next paragraph, we treat several examples of geometric codes. It is already clear that we find some *good* codes. E.g. from Theorem 10.6.3 we see that such codes over a curve of genus 0 (the projective line) are MDS codes (cf. §5.1). In fact, Theorem 10.6.3 says that $d \geq n - k + 1 - g$, so if g is small, we are close to the Singleton bound (cf. (5.2.2)).

§10.7. Some Geometric Codes

We know that to find good codes, we must find long codes. To use the methods from algebraic geometry, it is necessary to find rational points on a given curve. The number of these is a bound on the length of the code. A central problem in algebraic geometry is finding bounds for the number of rational points on a variety. In order to appreciate some of the examples in this paragraph, we mention without proof the *Hasse-Weil bound.*

(10.7.1) Theorem. *Let $\mathcal{X}$ be a curve of genus g over $\mathbb{F}_q$. If $N_q(\mathcal{X})$ denotes the number of rational points on $\mathcal{X}$, then*

$$|N_q(\mathcal{X}) - (q + 1)| \leq 2g\sqrt{q}.$$

We first give an example that does not yield anything new.

(10.7.2) Example. Let $\mathcal{X}$ be the projective line over $\mathbb{F}_{q^m}$. Let $n := q^m - 1$. We define $P_0 := (0 : 1)$, $P_\infty := (1 : 0)$ and we define the divisor D as $\sum_{j=1}^{n} P_j$, where $P_j := (\beta^j : 1)$, $(1 \leq j \leq n)$. We define $G := aP_0 + bP_\infty$, $a \geq 0, b \geq 0$. (Here β is a primitive nth root of unity.) By Theorem 10.5.1, $\mathcal{L}(G)$ has dimension $a + b + 1$ and one immediately sees that the functions $(x/y)^i$, $-a \leq i \leq b$ form a basis of $\mathcal{L}(G)$. Consider the code $C(D, G)$. A generator matrix for this code has as rows $(\beta^i, \beta^{2i}, \ldots, \beta^{ni})$ with $-a \leq i \leq b$. One easily checks that $(c_1, c_2, \ldots, c_n)$ is a codeword in $C(D, G)$ iff $\sum_{j=1}^{n} c_j(\beta^l)^j = 0$ for all l with $a < l < n - b$. It follows that $C(D, G)$ is a Reed-Solomon code. The subfield subcode with coordinates in $\mathbb{F}_q$ is a BCH code.

(10.7.3) Example. In this example we consider codes from *Hermitian* curves. Let $q = r^2 = 2^l$. A Hermitian curve $\mathcal{X}$ in $\mathbb{P}^2$ over $\mathbb{F}_q$ is defined by the equation

$$x^{r+1} + y^{r+1} + z^{r+1} = 0. \tag{10.7.4}$$

By Theorem 10.4.6, the genus g of $\mathcal{X}$ equals $\frac{1}{2}r(r-1) = \frac{1}{2}(q - \sqrt{q})$. We shall first show that $\mathcal{X}$ has the maximal number of rational points, i.e. by Theorem 10.7.1 exactly $1 + q\sqrt{q}$ rational points. If in (10.7.4) one of the coordinates is 0, then w.l.o.g. one of the others is 1 and the third one is one of the solutions of $\xi^{r+1} = 1$, which has $r+1$ solutions in $\mathbb{F}_q$. This shows that $\mathcal{X}$ has $3(r+1)$ points with $xyz = 0$. If $xyz \neq 0$, we may take $z = 1$ and y any element in $\mathbb{F}_q^*$ such that $y^{r+1} \neq 1$. For each choice of y, there are $r+1$ solutions x. This yields $(r-2)(r+1)^2$ pairs (x, y). It follows that $\mathcal{X}$ has $3(r+1) + (r-2)(r+1)^2 = 1 + q\sqrt{q}$ rational points. (We remark that this calculation could have been made for other primes than 2.)

We take $G := mQ$, where $Q := (0 : 1 : 1)$ and $q - \sqrt{q} < m < q\sqrt{q}$. The code $C(D, G)$ over $\mathbb{F}_q$ has length $n = q\sqrt{q}$, dimension $k = m - g + 1$, and distance $d \geq n - m$. To see how good these codes are, we take as example $q = 16$. A basis for $\mathcal{L}(G)$ is easily found. The functions $f_{i,j}(x, y, z) := x^i y^j/(y+z)^{i+j}$, $0 \leq i \leq 4$, $4i + 5j \leq m$ will do the job. First, observe that there are $m - 5 = m - g + 1$ pairs (i, j) satisfying these conditions. The functions $x/(y+z)$ and $y/(y+z)$ can be treated in exactly the same way as in Example 10.2.13, showing that $f_{i,j}$ has a pole of order $4i + 5j$ in Q. Hence, these functions are independent. Therefore, the code is easily constructed. Decoding is another question! Let us try to get some idea of the quality of this code. Suppose that we intend to send a long message (say 10^9 bits) over a channel with an error probability $p_e = 0.01$ (quite a bad channel). We compare coding using a rate $\frac{1}{2}$ Reed-Solomon code over $\mathbb{F}_{16}$ with using $C(D, G)$, where we take $m = 37$ to also have rate $\frac{1}{2}$. In this case, $C(D, G)$ has distance 27. The RS code has word length 16 (so 64 bits) and distance 9. If a word is received incorrectly, we assume that all the bits are wrong when we count the number of errors. For the RS code, the error probability after decoding is roughly $3 \cdot 10^{-4}$ (indeed a nice improvement); however, for the code $C(D, G)$, the error probability after decoding is less than $2 \cdot 10^{-7}$. In this example, it is important to keep in mind that we are fixing the alphabet (in this case $\mathbb{F}_{16}$). If we compare the code $C(D, G)$, for which the words are strings of 256 bits, with a rate $\frac{1}{2}$ RS code over $\mathbb{F}_{2^5}$ (words are 160 bits long), the latter will come close in performance (error probability $2 \cdot 10^{-6}$) and a rate $\frac{1}{2}$ RS code over $\mathbb{F}_{2^6}$ (words are 384 bits long) performs better (roughly 10^{-7}).

One could also compare our code with a binary BCH code of length 255 and rate about $\frac{1}{2}$. The BCH code wins when we are concerned with random errors. If we are using a bursty channel, then the code $C(D, G)$ can handle bursts of length up to 46 bits (which influence at most 13 letters of a codeword) while the BCH code would fail completely. Although one could argue about the question which of these comparisons really says something, it was this example (used by the author) that convinced several engineers, who believed firmly that RS codes were the only useful codes for them, to look more closely at codes from algebraic geometry. This has led to nice results on decoding, a problem we neglect in this book, but clearly central to applications.

(10.7.4) Example. Let $\mathcal{X}$ be the Klein quartic over $\mathbb{F}_8$ of Example 10.2.12. By Theorem 10.4.6, the genus is 3. By Theorem 10.7.1, $\mathcal{X}$ can have at most 25 rational points and as we saw in Example 10.2.12, it has 24 rational points; (in fact, this is optimal by an improvement of Theorem 10.7.1, due to J.-P.Serre [99]). Let $Q := (0:0:1)$ and let D be the sum of the other 23 rational points, $G := 10Q$. From Theorem 10.6.3, we find that $C(D, G)$ has dimension $10 - g + 1 = 8$ and minimum distance $d \geq 23 - 10 = 13$. We now concatenate this code with the [4,3,2] single parity check code as follows. The symbols in codewords of $C(D, G)$ are elements of $\mathbb{F}_8$ which we interpret as column vectors of length 3 over $\mathbb{F}_2$ and then we adjoin the parity check. The resulting code C is a binary $[92, 24, 26]$ code. The punctured code, a $[91, 24, 25]$ code (constructed by A. M. Barg et al. [82] in 1987) set a new world record for codes with $n = 91$, $d = 25$. Several other codes of this kind are given in the same paper.

(10.7.5) Example. We show how to construct a generator matrix for the code of the previous example. We consider the functions y/z, z/x, and x/y. The points where these functions can have zeros or poles are $P_1 := (1:0:0)$, $P_2 := (0:1:0)$, and $Q = (0:0:1)$. Since the line with equation $y = 0$ (in affine coordinates) is not a tangent at Q of the curve with affine equation $x^3y + y^3 + x = 0$, we see that y/z is a local parameter in Q (an idea that has been used in earlier examples). Similarly, z/x is a local parameter in P_1, and x/y is a local parameter in P_2. We analyze the behavior of y/z in P_1 and P_2. In P_1 we have

$$\frac{y}{z} = \left(\frac{z}{x}\right)^2 \frac{x^3}{x^3 + y^2z},$$

so y/z has a zero with multiplicity 2 in P_1. Similarly in P_2 we have

$$\frac{y}{z} = \left(\frac{y}{x}\right)^3 \frac{y^3 + z^2x}{y^3},$$

so P_2 is a pole of order 3 for the function y/z. Therefore $(\frac{y}{z}) = 2P_1 - 3P_2 + Q$. In the same way one calculates $(\frac{z}{x}) = P_1 + 2P_2 - 3Q$ and $(\frac{x}{y}) = -3P_1 + P_2 + 2Q$. From these divisors, we can deduce that the functions $(z/x)^i(y/x)^j$ with $0 \leq 3i + 2j \leq 10$, $0 \leq j \leq 2i$ are in $\mathcal{L}(10Q)$. We thus have eight functions in $\mathcal{L}(10Q)$ with poles in Q of order 0,3,5,6,7,8,9, and 10 respectively. Hence they are independent and since $l(10Q) = 8$, they are a basis of $\mathcal{L}(10Q)$. By substituting the coordinates of the rational points of $\mathcal{X}$ in these functions, we find the 8 by 23 generator matrix of the code.

(10.7.6) Example. Let $\mathbb{F}_4 = \{0, 1, \alpha, \overline{\alpha}\}$, where $\alpha^2 = \alpha + 1 = \overline{\alpha}$. Consider the curve $\mathcal{X}$ over $\mathbb{F}_4$ given by the equation $x^2y + \alpha y^2z + \overline{\alpha}z^2x = 0$. This is a nonsingular curve with genus 1. Its nine rational points are given by

	P_1	P_2	P_3	P_4	P_5	P_6	Q_1	Q_2	Q_3
x	1	0	0	1	1	1	α	1	1
y	0	1	0	α	$\overline{\alpha}$	1	1	α	1
z	0	0	1	$\overline{\alpha}$	α	1	1	1	α

Let $D := P_1 + P_2 + \ldots + P_6$, $G := 2Q_1 + Q_2$. We claim that the functions $x/(x+y+\overline{\alpha}z), y/(x+y+\overline{\alpha}z), \overline{\alpha}z/(x+y+\overline{\alpha}z)$ are a basis of $\mathcal{L}(G)$. To see this, note that the numerators in these fractions are not 0 in Q_1 and Q_2 and that the line with equation $x+y+\overline{\alpha}z = 0$ meets $\mathcal{X}$ in Q_2 and is tangent to $\mathcal{X}$ in Q_1. By Theorem 10.6.2, the code $C(D, G)$ of length 6 has minimum distance at least 3. However, the code is in fact an MDS code, namely the *hexacode* of §4.2.

§10.8. Improvement of the Gilbert-Varshamov Bound

We fix the alphabet $\mathbb{F}_q$. We consider codes $C(D, G)$ as defined in §10.6, with a curve $\mathcal{X}$ that has $n+1$ rational points $P_1, P_2, \ldots, P_n, Q$. We take $G = mQ$ with $2g - 2 < m < n$. We define $\gamma(\mathcal{X}) := g/n$. It was shown by Tsfasman, Vlăduţ, and Zink [99] that there exists a sequence of curves $\mathcal{X}$ such that the corresponding geometric codes are a sequence of codes that yield an improvement of Theorem 5.1.9. They proved the following theorem.

(10.8.1) Theorem. *Let q be a prime power and a square. There exists a sequence of curves $\mathcal{X}_i$ over $\mathbb{F}_q$ ($i \in \mathbb{N}$) such that $\mathcal{X}_i$ has $n_i + 1$ rational points, genus g_i, where $n_i \to \infty$ as $i \to \infty$, $\gamma(\mathcal{X}_i) \to (q^{\frac{1}{2}} - 1)^{-1} =: \overline{\gamma}$ for $i \to \infty$.*

As we saw in Theorem 10.6.3, the corresponding codes $C_i := C(D, G)$ over $\mathcal{X}_i$ have rate $R_i = (m_i - g_i + 1)/n_i$ and distance $d_i \geq n_i - m_i$. So, with the notation of §5.1, we have $R_i + \delta_i \geq 1 - \gamma(\mathcal{X}_i)$. From Theorem 10.8.1 we then find :

(10.8.2) Theorem. $\delta + \alpha(\delta) \geq 1 - \overline{\gamma}$.

It is an elementary exercise in calculus to determine whether or not the straight line in the (δ, α) plane, defined by the equation $\delta + \alpha = 1 - \overline{\gamma}$, intersects the curve of Theorem 5.1.9. For intersection, one finds $q \geq 43$ and since q must be a square, we have an improvement of the Gilbert-Varshamov bound for $q \geq 49$.

§10.9. Comments

The first interesting decoding methods were given in a paper by J. Justesen et al. [91]. The ideas were generalized by A.N. Skorobogatov and S.G. Vlăduţ [97]. Since then, the methods have been considerably improved (cf. [84]) and simplified by G.-L. Feng et al. and several others. For these results we refer to the survey paper [89].

As was mentioned in the introduction, many of the results presented in this chapter can be explained, more or less avoiding the deep theorems from algebraic geometry. The ideas, due to G.-L. Feng et al. [85,86] were recast and can be found in [90].

There are some textbooks on the material of this chapter. We recommend [98] that uses a purely algebraic approach by means of function fields.

The improvement of the Gilbert-Varshamov bound ([81]) uses the theory of modular curves. This is a central but very involved part of mathematics, much more so than the Riemann-Roch theorem. It needs the theory of schemes, i. e. curves over rings instead of fields, and the analytic and algebraic properties of curves.

The recent work of Garcia and Stichtenoth [87] gives an explicit description of sequences of curves proving Theorem 10.8.1 by means of more moderate tools from algebraic geometry.

§10.10. Problems

10.10.1. Consider the curve of Example 10.2.7. What is the behavior of the function x/z in the point (1:0:0)?

10.10.2. Show that if the Klein quartic is singular over $\mathbb{F}_p$, then $p = 7$. If $p = 7$, find a singular point.

10.10.3. Consider the parabola $\mathcal{X}$ of Example 10.2.7 over $\mathbb{F}_4$. Let $g = (z^3 + xyz)/x^3$. Determine the divisor of g.

10.10.4. Let $\mathcal{X}$ be the projective curve over the algebraic closure of $\mathbb{F}_2$ defined by the equation $x^4y + y^4z + z^4x = 0$. Determine the divisor of $f := x/z$.

10.10.5. Show that the code of Example 10.7.6 is indeed equivalent to the hexacode.

10.10.6. Consider the curve of Example 10.7.4. What does the Riemann-Roch theorem say about $l(3Q)$? Show that $l(3Q) = 2$.

CHAPTER 11

Asymptotically Good Algebraic Codes

§11.1. A Simple Nonconstructive Example

In the previous chapters we have described several constructions of codes. If we considered these codes from the point of view of Chapter 5, we would be in for a disappointment. The Hadamard codes of Section 4.1 have $\delta = \frac{1}{2}$ and if $n \to \infty$ their rate R tends to 0. For Hamming codes R tends to 1 but δ tends to 0. For BCH codes we also find $\delta \to 0$ if we fix the rate. For all examples of codes which we have treated, an explicitly defined sequence of these codes, either has $\delta \to 0$ or $R \to 0$.

As an introduction to the next section we shall now show that one can give a simple algebraic definition which yields good codes. However, the definition is not constructive and we are left at the point we reached with Theorem 2.2.3. We shall describe binary codes with $R = \frac{1}{2}$. Fix m and choose an element $\alpha_m \in \mathbb{F}_{2^m}$. How this element is to be chosen will be explained below. We interpret vectors $\mathbf{a} \in \mathbb{F}_2^m$ as elements of $\mathbb{F}_{2^m}$ and define

$$C_\alpha := \{(\mathbf{a}, \alpha\mathbf{a}) | \mathbf{a} \in \mathbb{F}_2^m\}.$$

Let $\lambda = \lambda_m$ be given. If C_α contains a nonzero word $(\mathbf{a}, \alpha\mathbf{a})$ of weight $< 2m\lambda$, then this word uniquely determines α as the quotient (in $\mathbb{F}_{2^m}$) of $\alpha\mathbf{a}$ and $\mathbf{a}$. It follows that at most $\sum_{i<2m\lambda} \binom{2m}{i}$ choices of α will lead to a code C_α which has minimum distance $< 2m\lambda$. Now we take $\lambda := H^{-}(\frac{1}{2} - (1/\log m))$. By Theorem 1.4.5 the number of "bad" choices for α is $o(2^m)$ $(m \to \infty)$. Therefore for almost all choices of α we have

$$d \geq 2mH^{-}\left(\frac{1}{2} - \frac{1}{\log m}\right),$$

where d denotes the minimum distance of C_α. Letting $m \to \infty$ and taking suitable choices for α_m we thus have a sequence of codes with rate $\frac{1}{2}$ such that the corresponding δ satisfies

$$\delta = H^{-}(\tfrac{1}{2}) + o(1), \qquad (m \to \infty).$$

So this sequence meets the Gilbert bound (5.1.9). If we could give an explicit way of choosing α_m, that would be a sensational result. For a long time there was serious doubt whether it is at all possible to give an explicit algebraic construction of a sequence of codes such that both the rate and d/n are bounded away from zero. In 1972, J. Justesen [37] succeeded in doing this. The essential idea is a variation of the construction described above. Instead of one (difficult to choose) value of α, all possible multipliers α occur within one codeword. The average effect is nearly as good as one smart choice of α.

§11.2. Justesen Codes

The codes we shall describe are a generalization of *concatenated codes* which were introduced by G. D. Forney [22] in 1966. The idea is to construct a code in two steps by starting with a code C_1 and interpreting words of C_1 as symbols of a new alphabet with which C_2 is constructed. We discuss this in more detail. Let C_2 be a code over $\mathbb{F}_{2^m}$. The symbols c_i of a codeword $(c_0, c_1, \ldots, c_{n-1})$ can be written as m-tuples over $\mathbb{F}_2$, i.e. $c_i = (c_{i1}, c_{i2}, \ldots, c_{im})$ $(i = 0, 1, \ldots, n-1)$, where $c_{ij} \in \mathbb{F}_2$. Such an m-tuple is the sequence of information symbols for a word in the so-called *inner code* C_1. Let us consider the simplest case where the rate is $\frac{1}{2}$. Corresponding to $c_i = (c_{i1}, c_{i2}, \ldots, c_{im})$ we have a word of length $2m$ in the inner code.

The rate of the concatenated code is half of the rate of C_2. Justesen's idea was to vary the inner code C_1, i.e. to let the choice of C_1 depend on i. As in the previous section, the inner codes are chosen in such a way that a word of length $2m$ starts with the symbols of c_i. For the *outer code* C_2 we take a Reed-Solomon code.

The details of the construction are as follows. Since we intend to let m tend to infinity, we must have a simple construction for $\mathbb{F}_{2^m}$. We use Theorem 1.1.28. So take $m = 2 \cdot 3^{l-1}$ and $\mathbb{F}_{2^m}$ in the representation as $\mathbb{F}_2[x]$ (mod $g(x)$), where $g(x) = x^m + x^{m/2} + 1$. The Reed-Solomon code C_2 which is our outer code is represented as follows (cf. Section 6.8). An m-tuple of information symbols $(i_0, i_1, \ldots, i_{m-1})$ is interpreted as the element $i_0 + i_1 x + \cdots + i_{m-1}x^{m-1} \in \mathbb{F}_{2^m}$. Take K successive m-tuples $a_0, a_1, \ldots, a_{K-1}$ and form the polynomial $a(Z) := a_0 + a_1 Z + \cdots + a_{K-1}Z^{K-1} \in \mathbb{F}_{2^m}[Z]$. For $j = 1, 2, \ldots, 2^m - 1 =: N$, define $j(x)$ by $j(x) = \sum_{i=0}^{m-1} \varepsilon_i x^i$ if $\sum_{i=0}^{m-1} \varepsilon_i 2^i$ is the binary representation of j. Then $j(x)$ runs through the nonzero elements of $\mathbb{F}_{2^m}$. We substitute these in $a(Z)$ and thus obtain a sequence of N elements of $\mathbb{F}_{2^m}$. This is a codeword in the linear code C_2, which has rate K/N. Since $a(Z)$ has degree $\le K-1$, it has at most $K-1$ zeros, i.e. C_2 has minimum weight $D \ge N-K+1$ (cf. Section 6.8). This is a completely constructive way of producing a sequence of Reed-Solomon codes. We proceed in a similar way to form the inner codes. If c_j is the jth symbol in a codeword of the outer code (still in the representation as polynomial over $\mathbb{F}_2$) we replace it by $(c_j, j(x)c_j)$, where multiplication is again to be taken mod $g(x)$. Finally, we interpret this as a $2m$-tuple of elements of $\mathbb{F}_2$.

(11.2.1) Definition. Let $m = 2 \cdot 3^{l-1}$, $N = 2^m - 1$. K will be chosen in a suitable way below; $D = N + 1 - K$. The binary code with word length $n := n_m := 2mN$ defined above will be denoted by $\mathscr{C}_m$. It is called a *Justesen code*. The dimension of $\mathscr{C}_m$ is $k := mK$ and the rate is $\frac{1}{2}K/N$.

In our analysis of $\mathscr{C}_m$ we use the same idea as in Section 11.1, namely the fact that a nonzero $2m$-tuple $(c_j, j(x)c_j)$ occurring in a codeword of $\mathscr{C}_m$ determines the value of j.

(11.2.2) Lemma. *Let $\gamma \in (0, 1)$, $\delta \in (0, 1)$. Let $(M_L)_{L \in \mathbb{N}}$ be a sequence of natural numbers with the property $M_L \cdot 2^{-L\delta} = \gamma + o(1)$ $(L \to \infty)$. Let W be the sum of the weights of M_L distinct words in $\mathbb{F}_2^L$. Then*

$$W \ge \gamma L 2^{L\delta}\{H^{-}(\delta) + o(1)\}, \quad (L \to \infty).$$

PROOF. For L sufficiently large we define

$$\lambda := H^{-}\left(\delta - \frac{1}{\log L}\right).$$

By Theorem 1.4.5 we have

$$\sum_{0 \le i \le \lambda L} \binom{L}{i} \le 2^{L(\delta - (1/\log L))}.$$

Hence

$$W \ge \left\{M_L - \sum_{0 \le i \le \lambda L} \binom{L}{i}\right\} \lambda L \ge \lambda L\{M_L - 2^{L(\delta - (1/\log L))}\}$$

$$= \lambda L 2^{L\delta}\{\gamma + o(1)\} = \gamma L 2^{L\delta}\{H^{-}(\delta) + o(1)\}, \quad (L \to \infty). \qquad \square$$

We choose a rate R, $0 < R < \frac{1}{2}$. The number K in (11.2.1) is taken to be the minimal value for which $R_m := \frac{1}{2}K/N \ge R$. This ensures that the sequence of codes $\mathscr{C}_m$ obtained by taking $l = 1, 2, \ldots$ in (11.2.1) has rate $R_m \to R$ $(l \to \infty)$ What about the minimum distance of $\mathscr{C}_m$? A nonzero word in the outer code has weight at least $N - K + 1 = D$.

Furthermore

(11.2.3) $$N - K + 1 > N - K = N(1 - 2R_m)$$
$$= (2^m - 1)\{1 - 2R + o(1)\}, \quad (m \to \infty).$$

Every nonzero symbol in a codeword of the outer code yields a $2m$-tuple $(c_j, j(x)c_j)$ in the corresponding codeword $\mathbf{c}$ of $\mathscr{C}_m$ and these must all be different (by the remark following (11.2.1)). Apply Lemma 11.2.2 to estimate the weight of $\mathbf{c}$. We take $L = 2m$, $\delta = \frac{1}{2}$, $\gamma = 1 - 2R$ and $M_L = D$. By (11.2.3) the conditions of the lemma are satisfied. Hence

$$w(\mathbf{c}) \geq (1 - 2R)\cdot 2m \cdot 2^m\{H^{\leftarrow}(\tfrac{1}{2}) + o(1)\}, \qquad (m \to \infty).$$

Therefore

$$d_m/n \geq (1 - 2R)\{H^{\leftarrow}(\tfrac{1}{2}) + o(1)\}, \qquad (m \to \infty),$$

We have proved the following theorem.

(11.2.4)Theorem. *Let $0 < R < \frac{1}{2}$. The Justesen codes $\mathscr{C}_m$ defined above have word length $n = 2m(2^m - 1)$, rate R_m and minimum distance d_m, where*

(i) $R_m \to R$, $(m \to \infty)$,
(ii) $\liminf_{m\to\infty} d_m/n \geq (1 - 2R)H^{\leftarrow}(\frac{1}{2})$.

Using the notation of Chapter 5, we now have $\delta \geq (1 - 2R)H^{\leftarrow}(\frac{1}{2})$ for values of R less than $\frac{1}{2}$. For the first time δ does not tend to 0 for $n \to \infty$.

A slight modification of the previous construction is necessary to achieve rates larger than $\frac{1}{2}$. Let $0 \leq s < m$ (we shall choose s later). Consider $\mathscr{C}_m$. For every $2m$-tuple $(c_j, j(x)c_j)$ in a codeword $\mathbf{c}$ we delete the last s symbols. The resulting code is denoted by $\mathscr{C}_{m,s}$. Let R be fixed, $0 < R < 1$. Given m and s, we choose for K the minimal integer such that $R_{m,s} := [m/(2m - s)](K/N) \geq R$ (this is possible if $m(2m - s) \geq R$). In the proof of Theorem 11.2.4 we used the fact that a codeword $\mathbf{c}$ contained at least D distinct nonzero $2m$-tuples $(c_j, j(x)c_j)$. By truncating, we have obtained $(2m - s)$-tuples which are no longer necessarily distinct but each possible value will occur at most 2^s times. So there are at least

$$M_s := 2^{-s}(N - K) = 2^{-s}N\left(1 - \frac{2m - s}{m}R_{m,s}\right)$$

distinct $(2m - s)$-tuples in a codeword $\mathbf{c}$ of $\mathscr{C}_{m,s}$.

Again we apply Lemma 11.2.2, this time with

$$L = 2m - s, \qquad \delta = \frac{m - s}{L}, \qquad \gamma = 1 - \frac{2m - S}{m}R, \qquad M_L = M_s.$$

Let $d_{m,s}$ be the minimum distance of $\mathscr{C}_{m,s}$. We find

$$d_{m,s} \geq \left(1 - \frac{2m - s}{m}R\right)(2m - s)2^{m-s}\left\{H^{\leftarrow}\left(\frac{m - s}{2m - s}\right) + o(1)\right\}2^s, \qquad (m \to \infty).$$

Therefore

$$(11.2.5)\quad \frac{d_{m,s}}{n} \geq \left(1 - \frac{2m-s}{m} R\right)\left\{H^{-}\left(\frac{m-s}{2m-s}\right) + o(1)\right\}, \qquad (m \to \infty).$$

We must now find a choice of s which produces the best result. Let r be fixed, $r \in (\frac{1}{2}, 1)$. Take $s := \lfloor m(2r-1)/r \rfloor + 1$. If $r \geq R$ then we also have $m/(2m-s) \geq R$. From (11.2.5) we find

$$(11.2.6)\qquad \frac{d_{m,s}}{n} \geq \left(1 - \frac{R}{r}\right)\{H^{-}(1-r) + o(1)\}, \qquad (m \to \infty).$$

The right-hand side of (11.2.6) is maximal if r satisfies

$$(11.2.7)\qquad R = \frac{r^2}{1 + \log\{1 - H^{-}(1-r)\}}.$$

If the solution of (11.2.7) is less than $\frac{1}{2}$, we take $r = \frac{1}{2}$ instead. The following theorem summarizes this construction.

(11.2.8) Theorem. *Let $0 < R < 1$ and let r be the maximum of $\frac{1}{2}$ and the solution of (11.2.7). Let $s = \lfloor m(2r-1)/r \rfloor + 1$.*

The Justesen codes $\mathscr{C}_{m,s}$ have word length n, rate $R_{m,s}$, and minimum distance $d_{m,s}$, where

$$\liminf \frac{d_{m,s}}{n} \geq \left(1 - \frac{R}{r}\right) H^{-}(1-r).$$

In Figure 3 we compare the Justesen codes with the Gilbert bound. For $r > \frac{1}{2}$, the curve is the envelope of the lines given by (11.2.6).

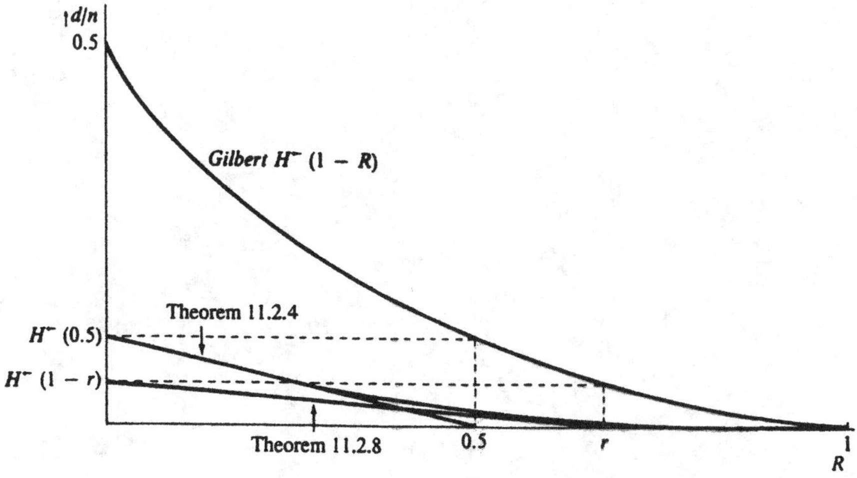

Figure 3

§11.3. Comments

The extremely simple idea of Section 11.1 has received very little attention up to now. A more serious attempt may lead to the discovery of explicit choices of α which yield relatively good codes (but see Problem 11.4.1). The discovery of the Justesen codes was one of the major developments in coding theory in the 1970s.

§11.4. Problems

11.4.1. Let $\mathbb{F}_{2^6}$ be represented as in Section 11.2. Show that there is no value of α for the construction os Section 11.1 such that C_α has distance greater than 3. Compare with other known [12, k] codes with rate $\geq \frac{1}{2}$.

11.4.2. Let $\alpha(x)$ be a polynomial of degree $< k$. A $[2k, k]$ binary *double circulant code* consists of all words of the form $(a(x), \alpha(x)a(x))$, where multiplication is $\bmod(x^k - 1)$. In this case the code is invariant under a simultaneous cyclic shift of both halves of the words. Construct a [12, 6] code of this type which has $d = 4$.

11.4.3. Use the truncation method of Section 11.2 to show that the idea of Section 11.1 leads to codes meeting the Gilbert bound for any rate R.

CHAPTER 12
Arithmetic Codes

§12.1. AN Codes

In this chapter we shall give a brief introduction to codes which are used to check and correct arithmetic operations performed by a computer. Operations are now ordinary arithmetic and as a result the theory is quite different from the preceding chapters. However, there is in several places a similarity to the theory of cyclic codes. In some cases we shall leave the details of proofs to the reader. For further information on this area see the references mentioned in Section 12.4.

The arithmetic operations in this chapter are carried out with numbers represented in the number system with base r ($r \in \mathbb{N}$, $r \geq 2$). For practical purposes the binary case ($r = 2$) and the decimal case ($r = 10$) are the most important. The first thing we have to do is to find a suitable distance function. In the previous chapters we have used Hamming distance but that is not a suitable distance function for the present purposes. One error in an addition can cause many incorrect digits in the answer because of carry. We need a distance function that corresponds to arithmetical errors in the same way as Hamming distance corresponds to misprints in words.

(12.1.1) Definition. The *arithmetic weight* $w(x)$ of an integer x is the minimal $t \geq 0$ such that there is a representation

$$x = \sum_{i=1}^{t} a_i r^{n(i)},$$

with integers a_i, $n(i)$ for which $|a_i| < r$, $n(i) \geq 0$ ($i = 1, 2, \ldots, t$). The *arithmetic distance* $d(x, y)$ of two integers is defined by

$$d(x, y) := w(x - y).$$

It is easily checked that this is indeed a metric on $\mathbb{Z}$. Arithmetic distance is translation invariant, i.e. $d(x, y) = d(x + z, y + z)$. This is not true for the Hamming distance of two integers (in r-ary representation). Arithmetic distance is at most equal to the Hamming distance.

We shall consider codes C of the form

$$C := \{AN | N \in \mathbb{Z}, 0 \leq N < B\},$$

where A and B are fixed positive integers. Such codes are called *AN codes*. These codes are used in the following way. Suppose we wish to add two integers N_1 and N_2 (both positive and small compared to B). These are encoded as AN_1 and AN_2 and then these two integers are added. Let S be the sum. If no errors have been made, then we find $N_1 + N_2$ by dividing by A. If S is not divisible by A, i.e. errors have been made, we look for the code word AN_3 such that $d(S, AN_3)$ is minimal. The most likely value of $N_1 + N_2$ is N_3. In order to be able to correct all possible patterns of at most e errors it is again necessary and sufficient that the code C has minimum distance $\geq 2e + 1$. As before, that is equivalent to requiring that C has minimum weight at least $2e + 1$. These properties of the code C are based on the resemblance of C to the subgroup $H := \{AN | N \in \mathbb{Z}\}$ of $\mathbb{Z}$. It would not be a good idea to take H as our code because H has minimum weight ≤ 2 (see Problem 12.5.1).

In order to avoid this difficulty we shall consider so-called *modular AN codes*. Define $m := AB$. Now we can consider C as a subgroup of $\mathbb{Z}/m\mathbb{Z}$. This makes it necessary to modify our distance function. Consider the elements of $\mathbb{Z}/m\mathbb{Z}$ as vertices of a graph Γ_m and let $x \pmod m$ and $x' \pmod m$ be joined by an edge iff

$$x - x' \equiv \pm c \cdot r^j \pmod m$$

for some integers c, j with $0 < c < r$, $j \geq 0$.

(12.1.2) Definition. The *modular distance* $d_m(x, y)$ of two integers x and y (considered as elements of $\mathbb{Z}/m\mathbb{Z}$) is the distance of x and y in the graph Γ_m. The *modular weight* $w_m(x)$ of x is $d_m(x, 0)$. Note that

$$w_m(x) = \min\{w(y) | y \in \mathbb{Z}, y \equiv x \pmod m\}.$$

Although we now have achieved a strong resemblance to linear codes there is another difficulty. Not every choice of m makes good sense. For example, if we take $r = 3$, $A = 5$, $B = 7$, i.e. $m = 35$, then by (12.1.2) we have $d_m(0, 4) = 1$ because $4 \equiv 3^{10} \pmod{35}$. But it is not very realistic to consider errors in the position corresponding to 3^{10} when adding integers less than 35. Restricting j in the definition of edges of Γ_m also has drawbacks. It turns out that we get an acceptable theory if we take $m = r^n - 1$ ($n \in \mathbb{Z}$, $n \geq 2$). In practice this is also a good choice because many computers do arithmetic operations mod $2^n - 1$.

Every integer x has a unique representation

$$x \equiv \sum_{i=0}^{n-1} c_i r^i \pmod{r^n - 1},$$

with $c_i \in \{0, 1, \ldots, r-1\}$ $(0 \le i < n)$, not all $c_i = 0$. Hence $\mathbb{Z}/(r^n - 1)$ can be interpreted as the set of nonzero words of length n over the alphabet $\{0, 1, \ldots, r-1\}$. Of course it would not have been necessary to exclude $\mathbf{0}$ if we had taken $m = r^n$, which is again a practical choice because many computers work mod 2^n. However, we cannot expect good codes for $r = 2$, $m = 2^n$. We would have to take $A = 2^k$ for some k and then the code C would consist of the integers $\sum_{i=0}^{n-1} c_i 2^i$, $c_i \in \{0, 1\}$, for which $c_0 = c_1 = \cdots = c_{k-1} = 0$. An integer $x \pmod B$ would be encoded by adding k 0s to its representation. This would serve no purpose. For arbitrary r there are similar objections. The reader should convince himself that in the case $AB = m = r^n - 1$ modular distance is a natural function for arithmetic in $\mathbb{Z}/m\mathbb{Z}$ and that C behaves as a linear code. In fact we have an even stronger analogy with earlier chapters.

(12.1.3) Definition. A *cyclic AN code* of length n and base r is a subgroup C of $\mathbb{Z}/(r^n - 1)$. Such a code is a principal ideal in this ring, i.e. there are integers A and B such that $AB = r^n - 1$ and

$$C = \{AN \mid N \in \mathbb{Z}, 0 \le N < B\}.$$

As in Section 6.1, we call A the *generator* of C. By now it will not be surprising to the reader that we are primarily interested in codes C with a large *rate* $(=(1/n)\log_r B)$ and a large minimum distance. The terminology of (12.1.3) is in accordance with (6.1.1). If $x \in C$ then $rx \pmod{r^n - 1}$ is also a codeword because C is a group and $rx \pmod{r^n - 1}$ is indeed a cyclic shift of x (both represented in base r). The integer B can be compared with the check polynomial of a cyclic code.

The idea of negacyclic codes can be generalized in the same way by taking $m = r^n + 1$ and then considering subgroups of $\mathbb{Z}/m\mathbb{Z}$.

(12.1.4) Example. Let $r = 2$, $n = 11$. Then $m = r^n - 1 = 2047$. We take $A = 23$, $B = 89$. We obtain the cyclic AN code consisting of 89 multiples of 23 up to 2047. There are 22 ways to make one error, corresponding to the integers $\pm 2^j$ $(0 \le j < 11)$. These are exactly the integers mod 23 except 0. Therefore every integer in $[1,2047]$ has modular distance 0 or 1 to exactly one codeword. This cyclic AN code is therefore *perfect*. It is a generalization of the Hamming codes.

§12.2. The Arithmetic and Modular Weight

In order to be able to construct AN codes that correct more than one error we need an easy way to calculate the arithmetic or modular weight of an integer.

By Definition 12.1.1, every integer x can be written as

$$x = \sum_{i=1}^{w(x)} a_i r^{n(i)}$$

with integers a_i, $n(i)$, $|a_i| < r$, $n(i) \geq 0$ ($i = 1, \ldots, w(x)$). It is easy to find examples which show that this representation is not unique. We shall put some more restrictions on the coefficients which will make the representation unique.

(12.2.1) Definition. Let $b \in \mathbb{Z}$, $c \in \mathbb{Z}$, $|b| < r$, $|c| < r$. The pair (b, c) is called *admissible* if one of the following holds

(i) $bc = 0$,
(ii) $bc > 0$ and $|b + c| < r$,
(iii) $bc < 0$ and $|b| > |c|$.

Note that if $r = 2$ we must have possibility (i). Therefore a representation $x = \sum_{i=0}^{\infty} c_i 2^i$ in which all pairs (c_{i+1}, c_i) are admissible has no two adjacent nonzero digits. This led to the name *nonadjacent form* (*NAF*) which we now generalize.

(12.2.2) Definition. A representation

$$x = \sum_{i=0}^{\infty} c_i r^i,$$

with $c_i \in \mathbb{Z}$, $|c_i| < r$ for all i and $c_i = 0$ for all large i is called an *NAF* for x if for every $i \geq 0$ the pair (c_{i+1}, c_i) is admissible.

(12.2.3) Theorem. *Every integer x has exactly one* NAF. *If this is*

$$x = \sum_{i=0}^{\infty} c_i r^i,$$

then

$$w(x) = |\{i \mid i \geq 0, c_i \neq 0\}|.$$

PROOF.

(a) Suppose x is represented as $\sum_{i=0}^{\infty} b_i r^i$, $|b_i| < r$. Let i be the minimal value such that the pair (b_{i+1}, b_i) is not admissible. W.l.o.g. $b_i > 0$ (otherwise consider $-x$). Replace b_i by $b_i' := b_i - r$ and replace b_{i+1} by $b_{i+1}' := b_{i+1} + 1$ (if $b_{i+1} + 1 = r$ we carry forward). If $b_{i+1} > 0$ we either have $b_{i+1}' = 0$ or $b_i' b_{i+1}' < 0$ and $b_{i+1}' = b_{i+1} + 1 > r - b_i = |b_i'|$ since (b_{i+1}, b_i) was not admissible. If $b_{i+1} < 0$ then $b_{i+1}' = 0$ or $b_i' b_{i+1}' > 0$ and $|b_i' + b_{i+1}'| = r - b_i - b_{i+1} - 1 < r$ because $-b_{i+1} \leq b_i$ as (b_{i+1}, b_i) is not admissible. So (b_{i+1}', b_i') is admissible and one checks in the same way that (b_i', b_{i-1}) is admissible. In this way we can construct an NAF and in the process the weight of the representation does not increase.

(b) It remains to show that the NAF is unique. Suppose some x has two such representations $x = \sum_{i=0}^{\infty} c_i r^i = \sum_{i=0}^{\infty} c_i' r^i$. W.l.o.g. we may assume $c_0 \neq c_0'$, $c_0 > 0$. Therefore $c_0' = c_0 - r$. It follows that $c_1' \in \{c_1 + 1 - r, c_1 + 1, c_1 + 1 + r\}$. If $c_1' = c_1 + 1 - r$ then $c_1 \geq 0$ and hence $c_0 + c_1 \leq r - 1$. Since $c_0' c_1' > 0$ we must have $-c_0' - c_1' < r$, i.e. $r - c_0 + r - c_1 - 1 < r$, so $c_0 + c_1 > r - 1$, a contradiction. In the same way the assumptions $c_1' = c_1 + 1$ resp. $c_1' = c_1 + 1 + r$ lead to a contradiction. Therefore the NAF is unique. □

A direct way to find the NAF of an integer x is provided in the next theorem.

(12.2.4) Theorem. *Let $x \in \mathbb{Z}$, $x \geq 0$. Let the r-ary representations of $(r + 1)x$ and x be*

$$(r + 1)x = \sum_{j=0}^{\infty} a_j r^j, \qquad x = \sum_{j=0}^{\infty} b_j r^j.$$

with a_j, $b_j \in \{0, 1, \ldots, r - 1\}$ for all j and $a_j = b_j = 0$ for j sufficiently large. Then the NAF *for x is*

$$x = \sum_{j=0}^{\infty} (a_{j+1} - b_{j+1}) r^j.$$

PROOF. We calculate the numbers a_j by adding $\sum_{j=0}^{\infty} b_j r^j$ and $\sum_{j=0}^{\infty} b_j r^{j+1}$. Let the carry sequence be $\varepsilon_0, \varepsilon_1, \ldots$, so $\varepsilon_0 = 0$ and $\varepsilon_i := \lfloor (\varepsilon_{i-1} + b_{i-1} + b_i)/r \rfloor$. We find that $a_i = \varepsilon_{i-1} + b_{i-1} + b_i - \varepsilon_i r$. If we denote $a_i - b_i$ by c_i then $c_i = \varepsilon_{i-1} + b_{i-1} - \varepsilon_i r$. We must now check whether (c_{i+1}, c_i) is an admissible pair. That $|c_{i+1} + c_i| < r$ is a trivial consequence of the definition of ε_i. Suppose $c_i > 0$, $c_{i+1} < 0$. Then $\varepsilon_i = 0$. We then have $c_i = \varepsilon_{i-1} + b_{i-1}$, $c_{i+1} = b_i - r$ and the condition $|c_{i+1}| > |c_i|$ is equivalent to $\varepsilon_{i-1} + b_{i-1} + b_i < r$, i.e. $\varepsilon_i = 0$. The final case is similar. □

The NAF for x provides us with a simple estimate for x as shown by the next theorem.

(12.2.5) Theorem. *If we denote the maximal value of i for which $c_i \neq 0$ in an* NAF *for x by $i(x)$ and define $i(0) := -1$ then*

$$i(x) \leq k \Leftrightarrow |x| < \frac{r^{k+2}}{r+1}.$$

We leave the completely elementary proof to the reader.

From Section 12.1, it will be clear that we must now generalize these ideas in some way to modular representations. We take $m = r^n - 1$, $n \geq 2$.

(12.2.6) Definition. A representation

$$x \equiv \sum_{i=0}^{n-1} c_i r^i \pmod{m},$$

with $c_i \in \mathbb{Z}$, $|c_i| < r$ is called a *CNAF* (=*cyclic NAF*) for x if (c_{i+1}, c_i) is admissible for $i = 0, 1, \ldots, n-1$; here $c_n := c_0$.

The next two theorems of CNAF's are straightforward generalizations of Theorem 12.2.3 and can be obtained from this theorem or by using Theorem 12.2.4. A little care is necessary because of the exception, but the reader should have no difficulty proving the theorems.

(12.2.7) Theorem. *Every integer x has a* CNAF mod m; *this* CNAF *is unique except if*

$$(r+1)x \equiv 0 \not\equiv x \pmod{m}$$

in which case there are two CNAFs *for x (mod m). If $x \equiv \sum_{i=0}^{n-1} c_i r^i \pmod{m}$ is a* CNAF *for x then*

$$w_m(x) = |\{i \mid 0 \le i < n, c_i \ne 0\}|.$$

(12.2.8) Theorem. *If $(r+1)x \equiv 0 \not\equiv x \pmod{m}$ then $w_m(x) = n$ except if $n \equiv 0$ (mod 2) and $x \equiv \pm[m/(r+1)] \pmod{m}$, in which case $w_m(x) = \frac{1}{2}n$.*

If we have an NAF for x for which $c_{n-1} = 0$ then the additional requirement for this to be a CNAF is satisfied. Therefore Theorem 12.2.5 implies the following theorem.

(12.2.9) Theorem. *An integer x has a* CNAF *with $c_{n-1} = 0$ iff there is a $y \in \mathbb{Z}$ with $x \equiv y \pmod{m}$, $|y| \le m/(r+1)$.*

This theorem leads to another way of finding the modular weight of an integer.

(12.2.10) Theorem. *For $x \in \mathbb{Z}$ we have $w_m(x) = |\{j \mid 0 \le j < n$, there is a $y \in \mathbb{Z}$ with*

$$m/(r+1) < y \le mr/(r+1),\ y \equiv r^j x \pmod{m}\}|.$$

Proof. Clearly a CNAF for rx is a cyclic shift of a CNAF for x, i.e. $w_m(rx) = w_m(x)$. Suppose $x \equiv \sum_{i=0}^{n-1} c_i r^i \pmod{m}$ is a CNAF and $c_{n-1-j} = 0$. Then $r^j x$ has a CNAF with 0 as coefficient of r^{n-1}. By Theorem 12.2.9 this is the case if there is a y with $y \equiv r^j x \pmod{m}$ and $|y| \le m/(r+1)$. Since the modular weight is the number of nonzero coefficients, the assertion now follows unless we are in one of the exceptional cases of Theorem 12.2.7, but then the result follows from Theorem 12.2.8. □

§12.3. Mandelbaum–Barrows Codes

We now introduce a class of cyclic multiple-error-correcting AN codes which is a generalization of codes introduced by J. T. Barrows and D. Mandelbaum. We first need a theorem on modular weights in cyclic AN codes.

(12.3.1) Theorem. *Let $C \subset \mathbb{Z}/(r^n - 1)$ be a cyclic* AN *code with generator* A *and let*

$$B := (r^n - 1)/A = |C|.$$

Then

$$\sum_{x \in C} w_m(x) = n\left(\left\lfloor \frac{rB}{r+1} \right\rfloor - \left\lfloor \frac{B}{r+1} \right\rfloor\right).$$

PROOF. We assume that every $x \in C$ has a unique CNAF

$$x \equiv \sum_{i=0}^{n-1} c_{i,x} r^i \pmod{r^n - 1}.$$

The case where C has an element with two CNAFs is slightly more difficult. We leave it to the reader. We must determine the number of nonzero coefficients $c_{i,x}$, where $0 \le i \le n-1$ and $x \in C$, which we consider as elements of a matrix. Since C is cyclic every column of this matrix has the same number of zeros. So the number to be determined is equal to $n|\{x \in C \mid c_{n-1,x} \neq 0\}|$. By Theorem 12.2.9, we have $c_{n-1,x} \neq 0$ iff there is a $y \in \mathbb{Z}$ with $y \equiv x \pmod{r^n - 1}$, $m/(r+1) < y \le mr/(r+1)$. Since x has the form AN $\pmod{r^n - 1}$ $(0 \le N < B)$, we must have $B/(r+1) < N \le Br/(r+1)$. □

The expression in Theorem 12.3.1 is nearly equal to $n|C|[(r-1)/(r+1)]$ and hence the theorem resembles our earlier result

$$\sum_{x \in C} w(x) = n|C| \cdot \frac{q-1}{q}$$

for a linear code C (cf. (3.7.5)).

The next theorem introduces the generalized *Mandelbaum–Barrows codes* and shows that these codes are *equidistant*.

(12.3.2) Theorem. *Let B be a prime number that does not divide r with the property that $(\mathbb{Z}/B\mathbb{Z})$ is generated by the elements r and -1. Let n be a positive integer with $r^n \equiv 1 \pmod{B}$ and let $A := (r^n - 1)/B$. Then the code $C \subset \mathbb{Z}/(r^n - 1)$ generated by A is an equidistant code with distance*

$$\frac{n}{(B-1)}\left(\left\lfloor \frac{rB}{r+1} \right\rfloor - \left\lfloor \frac{B}{r+1} \right\rfloor\right).$$

Proof. Let $x \in C$, $x \neq 0$. Then $x = AN \pmod{r^n - 1}$, with $N \not\equiv 0 \pmod{B}$. Our assumptions imply that there is a j such that $N \equiv \pm r^j \pmod{B}$. Therefore $w_m(x) = w_m(\pm r^j A) = w_m(A)$. This shows that C is equidistant and then the constant weight follows from Theorem 12.3.1. □

The Mandelbaum–Barrows codes correspond to the minimal cyclic codes M_i^- of Section 6.2. Notice that these codes have word length at least $\frac{1}{2}(B-1)$ which is large with respect to the number of codewords which is B. So, for practical purposes these codes do not seem to be important.

§12.4. Comments

The reader interested in more information about arithmetic codes is referred to W. W. Peterson and E. J. Weldon [53], J. L. Massey and O. N. Garcia [48], T. R. N. Rao [58]. Perfect single error-correcting cyclic AN codes have been studied extensively. We refer to M. Goto [28], M. Goto and T. Fukumara [29], and V. M. Gritsenko [31]. A perfect single error-correcting cyclic AN code with $r = 10$ or $r = 2^k$ ($k > 1$) does not exist.

For more details about the NAF and CNAF we refer to W. E. Clark and J. J. Liang [14], [15]. References for binary Mandelbaum–Barrows codes can be found in [48]. There is a class of cyclic AN codes which has some resemblance to BCH codes. These can be found in C. L. Chen, R. T. Chien and C. K. Liu [12].

For more information about perfect arithmetic codes we refer to a contribution with that title by H. W. Lenstra in the Séminaire Delange–Pisot–Poitou (Théorie des Nombres, 1977/78).

§12.5. Problems

12.5.1. Prove that $\min\{w(AN) | N \in \mathbb{Z}, N \neq 0\} \leq 2$ for every $A \in \mathbb{Z}$ if w is as defined in (12.1.1).

12.5.2. Generalize (10.1.4). Find an example with $r = 3$.

12.5.3. Consider ternary representations mod $3^6 - 1$. Find a CNAF for 455 using the method of the proof of Theorem 12.2.3..

12.5.4. Determine the words of the Mandelbaum–Barrows code with $B = 11$, $r = 3$, $n = 5$.

CHAPTER 13

Convolutional Codes

§13.1. Introduction

The codes which we consider in this chapter are quite different from those in previous chapters. They are not block codes, i.e., the words do not have constant length. Although there are analogies and connections to block codes, there is one big difference, namely that the mathematical theory of convolutional codes is not well developed. This is one of the reasons that mathematicians find it difficult to become interested in these codes.

However, in our introduction we gave communication with satellites as one of the most impressive examples of the use of coding theory and at present one of the main tools used in this area is convolutional coding! Therefore a short introduction to the subject seems appropriate. For a comparison of block and convolutional codes we refer to [51, Section 11.4]. After the introductory sections we treat a few of the more mathematical aspects of the subject. The main area of research of investigators of these codes is the reduction of decoding complexity. We do not touch on these aspects and we must refer the interested reader to the literature.

In this chapter we assume that the alphabet is binary. The generalization to $\mathbb{F}_q$ is straightforward. Every introduction to convolutional coding seems to use the same example. Adding one more instance to the list might strengthen the belief of some students that no other examples exist, but nevertheless we shall use this canonical example.

In Figure 4 we demonstrate the encoding device for our code. The three squares are storage elements (flip-flops) which can be in one of two possible states which we denote by 0 and 1. The system is governed by an external clock which produces a signal every t_0 seconds (for the sake of simplicity we choose our unit of time such that $t_0 = 1$). The effect of this signal is that the

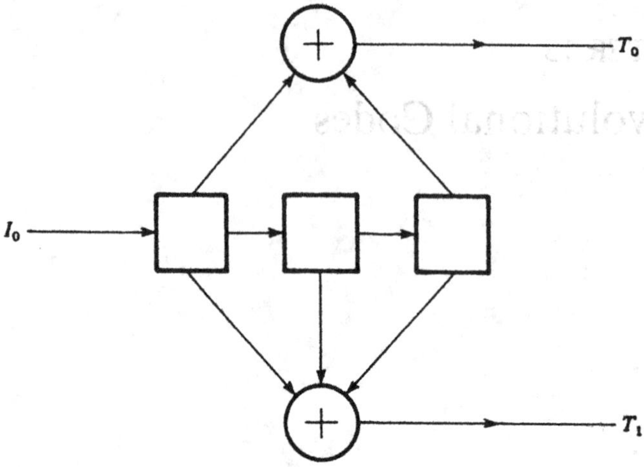

Figure 4

contents of the flip-flops move in the direction of the arrows to the next element of this so-called *shift register*. The elements ⊕ indicate mod 2 *adders*. For every pulse of the clock, we see that the contents of the first and third flip-flop are added and then leave the encoder via the stream T_0. The information to be processed enters on the left as stream I_0. Notice that the first flip-flop is actually superfluous since it only serves to divide the input stream into three directions. The second and third element of the shift register show the essential difference with block coding. They are *memory* elements which see to it that at time t the input signals for $t-1$ and $t-2$ are still available. The output depends on these three input symbols. In practice the output streams T_0 and T_1 are interlaced to produce one output stream. So, if we start with (0 0 0) in the register and have a message stream consisting of 1 followed by 0s, we see that the register first changes to (1 0 0) and yields the output (1 1); then it moves to (0 1 0) with output (0 1); subsequently (0 0 1) with output (1 1) and then back to the original state and a stream of 0s.

An efficient way of describing the action of this encoder is given by the *state diagram* of Figure 5. Here the contents of the second and third flip-flop are called the state of the register. We connect two states by a solid edge if the register goes from one to the other by an input 0. Similarly a dashed edge corresponds to an input 1. Along these edges we indicate in brackets the two outputs at T_0, T_1. An input stream I_0 corresponds to a walk through the graph of Figure 5.

A mathematical description of this encoding procedure can be given as follows. Describe the input stream $i_0, i_1, i_2, \ldots$ by the formal power series $I_0(x) := i_0 + i_1 x + i_2 x^2 + \cdots$ with coefficients in $\mathbb{F}_2$. Similarly describe the outputs at T_0 and T_1 by power series $T_0(x)$ resp. $T_1(x)$. We synchronize time in such a way that first input corresponds to first output. Then it is clear that

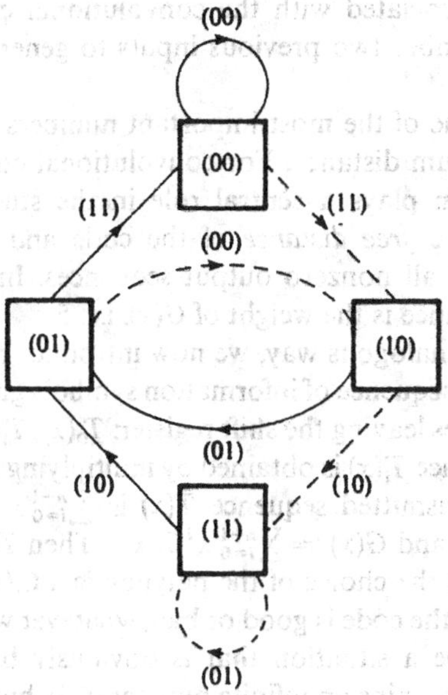

Figure 5

$$T_0(x) = (1 + x^2)I_0(x),$$
$$T_1(x) = (1 + x + x^2)I_0(x).$$

The interlacing of T_0 and T_1 then is described by

$$T(x) = T_0(x^2) + xT_1(x^2).$$

In our example we had input $I_0(x) = 1$ and the output sequence

$$11\ 01\ 11\ 00\ 00\ \ldots$$

which is

$$G(x) := 1 + x + x^3 + x^4 + x^5 = (1 + (x^2)^2) + x(1 + (x^2) + (x^2)^2).$$

So if we define $I(x) := I_0(x^2)$, then

(13.1.1) $$T(x) = G(x)I(x).$$

For obvious reasons we say that the rate of this convolutional code is $\frac{1}{2}$. As usual, the code is the set of possible output sequences $T(x)$. The polynomial $G(x)$ is sometimes called the *generator polynomial* of this code.

In the description given above, the information symbol i_ν entering the shift register influences six of the symbols of the transmitted sequence. This number is called the *constraint length* of the code. The constraint length equals $1 + \text{degree } G(x)$. The reader is warned that there are at least two other definitions of constraint length used by other authors (one of these is: the length of the shift register, e.g. in Figure 4 this is 3). In our example we say that the *memory* associated with the convolutional code is 2 because the encoder must remember two previous inputs to generate output when i_v is presented.

We know that one of the most important numbers connected to a block code C is its minimum distance. For convolutional codes there is a similar concept which again plays a central rôle in the study of decoding. This number is called the *free distance* of the code and is defined to be the minimum weight of all nonzero output sequences. In the example treated above, this free distance is the weight of $G(x)$, i.e. 5.

In a completely analogous way, we now introduce rate $1/n$ convolutional codes. We have one sequence of information symbols given by the series $I_0(x)$. There are n sequences leaving the shift register: $T_0(x), T_1(x), \ldots, T_{n-1}(x)$, where each encoded sequence $T_i(x)$ is obtained by multiplying $I_0(x)$ by some polynomial $G_i(x)$. The transmitted sequence $T(x)$ is $\sum_{i=0}^{n-1} x^i T_i(x^n)$. As before, we define $I(x) := I_0(x^n)$ and $G(x) := \sum_{i=0}^{n-1} x^i G_i(x^n)$. Then $T(x) = G(x)I(x)$.

It is obvious that the choice of the polynomials $G_i(x)$ $(i = 0, 1, \ldots, n-1)$ determines whether the code is good or bad, whatever we decide this to mean. Let us now describe a situation that is obviously bad. Suppose that the input stream $I_0(x)$ contains an infinite number of 1s but that the corresponding output stream $T(x)$ has only finitely many 1s. If the channel accidentally makes errors in the position of these 1s the resulting all-zero output stream will be interpreted by the receiver as coming from input $I_0(x) = 0$. Therefore a finite number of channel errors has then caused an infinite number of decoding errors! Such a code is called a *catastrophic code*. There is an easy way to avoid that the rate $1/n$ convolutional code is catastrophic, namely by requiring that

$$\text{g.c.d.}(G_0(x), G_1(x), \ldots, G_{n-1}(x)) = 1.$$

It is well known that this implies that there are polynomials $a_i(x)$ $(i = 0, \ldots, n-1)$ such that $\sum_{i=0}^{n-1} a_i(x)G_i(x) = 1$. From this we find

$$\sum_{i=0}^{n-1} a_i(x^n)T_i(x^n) = \sum_{i=0}^{n-1} a_i(x^n)G_i(x^n)I_0(x^n) = I(x),$$

i.e. the input can be determined from the output and furthermore a finite number of errors in $T(x)$ cannot cause an infinite number of decoding errors.

There are two ways of describing the generalization to rate k/n codes. We now have k shift registers with input streams $I_0(x), \ldots, I_{k-1}(x)$. There are n output streams $T_i(x)$ $(i = 0, \ldots, n-1)$, where $T_i(x)$ is formed using all the shift registers. We first use the method which was used above. Therefore we now need kn polynomials $G_{ij}(x)$ $(i = 0, \ldots, k-1;\ j = 0, \ldots, n-1)$ to describe the situation. We have

$$T_j(x) = \sum_{i=0}^{k-1} G_{ij}(x)I_i(x).$$

It is no longer possible to describe the encoding with one generator polynomial. We prefer the following way of describing rate k/n convolutional codes which makes them into block codes over a suitably chosen field.

Let $\mathscr{F}$ be the quotient field of $\mathbb{F}_2[x]$, i.e. the field of all Laurent series of the form

$$\sum_{i=r}^{\infty} a_i x^i, \qquad (r \in \mathbb{Z}, a_i \in \mathbb{F}_2).$$

We consider the k bits entering the different shift registers at time t as a vector in $\mathbb{F}_2^k$. This means that the input sequences are interpreted as a vector in $\mathscr{F}^k$ (like in Chapter 3 the vectors are row vectors). We now consider the kn polynomials $G_{ij}(x)$ as the elements of a generator matrix G. Of course the n output sequences can be seen as an element of $\mathscr{F}^n$. This leads to the following definition.

(13.1.2) Definition. A rate k/n binary *convolutional code* C is a k-dimensional subspace of $\mathscr{F}^n$ which has a basis of k vectors from $\mathbb{F}[x]^n$. These basis vectors are the rows of G.

Although this shows some analogy to block codes we are hiding the difficulty by using $\mathscr{F}$ and furthermore the restriction on the elements of G is quite severe. In practice all messages are finite and we can assume that there is silence for $t < 0$. This means that we do not really need $\mathscr{F}$ and can do everything with $\mathbb{F}[x]$. Since this is not a field, there will be other difficulties for that approach.

When discussing block codes, we pointed out that for a given code there are several choices of G, some of which may make it easier to analyze C. The same situation occurs for convolutional codes. For such a treatment of generator matrices we refer the reader to [23]. This is one of the few examples of a mathematical analysis of convolutional codes.

§13.2. Decoding of Convolutional Codes

There are several algorithms used in practice for decoding convolutional codes. They are more or less similar and all mathematically not very deep. In fact they resemble the decoding of block codes by simply comparing a received word with all codewords. Since the codewords of a convolutional code are infinitely long this comparison has to be truncated, i.e., one only considers the first l symbols of the received message. After the comparison one decides on say the first information symbol and then repeats the procedure for subsequent symbols. We sketch the so-called *Viterbi-algorithm* in more detail using the example of Figures 4 and 5.

Suppose the received message is 10 00 01 10 00.... The diagram of Figure 6 shows the possible states at time $t = 0, 1, 2, 3$, the arrows showing the path through Figure 5. Four of the lines in Figure 6 are dashed and these

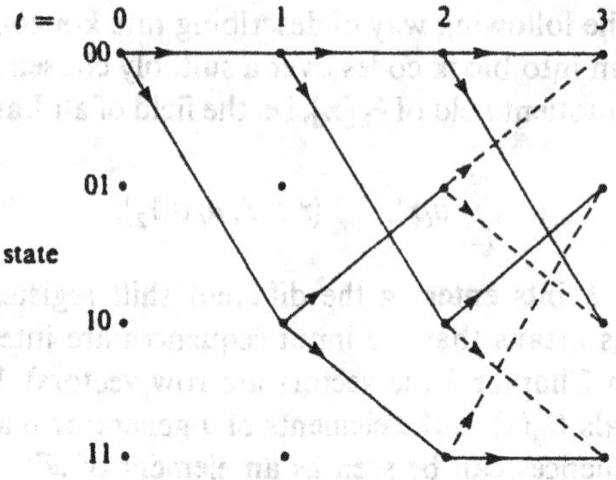

Figure 6

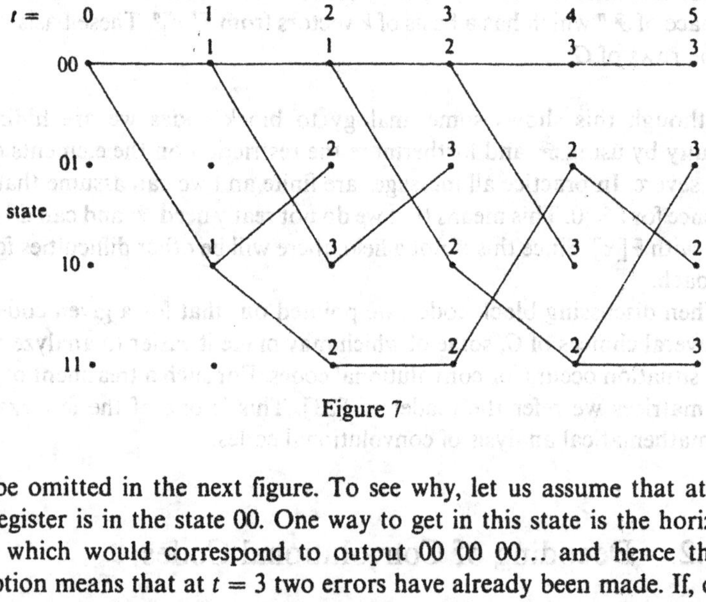

Figure 7

will be omitted in the next figure. To see why, let us assume that at $t = 3$ the register is in the state 00. One way to get in this state is the horizontal path which would correspond to output 00 00 00... and hence this assumption means that at $t = 3$ two errors have already been made. If, on the other hand the state 00 was reached via 10 and 01, there would have been output 11 01 11 and so there are already three errors. We do not know whether the register is in state 00 but if it is, then the horizontal path is the more likely way of having gotten there. Furthermore this path involves two errors. Let us extend Figure 6 to Figure 7 where $t = 4$ and $t = 5$ are included and we also show the number of errors involved in reaching the different stages.

Of course it can happen that there are two equally likely ways to reach a certain state in which case one chooses one of them and discards the other. Corresponding to each time and state, one can list the most likely outputs corresponding to that assumption. In this way Figure 7 would lead to the following list of outputs.

		$t = 1$	$t = 2$	$t = 3$	$t = 4$	$t = 5$
	00	00	00 00	00 00 00	00 00 00 00	00 00 00 00 00
state	01	—	11 01	00 11 01	11 10 01 10	00 00 11 10 10
	10	11	00 11	00 00 11	00 00 00 11	11 10 01 10 00
	11	—	11 10	11 10 01	00 00 11 10	00 00 11 10 01

Clearly at $t = 5$ a maximum likelihood decision would be that the register is in state 10, that two errors have been made and that the output was 11 10 01 10 00. The corresponding input can be found from Figure 7. The easiest way is to mark each edge with the input symbol corresponding to that edge. We leave it to the reader to invent several ways of continuing after truncation. It is clear that we have come upon problems of computing time and storage. A mathematically more interesting question is to calculate the effect of an incorrect decision on subsequent decoding decisions and also to find out what is the influence of the free distance of the code on the decoding accuracy. For a more detailed treatment of these questions we refer to [51].

§13.3. An Analog of the Gilbert Bound for Some Convolutional Codes

We consider a special class of rate k/n convolutional codes, namely those defined as follows. Take the k input sequences $I_0(x), \ldots, I_{k-1}(x)$ and form

$$I(x) := \sum_{i=0}^{k-1} x^i I_i(x^n).$$

This means that every n-tuple of information bits ends in $n - k$ zeros. Take a generator polynomial $G(x)$ and define output by $T(x) = G(x)I(x)$. This corresponds to a special choice of the polynomials $G_{ij}(x)$ in our earlier description of rate k/n codes. As before, we define the constraint length to be $l + 1 = 1 + \text{degree } G(x)$ and we consider only codes for which $l + 1 \leq mn$, where m is fixed. We are interested in the free distance d_f of such codes. Clearly $d_f \leq l + 1$. By shifting the time scale, we see that in studying the encoding procedure we may make the restriction that $i_0 = t_0 = G(0) = 1$. For every initial mn-tuple $1 = t_0, t_1, \ldots, t_{mn-1}$ of an output sequence and every initial mk-tuple $1 = i_0, i_1, \ldots, i_{mk-1}$, there is exactly one polynomial $G(x)$ of degree $\leq mn$ such that these initial sequences are in accordance with $T(x) = G(x)I(x)$. We wish to exclude all initial segments of $T(x)$ which have weight $< d$. This means that we exclude at most $2^{mk-1} \sum_{i=0}^{d-2} \binom{mm-1}{i}$ polynomials with $G(0) = 1$ as generators. Hence there is a choice of $G(x)$ yielding at least the required free distance if

$$2^{mk}\sum_{i=0}^{d}\binom{mn}{i} < 2^{mn}.$$

Taking $d = \lambda mn$ and writing $R := k/n$, we have from Theorem 1.4.5(i) by taking logarithms

$$\frac{1}{mn}\log\sum_{i=0}^{d}\binom{mn}{i} < H(\lambda) < 1 - R,$$

if

$$\lambda < H^{-}(1 - R).$$

Here λ is the quotient of the free distance and the constraint length. This bound should be compared with Theorem 5.1.9.

§13.4. Construction of Convolutional Codes from Cyclic Block Codes

Since quite a lot is known about block codes, it is not surprising that several authors have used good block codes to construct convolutional codes with desirable properties. In this section we describe one of the methods which were developed.

We use the notation of Section 4.5.

(13.4.1) Lemma. *Let $q = 2^r$, $P(x) \in \mathbb{F}_q[x]$, $c \in \mathbb{F}_q\backslash\{0\}$, $n \geq 0$, $N \geq 0$. Then*

$$w(P(x)(x^n - c)^N) \geq w((x - c)^N)\cdot w(P(x) \bmod (x^n - c)).$$

Proof. Write $P(x)$ as $\sum_{i=0}^{n-1} x^i Q_i(x^n)$. Then by Theorem 4.5.2 we have

$$\begin{aligned} w(P(x)(x^n - c)^N) &= \sum_{i=0}^{n-1} w(Q_i(x)(x - c)^N) \\ &\geq \sum_{i=0}^{n-1} w(Q_i(c))w((x - c)^N) \\ &= w((x - c)^N)\cdot w\left(\sum_{i=0}^{n-1} Q_i(c)x^i\right) \\ &= w((x - c)^N)\cdot w(P(x) \bmod (x^n - c)). \qquad \square \end{aligned}$$

Remark. It is not difficult to prove Lemma 13.4.1 for an arbitrary field $\mathbb{F}_q$.

(13.4.2) Theorem. *Let $g(x)$ be the generator of a cyclic code of length n (odd) over $\mathbb{F}_q$ ($q = 2^r$) with minimum distance d_g and let $h(x)$ be the parity check polynomial and d_h the minimum distance of the code with generator $h(x)$. Then* *the convolutional code with rate* $1/(2m)$ *over the same field and with generator* $G(x) = g(x)$ *is noncatastrophic and satisfies* $d_f \geq \min\{d_g, 2d_h\}$.

PROOF.

(i) Write $G(x) = \sum_{j=0}^{2m-1} x^j(\hat{G}_j(x^m))^2$. If we consider the representation of the convolutional code with polynomials $G_0(x), \ldots, G_{2m-1}(x)$ as in Section 13.1, then for any irreducible common factor $A(x)$ of these polynomials $G(x)$ would be divisible by $A^2(x)$ which is impossible since n is odd and $g(x)$ divides $x^n - 1$. So the code is noncatastrophic.

(ii) Consider an information sequence $I_0(x)$. We have $T(x) = G(x)I_0(x^{2m}) = G(x)(\hat{I}_0(x^m))^2$, and therefore $T(x)$ has the form

$$T(x) = P(x)(g(x))^{2i+1}(h(x))^{2j}$$

with $i \geq 0$, $j \geq 0$, $P(x) \neq 0$, $P(x)$ not divisible by $g(x)$ or $h(x)$. We consider two cases:

(a) Let $i \geq j$. Then we find

$$T(x) = P(x)(g(x))^{2(i-j)+1}(x^n - 1)^{2j}$$

and then Lemma 13.4.1 yields

$$w(T(x)) \geq w((x-1)^{2j}) \cdot w(P(x)(g(x))^{2(i-j)+1} \bmod (x^n - 1)) \geq d_g$$

since the second factor concerns a codeword in the cyclic code generated by $g(x)$.

(b) Let $i < j$. We then find

$$T(x) = P(x)(h(x))^{2(j-i)-1}(x^n - 1)^{2i+1}$$

and then Lemma 13.4.1 yields $w(T(x)) \geq 2d_h$ because

$$w((x-1)^{2i+1}) \geq 2. \qquad \square$$

Before looking at more examples of this argument, we discuss a bound for the free distance. Consider a convolutional code with rate $1/n$ over $\mathbb{F}_q$ with generator polynomial $G(x) = \sum_{i=0}^{n-1} x^i G_i(x^n)$. Let

$$L := n(1 + \max\{\text{degree } G_i(x) | 0 \leq i \leq n-1\}).$$

(Some authors call L the constraint length.) It is obvious that

$$d_f \leq L.$$

This trivial bound has a certain analogy with the Singleton bound $d \leq n - k + 1$ for block codes. We shall now describe a construction due to J. Justesen [38] of convolutional codes which meet this bound. First we show that this requirement yields a bound on L.

(13.4.3) Lemma. *If a rate* $1/n$ *convolutional code over* $\mathbb{F}_q$ *has* $d_f = L$ *then* $L \leq nq$.

PROOF. If $d_f = L$ then each of the polynomials $G_i(x)$ has weight L/n. We consider input sequences $I_0(x) = 1 + \alpha x$ where α runs through $\mathbb{F}_q\backslash\{0\}$ and determine the average weight $\bar{w}$ of the corresponding encoded sequences. We find

$$\bar{w} = (q-1)^{-1} \sum_{\alpha \in \mathbb{F}_q\backslash\{0\}} \sum_{i=0}^{n-1} w(G_i(x)(1+\alpha x))$$
$$= (q-1)^{-1} n \left\{ 2(q-1) + \left(\frac{L}{n} - 1\right)(q-2) \right\}.$$

Since $\bar{w} \geq L$, we must have $L \leq nq$. □

By using a method similar to the one of Theorem 13.4.2, we can give an easy example of a convolutional code with $d_f = L$.

(13.4.4) EXAMPLE. Let α be a primitive element of $\mathbb{F}_4$. Let $g_1(x) := x^2 + \alpha x + 1$, $g_2(x) = x^2 + \alpha^2 x + 1$. Then $(x^5 - 1) = (x-1)g_1(x)g_2(x)$ and $g_1(x)$ and $g_2(x)$ are relatively prime. We consider a rate $\frac{1}{2}$ convolutional code C over $\mathbb{F}_4$ with generator polynomial

$$G(x) := g_1(x^2) + x g_2(x^2).$$

The code is noncatastrophic. We write an information sequence as $I_0(x) = I_0'(x)(x^5 - 1)^N$ where N is maximal. By Lemma 13.4.1 we have

$$w(T(x)) = w(g_1(x)I_0(x)) + w(g_2(x)I_0(x))$$
$$\geq w((x-1)^N) \cdot \{w(g_1(x)I_0'(x) \bmod (x^5 - 1))$$
$$+ w(g_2(x)I_0'(x) \bmod (x^5 - 1))\}.$$

Now, if $I_0'(x)$ is not a multiple of $(x-1)g_1(x)$ or $(x-1)g_2(x)$, then the BCH bound shows that both terms in the second factor on the right are ≥ 3. If on the other hand $I_0'(x)$ is a multiple of $(x-1)g_1(x)$, then both terms on the right in the top line are even (because of the factor $x - 1$) and both are positive. If the second one is 2, then the first one is at least 4, again by the BCH bound. Therefore C has free distance at least 6. Since $L = 6$, we have $d_f = L$.

To generalize the idea behind Theorem 13.4.2 in order to construct rate $\frac{1}{2}$ codes over $\mathbb{F}_q$ we consider polynomials of the form

$$g_i(x) := (x - \alpha^m)(x - \alpha^{m+1}) \ldots (x - \alpha^{m+d-2}),$$

where α is a primitive element of $\mathbb{F}_q$. We choose $g_1(x)$ and $g_2(x)$ in such a way that they both have degree $\lfloor \frac{1}{3} q \rfloor$ and that they have no zeros in common. Then $G(x) := g_1(x^2) + x g_2(x^2)$ generates a noncatastrophic convolutional code C over $\mathbb{F}_q$. The two cyclic codes with generators $g_1(x)$ and $g_2(x)$ both have minimum distance $\geq 1 + \lfloor \frac{1}{3} q \rfloor =: d$. Let these codes have check polynomials $h_1(x)$ and $h_2(x)$. An information sequence $I_0(x)$ for C is written as

$$I_0(x) = (x^{q-1} - 1)^r (h_1(x))^s (h_2(x))^t p(x),$$

where $p(x)$ is not a multiple of $h_1(x)$ or $h_2(x)$ and s or t is 0. First assume that $s = t = 0$. By Lemma 13.4.1 we have

$$\begin{aligned} w(T(x)) &= w(g_1(x)I_0(x)) + w(g_2(x)I_0(x)) \\ &\geq \sum_{i=1}^{2} w((x-1)^r) w(p(x)g_i(x) \bmod (x^{q-1} - 1)) \\ &\geq 2d. \end{aligned}$$

If $s > 0$ we obtain in a similar way

$$\begin{aligned} w(T(x)) &= w((x^{q-1} - 1)^r (h_1(x))^s g_1(x) p(x)) + w((x^{q-1} - 1)^r (h_1(x))^s g_2(x) p(x)) \\ &\geq w((x-1)^{r+1}) w(p(x)(h_1(x))^{s-1} \bmod (x^{q-1} - 1)) \\ &\quad + w((x-1)^r) \cdot w(p(x)(h_1(x))^s g_2(x) \bmod (x^{q-1} - 1)) \\ &\geq 2 + (q - \lfloor \tfrac{1}{3} q \rfloor) \geq 2 + 2\lfloor \tfrac{1}{3} q \rfloor = 2d. \end{aligned}$$

From the construction we have $L = 2(1 + \lfloor \frac{1}{3} q \rfloor)$, so $d_f = L$. These examples illustrate that this is a good way to construct convolutional codes. The method generalizes to rate $1/n$. For details we refer to [38].

§13.5. Automorphisms of Convolutional Codes

We have seen, e.g., in the chapter on cyclic codes, that the requirement that a code is invariant under a certain group of permutations can lead to interesting developments. A lot of algebraic methods could be introduced and several good codes were found in this way. Therefore it is not surprising that attempts have been made to do something similar for convolutional codes. We shall sketch some of these ideas and define cyclic convolutional codes. We shall not give many details but hope that the treatment will be sufficient for the reader to decide whether he wishes to become interested in this area in which case he is referred to the work of Ph. Piret (see [54], [55]). This work was recast by C. Roos (see [59]). The ideas have led to some good codes and they are certainly worth further research.

We consider a convolutional code as defined in (13.1.2). If we call such a code cyclic if a simultaneous cyclic shift of the coefficients $\mathbf{a}_i$ of x^i ($\mathbf{a}_i \in \mathbb{F}_2^n$) leaves the code invariant, we do not get anything interesting. In fact the code is simply a block code. This shows that it is already difficult to define automorphisms in a sensible way. We do this as follows. Let K be the group of permutations acting on $\mathbb{F}_2^n$. Consider $K^{\mathbb{Z}}$, i.e. the set of all mappings $\varphi: \mathbb{Z} \to K$, which we make into a group by defining $\varphi_1\varphi_2(n) := \varphi_1(n)\varphi_2(n)$. We shall write φ_n instead of $\varphi(n)$. Note that $\varphi_n \in K$. We then define an action of φ on elements of $\mathscr{F}^n$ by

(13.5.1) $$\varphi\left(\sum_{i=r}^{\infty} \mathbf{a}_i x^i\right) := \sum_{i=r}^{\infty} \varphi_i(\mathbf{a}_i) x^i.$$

(13.5.2) Definition. If C is a convolutional code, then the set of all elements $\varphi \in K^{\mathbb{Z}}$ such that $\varphi(C) = C$ is called the *automorphism group* of C.

From the definition of convolutional codes, it is obvious that multiplication by x leaves a code C invariant. So, if φ is an automorphism of C, then $\varphi^x := x^{-1}\varphi x$ is also an automorphism. Furthermore it is clear that if we consider only the action on a fixed position, say only φ_i, we obtain a permutation group on $\mathbb{F}_2^n$ which is the projection of the automorphism group on the ith coordinate. By our remark above and the fact that $\varphi_i^x(\mathbf{a}_i) = \varphi_{i+1}(\mathbf{a}_i)$, we see that all projections are the same group. So it seems natural to try to find codes for which this projection is the cyclic group. We shall do this using the same algebraic approach as we did for block codes. We introduce a variable z and identify $\mathbb{F}_2^n$ with $\mathbb{F}_2[z] \bmod (z^n - 1)$. Let π be an integer such that $(\pi, n) = 1$ and let σ be the automorphism of $\mathbb{F}_2^n$ defined by $\sigma: f(z) \to f(z^\pi)$. The elements of $\mathscr{F}^n$ can now be written as $\sum_{i=r}^{\infty} a_i x^i$ where $a_i = a_i(z)$ is a polynomial of degree $< n$ ($i \in \mathbb{Z}$). In $\mathscr{F}^n$ we define addition in the obvious way and a multiplication (denoted by $*$) by

(13.5.3) $$\sum_i a_i x^i * \sum_j b_j x^j := \sum_i \sum_j \sigma^j(a_i) b_j x^{i+j}.$$

Suppose we take the factor on the left to be z (i.e. $a_0 = z$, $a_i = 0$ for $i \neq 0$). Then from (11.5.3) we find

(13.5.4) $$z * \sum_j b_j x^j = \sum_j (z^{\pi^j} b_j) x^j.$$

This means that the coefficient $\mathbf{b}_j$ of x^j is shifted cyclically over $\pi^j \pmod n$ positions. The main point of the definition above is that $(\mathscr{F}^n, +, *)$ is an algebra which we denote by $\mathscr{A}(n, \pi)$.

(13.5.5) Definition. A *cyclic convolutional code* (notation CCC) is a left ideal in the algebra $\mathscr{A}(n, \pi)$ which has a basis consisting of polynomials.

Observe that by (13.5.4), we now indeed have a group of automorphisms of the code. This group has the cyclic group as its projection on any coordinate. The difference with the trivial situation is the fact that the cyclic shifts are not the same for every position. We give one example to show that we have a class of nontrivial objects worth studying. We define a rate $\frac{4}{7}$ binary convolutional code with memory one by the matrix G given by

(13.5.6) $$G = \begin{bmatrix} 1 & 1 & 1 & 1 & 1 & 1 & 1 \\ 1+x & 1 & 1 & x & 1 & x & x \\ 0 & 1 & 1+x & 1 & x & 1+x & x \\ 0 & x & 1 & 1+x & 1+x & x & 1 \end{bmatrix}.$$

We shall now identify elements of $\mathbb{F}_2^7$ with polynomials in $\mathbb{F}_2[z] \bmod (z^7 - 1)$. Writing $z^7 - 1 = (1 + z)(1 + z + z^3)(1 + z^2 + z^3) = m_0(z)m_1(z)m_3(z)$ we can then abbreviate G as follows:

$$(13.5.7) \qquad G = \begin{bmatrix} m_1 m_3 \\ m_0 m_3 \\ z m_0 m_3 \\ z^2 m_0 m_3 \end{bmatrix} + \begin{bmatrix} 0 \\ m_0^3 m_1 \\ z^{-1} m_0^3 m_1 \\ z^{-2} m_0^3 m_1 \end{bmatrix} x.$$

We claim that G is the generator matrix for a CCC in $\mathscr{A}(7, -1)$. Since $\pi = -1$ we have from (13.5.4)

$$(13.5.8) \qquad z * \sum_i c_i x^i = \sum_i (z^{(-1)^i} c_i) x^i.$$

To show that the code is a CCC it is sufficient to consider the words $(1000)G$, $(0100)G$, $(0010)G$, $(0001)G$ and multiply these on the left by z and then show that the product is in the code. For the first three this is obvious from the form of (13.5.7) and from (13.5.8). For example,

$$\begin{aligned} z * (0100)G &= z * (m_0 m_3 + m_0^3 m_1 x) \\ &= z m_0 m_3 + z^{-1} m_0^3 m_1 x = (0010)G. \end{aligned}$$

Furthermore

$$\begin{aligned} z * (0001)G &= z * (z^2 m_0 m_3 + z^{-2} m_0^3 m_1 x) \\ &= z^3 m_0 m_3 + z^{-3} m_0^3 m_1 x \\ &= (1 + z) m_0 m_3 + (1 + z^6) m_0^3 m_1 x \\ &= (0110)G. \end{aligned}$$

The main point of Piret's theory is to show that a CCC always has a generator matrix with a "simple" form similar to (13.5.7). This makes it possible to construct examples in a relatively easy way and then study their properties, such as their free distance.

§13.6. Comments

Convolutional codes were introduced by P. Elias in 1955 (cf. [20]). There has been quite a lot of controversy about the question whether convolutional codes are "better" than block codes or not. Despite the lack of a deep mathematical theory, convolutional codes are used successfully in practice. Many of these codes were simply randomly chosen. For one of the deep space satellite missions ([1]) a scheme was proposed that combined block codes and convolutional codes. The idea was to take an information stream, divide it into blocks of twelve bits, and code these blocks with the [24, 12] extended Golay code. The resulting stream would be input for a convolutional encoder. This is the same idea as in the concatenated codes of Section 9.2.

For a connection between quasi-cyclic block codes and convolutional codes we refer to a paper by G. Solomon and H. C. A. van Tilborg [66]. They show, e.g. that the Golay code can be encoded and decoded convolutionally.

In Section 3.2 we saw that in decoding block codes the procedure for estimating the error pattern does not depend on the transmitted word. The idea was to introduce the syndrome (cf. (3.2.6)). The same idea has been used very successfully for convolutional codes. For more about this idea we refer to a paper by J. P. M. Schalkwijk, A. J. Vinck, and K. A. Post [60]. Some results concerning the error probability after decoding convolutional codes can be found in [51, Section 9.3].

The most complete treatment of convolutional codes can be found in a recent book by Ph. Piret [79].

§13.7. Problems

13.7.1. Suppose that in Figure 4 we remove the connection of the third flip-flop with the adder for T_1. Show that the resulting code is catastrophic.

13.7.2. Let $g(x)$ be the generator polynomial of a cyclic code with minimum distance d. As in Section 13.1 we take two polynomials $G_0(x)$ and $G_1(x)$ to generate a rate 1/2 convolutional code. If we take these such that $g(x) = G_0(x^2) + xG_1(x^2)$ then by (13.1.1) all encoded sequences are multiples of $g(x)$. Give an example where nevertheless the free distance is less than d. Check the result by constructing the encoder as in Figure 4.

13.7.3. Determine the free distance of the CCC given by (13.5.7).

Hints and Solutions to Problems

Chapter 2

2.5.1.
$$\sum_{0 \le k < N/2} \binom{N}{k} q^k p^{N-k} < (pq)^{N/2} \sum_{0 \le k < N/2} \binom{N}{k}$$
$$= 2^{N-1}(pq)^{N/2} < (0.07)^N.$$

2.5.2. There are 64 possible error patterns. We know that 8 of these lead to 3 correct information symbols after decoding. To analyse the remainder one should realize that there are only 4 essentially different 3-tuples (s_1, s_2, s_3). Consider one possibility, e.g. $(s_1, s_2, s_3) = (1, 1, 0)$. This can be caused by the error patterns (101011), (011101), (110000), (010011), (100101), (000110), (111110) and of course by (001000) which is the most likely one. Our decision is to assume that $e_3 = 1$. So here we obtain two correct information symbols with probability $p^2q^4 + 2p^4q^2$ and we have one correct information symbol with probability $2p^3q^3 + p^5q$.

By analysing the other cases in a similar way one finds as symbol error probability

$$\tfrac{1}{3}(22p^2q^4 + 36p^3q^3 + 24p^4q^2 + 12p^5q + 2p^6)$$
$$= \tfrac{1}{3}(22p^2 - 52p^3 + 48p^4 - 16p^5).$$

In our example this is 0.000007 compared to 0.001 without coding.

2.5.3. Take as codewords all possible 8-tuples of the form

$$(a_1, a_2, a_3, \quad a_2 + a_3, \quad a_1 + a_3, \quad a_1 + a_2, \quad a_1 + a_2 + a_3).$$

This code is obtained by taking the eight words of the code of the previous problem and adding an extra symbol which is the sum of the first 6 symbols.

This has the effect that any two distinct codewords differ in an even number of places, i.e. $d(\mathbf{x}, \mathbf{y}) \geq 4$ for any two distinct codewords $\mathbf{x}$, $\mathbf{y}$.

The analysis of the error patterns is similar to the one which was treated in Section 2.2. For $(e_1, e_2, \ldots, e_7)$ we find

$$e_2 + e_3 + e_4 = s_1,$$

$$e_1 + e_3 + e_5 = s_2,$$

$$e_1 + e_2 + e_6 = s_3,$$

$$e_1 + e_2 + e_3 + e_7 = s_4.$$

There are 16 possible outcomes (s_1, s_2, s_3, s_4). Eight of these can be explained by an error pattern with no errors or one error. Of the remainder there are seven, each of which can be explained by three different error patterns with two errors, e.g. $(s_1, s_2, s_3, s_4) = (1, 1, 0, 0)$ corresponds to $(e_1, e_2, \ldots, e_7)$ being (0010001), (1100000) or (0001100). The most likely explanation of (1, 1, 1, 0) is the occurrence of three errors. Hence the probability of correct decoding is

$$q^7 + 7q^6p + 7q^5p^2 + q^4p^3.$$

This is about $1 - 14p^2$, i.e. the code is not much better than the previous one even though it has smaller rate.

2.5.4. For the code using repetition of symbols the probability of correct reception of a repeated symbol is $1 - p^2$. Therefore the code of length 6 with codewords $(a_1, a_2, a_3, a_1, a_2, a_3)$ has probability $(1 - p^2)^3 = 0.97$ of correct reception. The code of Problem 2.5.2 has the property that any two codewords differ in three places and therefore two erasures can do no harm. In fact an analysis of all possible erasure patterns with three erasures shows that 16 of these do no harm either. This leads to a probability $(1 - p)^3(1 + 3p + 6p^2 + 6p^3) = 0.996$ of correct reception. This is a remarkable improvement considering the fact that the two codes are very similar.

2.5.5. Treat the erasures as zeros. The inner products are changed by at most $2e_1 + e_2$.

2.5.6. Replace a 1 in C by -1, a 0 by $+1$. The two conditions (i) and (ii) imply that the images of codewords are orthogonal vectors in $\mathbb{R}^{16}$. Hence $|C| \leq 16$. To construct such a code, we need a Hadamard matrix of order 16 with six -1s in every row. There is a well known construction. It yields a binary code that is most easily described by writing codewords as 4 by 4 matrices. Fix a row and a column; put 1s in this row and in this column, except where they meet. In this way, we find 16 words of weight 6 that indeed pairwise have distance 8.

2.5.7. For any $\mathbf{x}$, there are at most $n/2$ codewords that differ from $\mathbf{x}$ in two places. If there exists a codeword that differs from $\mathbf{x}$ in exactly one place, then there are at most $(n - 2)/2$ other codewords that differ from $\mathbf{x}$ in two places (because n is even). For any codeword **c**, there are exactly n words that differ from **c** in one place and $\binom{n}{2}$ words that differ in two places.

Counting pairs (**x**, **c**) in two ways, we find

$$|C| \cdot \binom{n}{2} \leq |C| \cdot n \cdot \frac{n-2}{2} + (2^n - |C| \cdot (n+1)) \cdot \frac{n}{2},$$

from which the result follows. The code of §2.1 is an example where equality holds.

Chapter 3

3.8.1. By (3.1.6) we have $\sum_{i=0}^{3} \binom{n}{i} = 2^l$ for some integer l. This equation reduces to $(n+1)(n^2 - n + 6) = 3.2^{l+1}$, i.e. $(n+1)\{(n+1)^2 - 3(n+1) + 8\} = 3.2^{l+1}$. If $n+1$ is divisible by 16 then the second factor on the left is divisible by 8 but not by 16, i.e. it is 8 or 24 which yields a contradiction. Therefore $n+1$ is a divisor of 24. Since $n \geq 7$ we see that $n = 7$, 11, or 23 but $n = 11$ does not satisfy the equation. For $n = 7$ the code $M = \{\mathbf{0}, \mathbf{1}\}$ is an example. For $n = 23$ see Section 4.2.

3.8.2. Let $\mathbf{c} \in C$, $w(\mathbf{c}) \leq n - k$. There is a set of k positions where **c** has a coordinate equal to 0. Since C is systematic on these k positions we have $\mathbf{c} = \mathbf{0}$. Hence $d > n - k$. Given k positions, there are codewords which have $d - 1$ zeros on these positions, i.e. $d \leq n - k + 1$. In accordance with the definition of separable given in Section 3.1 an $[n, k, n-k+1]$ code is called a *maximum distance separable* code (MDS code).

3.8.3. Since $C \subset C^\perp$ every $\mathbf{c} \in C$ has the property $\langle \mathbf{c}, \mathbf{c} \rangle = 0$, i.e. $w(\mathbf{c})$ is even and hence $\langle \mathbf{c}, \mathbf{1} \rangle = 0$. However, $\langle \mathbf{1}, \mathbf{1} \rangle = 1$ since the word length is odd. Therefore $C^\perp \backslash C$ is obtained by adding **1** to all the words of C.

3.8.4. $|B_1(\mathbf{x})| = 1 + 6 = 7$. Since $7|C| = 63 < 2^6$ one might think that such a code C exists. However, if such a C exists then by the pigeon hole principle some 3-tuple of words of C would have the same symbols in the last two positions. Omitting these symbols yields a binary code C' with three words of length 4 and minimum distance 3. W.l.o.g. one of these words is **0** and then the other two would have weight ≥ 3 and hence distance ≤ 2, a contradiction.

3.8.5. By elementary linear algebra, for every i it is possible to find a basis for C such that $k-1$ basis vectors have a 0 in position i and the remaining one has a 1. Hence exactly q^{k-1} code words have a 0 in position i.

3.8.6. The even weight subcode of C is determined by adding the row **1** to the parity check matrix of C. This decreases the dimension of the code by 1.

3.8.7. From the generator matrix we find for $\mathbf{c} \in C$

$$c_1 + c_2 + c_5 = c_3 + c_4 + c_6 = c_1 + c_2 + c_3 + c_4 + c_7 = 0.$$

Hence the syndromes

$$(s_1, s_2, s_3) = (e_1 + e_2 + e_5, e_3 + e_4 + e_6, e_1 + e_2 + e_3 + e_4 + e_7),$$

for the three received words are resp. (0, 0, 0), (0, 0, 1), (1, 0, 1). Hence (a) is a code word; by maximum likelihood decoding (b) has an error in position 7; (c) has an error in position 1 or an error in position 2, so here we have a choice.

3.8.8. (i) If $p \equiv 1 \pmod 4$ then there is an $\alpha \in \mathbb{F}_p$ such that $\alpha^2 = -1$. Then $G = (I_4, \alpha I_4)$ is the generator matrix of the required code.

(ii) If $p \equiv 3 \pmod 4$ we use the fact that not all the elements of $\mathbb{F}_p$ are squares and hence there is an α which is a square, say $\alpha = \beta^2$, such that $\alpha + 1$ is not a square, i.e. $\alpha + 1 = -\gamma^2$. Hence $\beta^2 + \gamma^2 = -1$. Then

$$G = \begin{bmatrix} 1 & 0 & 0 & 0 & \beta & \gamma & 0 & 0 \\ 0 & 1 & 0 & 0 & -\gamma & \beta & 0 & 0 \\ 0 & 0 & 1 & 0 & 0 & 0 & \beta & \gamma \\ 0 & 0 & 0 & 1 & 0 & 0 & -\gamma & \beta \end{bmatrix}$$

does the job.

(iii) If $p = 2$, see (3.3.3).

3.8.9. $$R_k = \frac{n-k}{n} = \frac{2^k - 1 - k}{2^k - 1} \to 1, \quad \text{as } k \to \infty.$$

3.8.10. (i) Let $(\bar{A}_0, \bar{A}_1, \ldots, \bar{A}_n, \bar{A}_{n+1})$ be the weight distribution of $\bar{C}$. Then $\bar{A}_{2k-1} = 0$ and $\bar{A}_{2k} = A_{2k-1} + A_{2k}$. Since $\sum A_{2k} z^{2k} = \frac{1}{2}\{A(z) + A(-z)\}$ and $\sum A_{2k-1} z^{2k-1} = \frac{1}{2}\{A(z) - A(-z)\}$, we find $\bar{A}(z) = \frac{1}{2}\{(1+z)A(z) + (1-z)A(-z)\}$.

(ii) From (i) and (3.5.2) we find the weight enumerator of the extended Hamming code of length $n + 1 = 2^k$ to be

$$\frac{1}{2}\left\{\frac{(1+z)^{n+1} + (1-z)^{n+1}}{n+1}\right\} + \frac{n}{n+1}(1 - z^2)^{(n+1)/2}.$$

Now apply Theorem 3.5.3. The weight enumerator of the dual code is $1 + 2nz^{(n+1)/2} + z^{n+1}$, i.e. all the words of this code, except $\mathbf{0}$ and $\mathbf{1}$, have weight 2^{k-1}.

3.8.11. The error pattern is a nonzero codeword $\mathbf{c}$. If $w(\mathbf{c}) = i$ then the probability of this error pattern is $p^i(1-p)^{n-i}$. Therefore the probability of an undetected error is $(1-p)^n\{-1 + A(p/(1-p))\}$.

3.8.12. Let G_i $(i = 1, 2)$ be a generator matrix of C_i in standard form. Define $A_{ij} \in \mathcal{R}$ $(1 \le i \le k_1, 1 \le j \le k_2)$ as follows. The first k_1 rows of A_{ij} are $\mathbf{0}$ except for row i which is equal to the ith row of G_1 and similarly for the first k_2 columns except for column j which is the transpose of the jth row of G_2. It is easy to see that this uniquely determines an element A_{ij} in $\mathcal{R}$. The A_{ij} are $k_1 k_2$ linearly independent elements of $\mathcal{R}$ which generate the code C. If $A \in \mathcal{R}$ has a nonzero row then this row has weight $\geq d_1$, i.e. A has at least d_1 columns with weight $\geq d_2$. So C has minimum distance $\geq d_1 d_2$. In fact equality holds.

3.8.13. The subcode on positions 1, 9 and 10 is the repetition code which is perfect and single-error correcting. The subcode on the remaining seven positions is the [7, 4] Hamming code which is also perfect. So we have a unique way of correcting at most one error on each of these two subsets of positions. C has minimum distance 3 and covering radius 2.

3.8.14. (i) Consider the assertion $A_k :=$ "after 2^k choices we have a linear code and for $i < k$ the word length increases after 2^i steps". For $k = 1$ the assertion is true. Suppose A_k is true for some value of k. Consider the codeword $\mathbf{c}_{2^k}$. The list of chosen words looks like

$$A\begin{cases} \mathbf{c}_0 & = 0 \quad 0 \quad \dots \quad 0 \quad 0 \quad \dots \quad 0 \\ \vdots & \\ \mathbf{c}_{2^{k-1}-1} & = * \quad * \quad \dots \quad 1 \quad 0 \quad \dots \quad 0 \end{cases}$$

$$B\begin{cases} \mathbf{c}_{2^{k-1}} & = * \quad * \quad \dots \quad * \quad 11 \quad \dots \quad 1 \\ \vdots & \\ \mathbf{c}_{2^k-1} & = * \quad * \quad \dots \quad * \quad 11 \quad \dots \quad 1 \end{cases}$$

where the words in B are obtained by adding $\mathbf{c}_{2^{k-1}}$ to the words of A. If $\mathbf{c}_{2^k}$ has the same length as the words of B, since it is lexicographically larger, $\mathbf{c}_{2^k}$ must be of the form $\mathbf{c}_{2^{k-1}} + \mathbf{x}$ where $\mathbf{x}$ has 0s in the last positions. However $d \leq d(\mathbf{c}_{2^{k-1}} + \mathbf{x}, \mathbf{c}_{2^{k-1}} + \mathbf{c}_i) = d(\mathbf{x}, \mathbf{c}_i)$, where $0 \leq i < 2^{k-1}$, shows that we should have chosen $\mathbf{x}$ instead of $\mathbf{c}_{2^{k-1}}$, a contradiction. So the length of the code increases when we choose $\mathbf{c}_{2^k}$. Now suppose that we have shown that $\mathbf{c}_{2^k+i} = \mathbf{c}_{2^k} + \mathbf{c}_i$ for $0 \leq i < j$ (it is true for $i = 0$). We have $d(\mathbf{c}_{2^k} + \mathbf{c}_j, \mathbf{c}_{2^k} + \mathbf{c}_i) = d(\mathbf{c}_j, \mathbf{c}_i) \geq d$, $d(\mathbf{c}_{2^k} + \mathbf{c}_j, \mathbf{c}_i) = d(\mathbf{c}_{2^k}, \mathbf{c}_i + \mathbf{c}_j) \geq d$ since by the assumption on linearity $\mathbf{c}_i + \mathbf{c}_j = \mathbf{c}_v$ for some v. This proves that $\mathbf{c}_{2^k} + \mathbf{c}_j$ is a possible choice for the codeword $\mathbf{c}_{2^k+j}$. The difficult part is to show that it is the least choice. On the contrary, suppose the choice should be $\mathbf{c}_{2^k} + \mathbf{x}$ where $\mathbf{c}_{2^k} + \mathbf{x} \prec \mathbf{c}_{2^k} + \mathbf{c}_j$ (we use $\prec$ for the lexicographic ordering). By the induction hypothesis $\mathbf{x} \succ \mathbf{c}_j$. These inequalities imply that $\mathbf{c}_j$, $\mathbf{x}$, and $\mathbf{c}_{2^k}$ look like

$$\begin{aligned} \mathbf{c}_j &= * \quad \dots \quad * \quad 0 \quad a_1 a_2 \quad \dots \quad a_t \quad 0 \quad 0 \quad \dots \quad 0 \quad 0 \\ \mathbf{x} &= * \quad \dots \quad * \quad 1 \quad a_1 a_2 \quad \dots \quad a_t \quad 0 \quad 0 \quad \dots \quad 0 \quad 0 \\ \mathbf{c}_{2^k} &= * \quad \dots \quad * \quad 1 \quad ** \quad \dots\dots\dots\dots\dots\dots \quad *1. \end{aligned}$$

The assumption that $\mathbf{c}_{2^k} + \mathbf{x}$ is an admissible choice implies (again using linearity)

$$d(\mathbf{c}_{2^k} + \mathbf{x}, \mathbf{c}_j + \mathbf{c}_i) \geq d, \qquad \text{for } 0 \leq i < 2^k,$$

i.e.

$$d(\mathbf{c}_{2^k} + \mathbf{x} + \mathbf{c}_j, \mathbf{c}_i) \geq d, \quad \text{for } 0 \leq i < 2^k.$$

But $\mathbf{c}_{2^k} + \mathbf{x} + \mathbf{c}_j \prec \mathbf{c}_{2^k}$, i.e. the choice for $\mathbf{c}_{2^k}$ was not the least, a contradiction. This completes the proof by induction of assertion A_k.

(ii) Now consider the case $d = 3$. Let n_k be the length of the code after 2^k vectors have been chosen. So $n_1 = 3$. Let C_k be the linear code after 2^k choices. If C_k is not perfect then there is a vector $\mathbf{x}$ of length n_k which has distance ≥ 2 to every word of C_k. So $(\mathbf{x}, 1)$ is a possible choice for $\mathbf{c}_{2^k}$. This gives us $n_{k+1} = n_k + 1$. If, on the other hand, C_k is perfect it is clear that $n_{k+1} = n_k + 2$ and $\mathbf{c}_{2^k} = (1, 0, 0, \ldots, 0, 1, 1)$. The assertion $B_a :=$ "the length n_k equals $2^a + i$ for $k = 2^a - a - 1 + i$ and $1 \leq i < 2^a$" now follows by induction from the observations above. In each of the sequences mentioned in B_a the final code is a Hamming code.

3.8.15. If C is a [15,8,5] code, then C must contain a word of weight 5 because otherwise we could puncture to a [14,8,5] code which cannot exist because 2^8 spheres of radius 2 contain too many points (see (5.2.7)). Take such a word as first row of a generator matrix. On the positions where this row has zeros, the other seven rows generate a [10,7] code which must have minimum distance 3. Again, this violates the sphere-packing condition. (See the part on residual codes in §4.4.)

Chapter 4

4.8.1. By (4.5.6) $\mathscr{R}(1, m)$ has dimension $m + 1$, i.e. it contains 2^{m+1} words of length $n = 2^m$. By Theorem 4.5.9 each of the $2(2^m - 1)$ hyperplanes of $\mathrm{AG}(m, 2)$ yields a codeword of $\mathscr{R}(1, m)$, i.e. except for $\mathbf{0}$ and $\mathbf{1}$ every codeword is the characteristic function of a hyperplane. Take $\mathbf{0}$ and the codewords corresponding to hyperplanes through the origin. Replace 0 by -1 in each of these codewords. Since two hyperplanes intersect in 2^{m-1} points the n vectors are pairwise orthogonal.

4.8.2. Since the code is perfect every word of weight 3 in $\mathbb{F}_3^{11}$ has distance 2 to a codeword of weight 5. Therefore $A_5 = 2^3\binom{11}{3}/\binom{5}{2} = 132$. Denote the 5-set corresponding to $\mathbf{x}$ by $B(\mathbf{x})$. Suppose $\mathbf{x}$ and $\mathbf{y} \notin \{\mathbf{x}, 2\mathbf{x}\}$ are codewords of weight 5 such that $|B(\mathbf{x}) \cap B(\mathbf{y})| > 3$. Then $w(\mathbf{x} + \mathbf{y}) + w(2\mathbf{x} + \mathbf{y}) \leq 8$ which is impossible because $\mathbf{x} + \mathbf{y}$ and $2\mathbf{x} + \mathbf{y}$ are codewords. Therefore the 66 sets $B(\mathbf{x})$ cover $66 \cdot \binom{5}{4} = \binom{11}{4}$ 4-subsets, i.e. all 4-subsets.

4.8.3. By Theorem 1.3.8 any two rows of A have distance 6 and after a permutation of symbols they are (111, 111, 000, 00) and (111, 000, 111, 00). Any other row of A then must be of type (100, 110, 110, 10) or (110, 100, 100, 11). Consider the 66 words $\mathbf{x}_i$ resp. $\mathbf{x}_j + \mathbf{x}_k$. We have $d(\mathbf{x}_i, \mathbf{x}_i + \mathbf{x}_k) = w(\mathbf{x}_k) = 6$, $d(\mathbf{x}_i, \mathbf{x}_j + \mathbf{x}_k) = w(\mathbf{x}_i + \mathbf{x}_j + \mathbf{x}_k) = 4$ or 8 by the above standard representation. Finally, we have $d(\mathbf{x}_i + \mathbf{x}_j, \mathbf{x}_k + \mathbf{x}_l) \geq 4$ by two applications of the above standard form (for any triple $\mathbf{x}_i, \mathbf{x}_j, \mathbf{x}_k$). Since

$d(\mathbf{x}, \mathbf{1} + \mathbf{y}) = 11 - d(\mathbf{x}, \mathbf{y})$, adding the complements of the 66 words to the set decreases the minimum distance to 3.

4.8.4. Apply the construction of Section 4.1 to the Paley matrix of order 17.

4.8.5. a) Show that the subcode corresponds to the subcode of the hexacode generated by $(1, \omega, 1, \omega, 1, \omega)$.

b) As in the proof for $\mathscr{G}_{24}$, show that the subcode has dimension 4.

c) Show that $d = 4$.

4.8.6. Let C be an (n, M, d) code, d even. Puncture C. The new code C' is an $(n - 1, M, d - 1)$ code (if we puncture in a suitable way). The code $\bar{C}'$ is an (n, M, d) code because all words in $\bar{C}'$ have even weight.

4.8.7. If R and S are 3 by 3 matrices then write

$$\begin{bmatrix} R & S & S & S \\ S & R & S & S \\ S & S & R & S \\ S & S & S & R \end{bmatrix} := M(R, S).$$

The rows of A, B, C, D have weight 5, 6, 8, 9, respectively. For two words $\mathbf{a}$, $\mathbf{b}$ we have $d(\mathbf{a}, \mathbf{b}) = w(\mathbf{a}) + w(\mathbf{b}) - 2\langle \mathbf{a}, \mathbf{b} \rangle$ where $\langle \mathbf{a}, \mathbf{b} \rangle$ is calculated in $\mathbb{Z}$. By block multiplication we find

$$AA^T = M(4I + J, 2J), \qquad BB^T = M(3I + 3J, 3J - I),$$
$$AB^T = M(5J, 3J) - 2B^T, \qquad \text{a matrix with entries 3 or 1.}$$

It follows that two rows of A resp. B have distance 6 or 8 and that a row of A and a row of B have distance 5 or 9. In the same way the fact that the remaining distances are at least 5 follows from

$$CA^T = (4J - 2I, 4J - 2I, 4J - 2I, 4J - 2I),$$
$$CB^T = (4J, 4J, 4J, 4J),$$
$$DA^T = 3J + D, \qquad DB^T = 3J + 2D,$$
$$CC^T = 4J + 4I, \qquad DD^T = 3I + 6J, DC^T = 6J.$$

(This construction is due to J. H. van Lint, cf. [43].)

4.8.8. From Theorem 1.3.8 we have $A^T = -A$ and $AA^T = 11I$. It follows that over $\mathbb{F}_3$ any two rows of G have inner product 0, i.e. G generates a [24, 12] self-dual code C. Therefore $(A \quad I)$ is also a generator matrix for C. So, when looking for words of minimum weight we may assume that the first twelve positions contribute at most one half of the total weight. Since C is self-dual all weights are divisible by 3. Every row of G has weight $1 + 11 = 12$ and a linear combination of two rows has weight $2 + 7 = 9$ (this follows from $AA^T = 11I$). Therefore a linear combination of three rows has weight at least $3 + (11 - 7)$ and hence at least 9. This shows that C has minimum weight 9.

Remark. This code and the ternary Golay code are examples of *symmetry codes*. These codes were introduced by V. Pless (1972). The words of fixed weight in such codes often yield t-designs (even with $t = 5$) as in Problem 4.8.2. The interested reader is referred to [11].

4.8.9. Let $\mathbf{1}, \mathbf{v}_0, \ldots, \mathbf{v}_{m-1}$ be the basis vectors of $\mathscr{R}(1, m)$. From (4.5.3)(i) and (ii) we then see that $\mathbf{1} = (\mathbf{1}, \mathbf{1})$, $\mathbf{w}_0 = (\mathbf{v}_0, \mathbf{v}_0), \ldots, \mathbf{w}_{m-1} = (\mathbf{v}_{m-1}, \mathbf{v}_{m-1})$, $\mathbf{w}_m = (\mathbf{0}, \mathbf{1})$ are the basis vectors (of length 2^{m+1}) of $\mathscr{R}(1, m+1)$. Therefore a basis vector of $\mathscr{R}(r+1, m+1)$ of the form $\mathbf{w}_{i_1} \ldots \mathbf{w}_{i_s}$ is of type $(\mathbf{u}, \mathbf{u})$ with $\mathbf{u}$ a basis vector of $\mathscr{R}(r+1, m)$ if $\mathbf{w}_m$ does not occur in the product and it is of type $(\mathbf{0}, \mathbf{v})$, where $\mathbf{v}$ is a basis vector of $\mathscr{R}(r, m)$ if $\mathbf{w}_m$ does occur in the product. If $d(r, m)$ is the minimum distance of $\mathscr{R}(r, m)$ then from the $(\mathbf{u}, \mathbf{u}+\mathbf{v})$-construction we know that $d(r+1, m+1) = \min\{2d(r+1, m), d(r, m)\}$ and then Theorem 4.5.7 follows by induction.

4.8.10. (i) We consider the vectors $\mathbf{x}^*$ and $\mathbf{c}^* = \pm\mathbf{a}_i$ as vectors in $\mathbb{R}^n$. Clearly $\langle \mathbf{x}^*, \mathbf{c}^* \rangle = n - 2d(\mathbf{x}, \mathbf{c})$. We may assume that the $\mathbf{a}_i$ are chosen in such a way that $\langle \mathbf{x}^*, \mathbf{a}_i \rangle$ is positive $(1 \le i \le n)$. The vectors $\mathbf{x}^*$ and all $\mathbf{a}_i$ have length $\sqrt{n}$. Therefore $\sum_{i=1}^n \langle \mathbf{x}^*, \mathbf{a}_i \rangle^2 = n^2$, because the $\mathbf{a}_i$ are pairwise orthogonal. It follows that there is an i such that $\langle \mathbf{x}^*, \mathbf{a}_i \rangle$ is at least $\sqrt{n}$.

(ii) Now let $m = 2k$ and $\mathbf{c} \in \mathscr{R}(1, m)$. By definition $\mathbf{c}$ is the list of values of a linear function on $\mathbb{F}_2^m$ and $d(\mathbf{x}, \mathbf{c})$ is therefore the number of points in $\mathbb{F}_2^m$ where the sum of $x_1x_2 + x_3x_4 + \cdots + x_{2k-1}x_{2k}$ and this linear function takes the value 1. Observe that

$$x_1x_2 + x_3x_4 + \cdots + x_{2k-1}x_{2k} + x_1$$
$$= x_1\bar{x}_2 + x_3x_4 + \cdots + x_{2k-1}x_{2k}$$

where $\bar{x}_2 := x_2 + 1$. Therefore a sequence of coordinate transformations $\bar{x}_i := x_i + 1$ (for $\mathbb{F}_2^m$) changes the sum into an expression equivalent to $x_1x_2 + \cdots + x_{2k-1}x_{2k}$ or to its complement (if the term $\mathbf{1}$ occurs in the linear function). We count the points $\mathbf{x}$, for which $x_1x_2 + \cdots + x_{2k-1}x_{2k} = 1$. Call this number n_k. Clearly $n_k = 3n_{k-1} + (2^{2k-2} - n_k)$, from which we find $n_k = 2^{2k-1} - 2^{k-1}$. This can also be calculated by considering the vector $(x_1, x_3, \ldots, x_{2k-1})$. If this is not $\mathbf{0}$ then 2^{k-1} choices for $(x_2, x_4, \ldots, x_{2k})$ are possible. Hence $n_k = (2^k - 1)2^{k-1}$.

4.8.11. Since the ternary Hamming code is self-dual it follows that C is self-dual; ($J + I$ has rank 4 and hence C has dimension 6). Therefore the minimum distance of C is either 3 or 6. Clearly a linear combination of the first four rows of G has weight > 4. On the other hand a linear combination of the last two rows has weight 6. Again since $J + I$ has rank 4 a combination involving rows of both kinds cannot have weight 3.

Chapter 5

5.5.1. We construct a suitable parity check matrix for the desired code C by successively picking columns. Any nonzero column can be our first choice. If m columns have been chosen we try to find a next column which is not a linear combination of i previous columns for any $i \leq d - 2$. This ensures that no $(d - 1)$-tuple of columns of the parity check matrix is linearly dependent, i.e. the code has minimum distance at least d. The method works if the number of linear combinations of at most $d - 2$ columns out of m chosen columns is less than q^{n-k} (for every $m \leq n - 1$). So a sufficient condition is

$$\sum_{i=0}^{d-2} \binom{n-1}{i}(q-1)^i < q^{n-k}.$$

The left-hand side of the inequality in (5.1.8) is at least $n(q - 1)/(d - 1)$ times as large, the right-hand side only q times as large, i.e. Problem 5.5.1 is a stronger result in general.

5.5.2. By (5.1.3) we have $A(10, 5) = A(11, 6)$. The best bound is obtained by using (5.2.4) (also see the example following (5.3.5)). It is $A(11, 6) \leq 12$. From (1.3.9) (also see the solution of Problem 4.8.3) we find an (11, 12, 6) code. Hence $A(10, 5) = 12$.

5.5.3. Equality in (5.2.4) can hold only if this is also the case for the inequalities used in the proof and that is possible only if $\frac{1}{2}M^2$ is an integer. So $M = l$ is impossible.

5.5.4. By Problem 4.8.4 we have $A(17, 8) \geq 36$. The best estimate from Section 5.2 is obtained by using the Plotkin bound. We find $A(17, 8) \leq 4A(15, 8) \leq 64$. A much better result is obtained by applying Theorem 5.3.4. The reader can check that the result is $A(17, 8) \leq 50$. The best known bound, obtained by refining the method of (5.3.4), is $A(17, 8) \leq 37$ (cf. [6]).

5.5.5. The columns of the generator matrix are the points $(x_1, x_2, x_3, x_4, x_5)$ of PG(4, 2). We know (cf. Problem 3.7.10) that all the nonzero codewords of the [31, 5] code have weight 16. By the same result the positions corresponding to $x_1 = x_2 = 0$ yield a subcode of length 7 with all nonzero weights equal to 4 and the positions with $x_3 = x_4 = x_5 = 0$ give a subcode of length 3 with all nonzero weights equal to 2. If we puncture by these ten positions the remaining [21, 5] code therefore has $d = 16 - 4 - 2 = 10$. From (5.2.6) we find $n \geq 10 + 5 + 3 + 2 + 1 = 21$, i.e. the punctured code meets the Griesmer bound.

5.5.6. This is a direct consequence of the proof of Lemma 5.2.14 (the average number of occurrences of a pair of ones is $|C| \cdot \binom{w}{2} \Big/ \binom{n}{2} = A(n - 2, 2k, w - 2)$ and no pair can occur more than this many times).

5.5.7. Let $n = 2^k - 2$. By Lemma 5.2.14 we have $A(n, 3, 3) \leq \frac{1}{6}n(n - 2)$. Hence Theorem 5.2.15 yields

$$A(n, 3) \le 2^n \Bigg/ \left\{1 + n + \left(\binom{n}{2} - \frac{3n(n-2)}{6}\right) \Bigg/ \binom{n}{2}\right\} = 2^{n-k}.$$

Hence the $[n, n-k, 3]$ shortened Hamming code is optimal.

5.5.8. If two words $\mathbf{c}$ and $\mathbf{c}'$ of weight w have distance 2, say $c_j = c'_k = 1$, $c'_j = c_k = 0$, then $\sum_{i=0}^{n-1} ic_i - \sum_{i=0}^{n-1} ic'_i \equiv j - k \pmod{n}$. It follows that each of the codes C_l $(0 \le l \le n-1)$ has minimum distance 4. Therefore $A(n, 4, w) \ge 1/n\binom{n}{w}$, since $\sum_{l=0}^{n-1} |C_l| = \binom{n}{w}$. By Lemma 5.2.14 we have $A(n, 4, w) \le \binom{n}{w}/(n - w + 1)$. Combining these inequalities the result follows. (For generalizations of this idea cf. [30].)

5.5.9. Let C be a binary $(n, M, 2k)$ code. Define

$$S := \{(\mathbf{c}, \mathbf{x}) | \mathbf{c} \in C, \mathbf{x} \in \mathbb{F}_2^n, d(\mathbf{c}, \mathbf{x}) = w\}.$$

Clearly $|S| = \binom{n}{w} M$. For a fixed $\mathbf{x}$ there are at most $A(n, 2k, w)$ words $\mathbf{c}$ in C such that $d(\mathbf{c}, \mathbf{x}) = w$. Therefore $\binom{n}{w} M \le 2^n A(n, 2k, w)$.

5.5.10. (i) The proof of this inequality is essentially the same as the proof of Lemma 5.2.10. If a constant weight code has m_i words with a 1 in position i then, in our usual notation,

$$2k\binom{M}{2} \le \sum_{i=1}^{n} m_i(M - m_i) \le M^2 w - n\left(\frac{Mw}{n}\right)^2.$$

(ii) Let $2k/n \to \delta$ as $n \to \infty$. Let $w/n \to \omega$ as $n \to \infty$. Then $A(n, 2k, w)$ is bounded as $n \to \infty$ and Problem 5.5.9 yields $\alpha(\delta) \le 1 - H_2(\omega)$. We must still satisfy the requirement $1 - (w/k)(1 - (w/n)) > 0$. So the best result is obtained if $1 - (2\omega/\delta)(1 - \omega) = 0$, i.e. $\omega = \frac{1}{2} - \frac{1}{2}\sqrt{1 - 2\delta}$, which is (5.2.12).

5.5.11. We have $K_2(i) = 2i^2 - 2ni + \binom{n}{2}$ and this is less than $2d^2 - 2nd + \binom{n}{2}$ for $d \le i \le n - d$. Since $A_0 = A_n = 1$ (w.l.o.g.) and $A_i = 0$ for all other values of i outside $[d, n-d]$, we find from (5.3.3)

$$2\binom{n}{2} + \left(2d^2 - 2nd + \binom{n}{2}\right) \sum_{i=d}^{n-d} A_i \ge 0.$$

This yields the required inequality.

5.5.12. Consider the problem of determining $A(9, 4)$ with Theorem 5.3.4. We find four inequalities for A_4, A_6, A_8. To this system we may add the obvious inequality $A_8 \le 1$. A rather tedious calculation yields the optimal solution $A_4 + A_6 + A_8 \le 20\frac{1}{3}$. So we have to consider the possibility of a

(9, 21, 4) code. Taking $\omega = -1$ in the proof of Lemma 5.3.3 yields the inequalities

$$21 \sum_{i=0}^{9} A_i K_k(i) \geq \binom{9}{k},$$

i.e. using

$$K_k(0) = \binom{9}{k},$$

$$\frac{20}{21} K_k(0) + \sum_{i=1}^{9} A_i K_k(i) \geq 0.$$

Since the code has 21 words, there are at most 10 pairs of codewords with distance 8, i.e. $A_8 \leq \frac{20}{21}$. Therefore the numbers $\frac{21}{20} A_i$ must satisfy the same inequalities which we solved to start with. This implies $A_4 + A_6 + A_8 \leq \frac{20}{21} \cdot \frac{61}{3} < 20$, a contradiction.

Chapter 6

6.13.1. We generalize Sections 6.1 and 6.2. Over $\mathbb{F}_3$ we have $x^4 + 1 = (x^2 + x + 2)(x^2 + 2x + 2)$. Hence $x^2 + x + 2$ is the generator of a negacyclic [4, 2] code which has generator matrix $\begin{pmatrix} 2 & 1 & 1 & 0 \\ 0 & 2 & 1 & 1 \end{pmatrix}$. By Definition 3.3.1 this is a [4, 2] Hamming code.

6.13.2. Since $x^4 + x + 1$ is primitive we can take it as generator for the [15, 11] Hamming code. Now follow the procedure of the proof of Theorem 6.4.1 to find $a(x)(x^4 + x + 1)$ which turns out to be $1 + (x + x^2 + x^4 + x^8) + (x^3 + x^6 + x^9 + x^{12})$, a sum of three idempotent corresponding to cyclotomic cosets. The method described after Theorem 6.4.4 and the example given there provide a second solution.

6.13.3. Consider the matrix E introduced in Section 4.5 and leave out the initial column. We now have a list of the points $\neq (0, 0, \ldots, 0)$ in AG(m, 2), written as column vectors. We can also consider this as a list of elements of $\mathbb{F}_{2^m}^*$. This is a cyclic group generated by a primitive element ξ of $\mathbb{F}_{2^m}$. The mapping $A: \mathbb{F}_{2^m} \to \mathbb{F}_{2^m}$ defined by $A(x) := \xi x$ is clearly a nonsingular linear transformation of AG(m, 2) into itself and as a permutation of the points of AG(m, 2)$\backslash\{0\}$ it has order $2^m - 1$. The mapping A maps flats into flats. It now follows from Lemma 4.5.5(i), (4.5.6) and Theorem 4.5.9 that the permutation of coordinate places corresponding to A yields a cyclic representation of the shortened code.

6.13.4. Use $x^3 + 2x + 1$ to generate $\mathbb{F}_3^3$. If β is a primitive element $x^3 + 2x + 1$ is its minimal polynomial. Using a table of the field we find the minimal polynomial of β^2 to be $(x - \beta^2)(x - \beta^6)(x - \beta^{18}) = x^3 + x^2 + x + 2$ and the minimal polynomial of β^4 is $(x^3 + x^2 + 2)$. The product of these functions has $\beta, \beta^2, \beta^3, \beta^4$ among its zeros. So it generates the required code. The generator polynomial is $1 + x + 2x^2 + 2x^3 + 2x^4 + x^5 + x^6 + x^7 + 2x^8 + x^9$. The dimension of the code is 17.

6.13.5. First make a table of $\mathbb{F}_{2^5}$. By substitution we find $E(\alpha^i) = R(\alpha^i)$ for $i = 1, 2, 3, 4$. These are respectively $\alpha^{28}, \alpha^{25}, 1, \alpha^{19}$. We must determine $\sigma(z) = 1 + \sigma_1 z + \sigma_2 z^2$ from the equations $1 + \sigma_1\alpha^{25} + \sigma_2\alpha^{28} = 0$, $\alpha^{19} + \sigma_1 + \sigma_2\alpha^{25} = 0$. We find $\sigma_1 = \alpha^{28}$, $\sigma_2 = \alpha^{10}$, i.e.

$$\sigma(z) = (1 - \alpha^{14}z)(1 - \alpha^{27}z).$$

So the codeword was

$$(1\,0\,0\,1\,0\,1\,1\,0\,1\,1\,1\,1\,0\,0\,1\,0\,1\,1\,0\,1\,0\,1\,0\,1\,0\,1\,1\,0\,1\,1\,1).$$

The generator of the code is $(1 + x^2 + x^5)(1 + x^2 + x^3 + x^4 + x^5)$, i.e. $g(x) = 1 + x^3 + x^5 + x^6 + x^8 + x^9 + x^{10}$. The codeword is

$$g(x)(1 + x^{11} + x^{20}).$$

6.13.6. Since the defining set of C contains $\{\beta^j \mid j = -2, -1, 1, 2\}$, it follows from Example 6.6.10 (with $d_A = 3$ and $|B| = 2$) that $d \geq 4$. Consider the even weight subcode C' of C. The words of this code have $\beta^{-2}, \beta^{-1}, \beta^0, \beta^1, \beta^2$ as zeros. Hence C' has minimum distance at least 6 by the BCH bound. It follows that $d \geq 5$.

6.13.7. Consider any $[q + 1, 2, q]$ code C'. This code is systematic on any two positions (cf. (3.8.2)). For the coefficients of the weight enumerator this implies

$$(q + 1)A_{q+1} + aAq = (q + 1)(q^2 - q).$$

Since $A_{q+1} + A_q = q^2 - 1$ we find that $A_{q+1} = 0$, i.e. every nonzero codeword has weight q. There is a unique codeword $\mathbf{c} = (c_0, c_1, \ldots, c_q)$ with $c_0 = c_{(q+1)/2} = 1$. Since exactly one coordinate of $\mathbf{c}$ is 0, a cyclic shift of $\mathbf{c}$ over $\frac{1}{2}(q + 1)$ positions does not yield the same word $\mathbf{c}$. Hence C' is not cyclic.

6.13.8. Over $\mathbb{F}_3$ we have

$$x^{11} - 1 = (x - 1)(x^5 - x^3 + x^2 - x - 1)(x^5 + x^4 - x^3 + x^2 - 1),$$

where the factors are irreducible. So, as in (6.9.1), we can take $g_0(x) = (x^5 - x^3 + x^2 - x - 1)$ as generator of the ternary $[11, 6]$ QR code C. Both the BCH bound and Theorem 6.9.2 yield $d \geq 4$; in the latter case there is the restriction $c(1) \neq 0$. The code $C^\perp$ has generator $(x - 1)g_0(x)$ (cf. Section 6.2). Now consider the code $\overline{C}$ obtained by adding an overall parity check in the usual way. If G is a generator matrix for $C^\perp$, then a generator matrix for C is obtained by adding the row $\mathbf{1}$ and a generator matrix for $\overline{C}$ is given by

$$\left[\begin{array}{c|c} G & \begin{matrix}0\\0\\ \vdots \\0\end{matrix} \\ \hline 1\ 1\ \cdots\ 1 & 1 \end{array}\right]$$

We see that $\bar{C}$ is self-dual. The inner product of a codeword with itself (over $\mathbb{R}$) is the number of nonzero coordinates. Hence every codeword in C has a weight divisible by 3. This proves $d \geq 5$. That C is perfect follows from the definition by using $1 + \binom{11}{1}\cdot 2 + \binom{11}{2}\cdot 2^2 = 3^5$.

6.13.9. By (6.9.5) d is odd. By Theorem 6.9.2(ii) and (iii) we have $d^2 - d + 1 \geq 47$ (i.e. $d \geq 8$) and $d \equiv 3 \pmod 4$. Hence $d \geq 11$. By the Hamming bound (Theorem 5.2.7) we have, with $d = 2e + 1$,

$$\sum_{i=0}^{e} \binom{47}{i} \leq 2^{47}/|C|.$$

Since C has dimension 24, we find $e \leq 5$. It follows that $d = 11$.

6.13.10. In the example of Section 6.9 and in Problem 6.12.8 we have already found the [7, 4] Hamming code and the two Golay codes as examples of perfect QR codes. There is one other perfect QR code. That there are no others with $e > 1$ is a consequence of the results of Chapter 7. Suppose C is a QR code of length n over $\mathbb{F}_q$ and C is a perfect code with $d = 3$. Then by (3.1.6) we have

$$1 + n(q - 1) = q^{(n-1)/2}, \qquad (\text{or } q^{(n+1)/2}).$$

The two cases are similar. In the first we have

$$n = 1 + q + q^2 + \cdots + q^{(n-3)/2}.$$

If $n > 5$ then the right-hand side is at least

$$1 + 2 + \frac{n-5}{2}\cdot 4 = 2n - 7,$$

i.e. $n = 7$ and $q = 2$ and C is the [7, 4] Hamming code. It remains to try $n = 3$ and $n = 5$. We find

$$1 + 3(q - 1) = q \qquad \text{resp. } 1 + 5(q - 1) = q^2.$$

So the only solution is $n = 5$, $q = 4$.

6.13.11. Let β be a primitive element of $\mathbb{F}_n$. Define $R_\nu := \{\beta^i \in \mathbb{F}_n \mid i \equiv \nu \pmod e\}$, $0 \leq \nu < e$. Let α be a primitive nth root of unity in an extension field of $\mathbb{F}_q$. We define

$$g_\nu(x) := \prod_{r \in R_\nu} (x - \alpha^r), \qquad 0 \leq \nu < e.$$

Since $q \in R_0$ each of the g_ν has coefficients in $\mathbb{F}_q$. Furthermore these polynomials all have degree $(n - 1)/e$ and

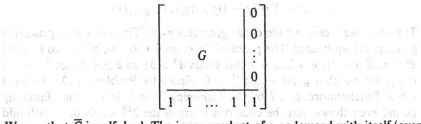

$$x^n - 1 = (x-1)g_0(x)g_1(x)\dots g_{e-1}(x).$$

The eth power residue code C has generator $g_0(x)$. The codes with generator $g_v(x)$ are all equivalent. The proof of Theorem 6.9.2(i) can be copied to yield $d^e > n$. If $n = 31$, $e = 3$, $q = 2$ this yields $d^3 > 31$ so $d \geq 4$. Since $5^3 = -1$ in $\mathbb{F}_{31}$ we see that $g_0(\alpha) = g_0(\alpha^{-1}) = 0$. Hence by Problem 6.13.6 we have $d \geq 5$. Furthermore, $d < 7$ by the Hamming bound. In fact the Hamming bound even shows that the code consisting of the 2^{20} words of C with odd weight cannot have $d = 7$. Hence $d = 5$.

6.13.12. (a) Since α, α^2, α^4, α^5 are zeros of codewords $d \geq 4$ is a direct consequence of Example (6.6.10).

(b) By the BCH bound we have $d \geq 3$. If $d = 3$ there would be a codeword with coordinates 1 in positions 0, i, j. Let $\xi = \alpha^i$, $\eta = \alpha^j$. Then $1 + \xi + \eta = 0$, $1 + \xi^5 + \eta^5 = 0$. We then find $1 = (\xi + \eta)^5 = (\xi^5 + \eta^5) + \xi\eta(\xi^3 + \eta^3)$, i.e. $\xi^3 + \eta^3 = 0$. This is a contradiction because $2^m - 1 \not\equiv 0 \pmod 3$, so $x^3 = 1$ has the unique solution $x = 1$, whereas $\xi \neq \eta$.

(c) If there is a word of weight 4, then by a cyclic shift, there is a word of weight 4 with its nonzero coordinates in positions ξ, $\xi + 1$, η, and $\eta + 1$. The sum of the fifth powers of these elements is 0. This yields $(\xi + \eta)^3 = 1$, i.e. $\xi + \eta = 1$, a contradiction.

6.13.13. Consider the representation of (6.13.8). Let α be a primitive element of $\mathbb{F}_{3^5}$. Then α^{22} is a primitive 11th root of unity, i.e. a zero of $g_0(x)$ or $g_1(x)$. So the representations of 1, α^{22}, α^{44}, ..., α^{220} as elements of $(\mathbb{F}_3)^5$ are the columns of a parity check matrix of C. Multiply the columns corresponding to α^{22i} with $i > 5$ by $-1 = \alpha^{121}$ and reorder to obtain 1, α^{11}, α^{22}, ..., α^{110} corresponding to a zero of $x^{11} + 1$. This code is equivalent to C which is known to be unique and this representation is negacyclic.

6.13.14. The defining set of this code is

$$R = \{\alpha^i \mid i = 1, 2, 4, 5, 7, 8, 9, 10, 14, 16, 18, 19, 20, 25, 28\}.$$

Show that the set R contains AB, where $A = \{\alpha^i \mid i = 8, 9, 10\}$ and $B = \{\beta^j \mid j = 0, 1, 3\}$, $\beta = \alpha^{10}$. This implies $d \geq 6$. For the even weight subcode take $A = \{\alpha^i \mid i = 4, 7, 8, 9\}$, $B = \{\beta^j \mid j = 0, 3, 4\}$, $\beta = \alpha^8$. Apply Theorem 6.6.9 with $|I| = 6$.

Chapter 7

7.7.1. Let C be the code. Since the Hamming code of length $n + 1 = 2^m - 1$ is perfect with $e = 1$, C has covering radius 2. Substitute $n = 2^m - 2$, $|C| = 2^{n-m}$, $e = 1$ in (5.2.16). We find equality. It follows that C is nearly perfect.

7.7.2. Suppose $\rho(\mathbf{u}, C) = 2$, i.e. $\mathbf{u} = \mathbf{c} + \mathbf{e}$, where $\mathbf{c} \in C$ and $\mathbf{e}$ has weight 2. So $c(\alpha) = c(\alpha^3) = 0$ and $e(x) = x^i + x^j$ for some i and j. We calculate in how many ways we can change three positions, labelled x_1, x_2, x_3, and thus find a codeword. So we are calculating the number of solutions of

$$x_1 + x_2 + x_3 + \alpha^i + \alpha^j = 0,$$
$$x_1^3 + x_2^3 + x_3^3 + \alpha^{3i} + \alpha^{3j} = 0.$$

Substitute $y_i := x_i + e(\alpha)$. We find

$$y_1 + y_2 + y_3 = 0$$
$$y_1^3 + y_2^3 + y_3^3 = s := \alpha^{i+j}(\alpha^i + \alpha^j),$$

where $s \neq 0$ and $y_k \notin \{\alpha^i, \alpha^j\}$.

From the first equation we have $y_3 = y_1 + y_2$ which we substitute in the second equation. The result is

$$y_1 y_2 (y_1 + y_2) = s.$$

Since $s \neq 0$ we have $y_2 \neq 0$. Define $y := y_1/y_2$. The equation is reduced to

$$y(1 + y) = s/y_2^3.$$

Since $(3, n) = (3, 2^{2m+1} - 1) = 1$ it follows that for every value of y, except $y = 0$, $y = 1$, this equation has a unique solution y_2 (in $\mathbb{F}_{2^{2m+1}}$). Hence we find $n - 1$ solutions $\{y_1, y_2\}$ and then y_3 follows. Clearly, each triple is found six times. Furthermore, we must reject the solution with $y_1 = \alpha^i$, $y_2 = \alpha^j$ because these correspond to $x_1 = \alpha^j$, $x_2 = \alpha^i$, $x_3 = 0$ (or any permutation). So $\rho(\mathbf{u}, C) = 2$ implies that there are $\frac{1}{6}(n - 1) - 1$ codewords with distance 3 to $\mathbf{u}$. In a similar way one can treat the case $\rho(\mathbf{u}, C) > 2$.

7.7.3. The code C is the Preparata code of length 15. However, we do not need to use this fact. Start with the (16, 256, 6) Nordstrom-Robinson code $\bar{C}$ and puncture to obtain a (15, 256, 5) code C. These parameters satisfy (5.2.16) with equality.

7.7.4. That C is not equivalent to a linear code is easily seen. If it were, then C would in fact be a linear code since $\mathbf{0} \in C$. Then the sum of two words of C would again be in C. This is obviously not true. To show that C is perfect we must consider two codewords $\mathbf{a} = (\mathbf{x}, \mathbf{x} + \mathbf{c}, \sum x_i + f(\mathbf{c}))$ and $\mathbf{b} = (\mathbf{y}, \mathbf{y} + \mathbf{c}', \sum y_i + f(\mathbf{c}'))$. If $\mathbf{c} = \mathbf{c}'$ and $\mathbf{x} \neq \mathbf{y}$ it is obvious that $d(\mathbf{a}, \mathbf{b}) \geq 3$. If $\mathbf{c} \neq \mathbf{c}'$ then $d(\mathbf{a}, \mathbf{b}) \geq w(\mathbf{x} - \mathbf{y}) + w(\mathbf{x} + \mathbf{c} - \mathbf{y} - \mathbf{c}') \geq w(\mathbf{c} - \mathbf{c}') \geq 3$. Since $|C| = 2^{11}$ and $d = 3$ we have equality in (3.1.6), i.e. C is perfect.

7.7.5. For the two zeros of Ψ_2 we find from (7.5.2) and (7.5.6)

$$x_1 + x_2 = n + 1 \qquad \text{and} \qquad x_1 x_2 = 2^{l-1}.$$

Hence $x_1 = 2^a$, $x_2 = 2^b$ $(a < b)$. From (3.1.6) we find $n^2 + n + 2 = 2^c$ (where $c \geq 3$ since $n \geq 2$).

So

$$(2^a + 2^b)(2^a + 2^b - 1) + 2 = 2^c.$$

Since on the left-hand side we have a term 2, the other term is not divisible by 4. Therefore $a = 0$ or $a = 1$. If $a = 0$ we find $2^b(2^b + 1) + 2 = 2^c$, i.e. $b = 1$ and $n = 2$ corresponding to the code $C = \{(00)\}$. If $a = 1$ we find $2^{2b} + 3 \cdot 2^b + 4 = 2^c$, i.e. $b = 2$ and then $n = 5$ corresponding to the repetition code $C = \{(00000), (11111)\}$.

7.7.6. First use Theorem 7.3.5 and (1.2.7). We find the equation

$$4x^2 - 4(n+1)x + (n^2 + n + 12) = 0,$$

with zeros $x_{1,2} = \frac{1}{2}(n + 1 \pm \sqrt{n - 11})$.

It follows that $n - 11 = m^2$ for some integer m. From (7.3.6) we find

$$12 \cdot 2^n = |C| \cdot (n^2 + n + 12) = |C|(n + 1 + m)(n + 1 - m).$$

So $n + 1 + m = a \cdot 2^{\alpha+1}$, $n + 1 - m = b \cdot 2^{\beta+1}$ with $ab = 1$ or 3. First try $a = b = 1$. We find $n + 1 = 2^\alpha + 2^\beta$, $m = 2^\alpha - 2^\beta$ $(\alpha > \beta)$ and hence

$$2^\alpha + 2^\beta - 12 = 2^{2\alpha} - 2^{\alpha+\beta+1} + 2^{2\beta},$$

i.e.

$$-12 = 2^\alpha(2^\alpha - 2^{\beta+1} - 1) + 2^\beta(2^\beta - 1),$$

an obvious contradiction.

Next, try $b = 3$. This leads to $n + 1 = a \cdot 2^\alpha + 3 \cdot 2^\beta$, $m = a \cdot 2^\alpha - 3 \cdot 2^\beta$ and hence

$$3 \cdot 2^\beta(3 \cdot 2^\beta - 2^{\alpha+1} - 1) + 2^\alpha(2^\alpha - 1) + 12 = 0.$$

If $\alpha > 2$ then we must have $\beta = 2$ and it follows that $\alpha = 4$. Since $\alpha \le 2$ does not give a solution and the final case $a = 3$ also does not yield anything we have proved that $n + 1 = 2^4 + 3 \cdot 2^2$, i.e. $n = 27$.

The construction of such a code is similar to (7.4.2). Replace the form used in (7.4.2) by $x_1x_2 + x_3x_4 + x_5x_6 + x_5 + x_6 = 0$. The rest of the argument is the same. We find a two-weight code of length 27 with weights 12 and 16 and then apply Theorem 7.3.7.

7.7.7. (i) Let N denote the number of pairs $(\mathbf{x}, \mathbf{c})$ with $\mathbf{x} \in \mathbb{F}_3^{14}$, $\mathbf{c} \in C$, $d(\mathbf{x}, \mathbf{c}) = 2$. By first choosing $\mathbf{c}$, we find $N = |C| \cdot \binom{14}{2} \cdot 2^2$. For any $\mathbf{x}$ with $d(\mathbf{x}, C) = 2$, there are at most seven possible codewords with $d(\mathbf{x}, \mathbf{c}) = 2$. Hence

$$N \le (3^{14} - |C|(1 + 2 \cdot 14)).$$

Comparing these results, we see that in the second one, equality must hold. Every $\mathbf{x} \in \mathbb{F}_3^{14}$ either has distance at most one to C or distance 2 to exactly seven codewords.

(ii) There are $\binom{14}{2} \cdot 2^2$ words of weight 2 in $\mathbb{F}_3^{14}$. Each has distance 2 to $\mathbf{0}$ and hence distance 2 to six codewords of weight 4. Since each codeword of weight 4 has distance 2 to six words of weight 2, we find $A_4 = 364$.

(iii) From (ii) we see that $4A_4 = 1456$ words of weight 3 have distance 1 to C. We count pairs $(\mathbf{x}, \mathbf{c})$ with $w(\mathbf{x}) = 3$, $\mathbf{c} \in C$, and $d(\mathbf{x}, \mathbf{c}) = 2$. Apparently there are $\binom{14}{3} \cdot 2^3 - 1456$ choices for $\mathbf{x}$. Starting with a codeword $\mathbf{c}$ with weight 4, there are 12 choices for $\mathbf{x}$. If $w(\mathbf{c}) = 5$, there are 10 choices for $\mathbf{x}$. So we find

$$12A_4 + 10A_5 = 7 \cdot 1456,$$

i.e. $A_5 = 582\frac{2}{5}$, which is absurd. C does not exist!

7.7.8. In the Lee metric, the volume of a sphere with radius one is $V_1 = 1+2n = m^2$. Split the words in the set $\mathbb{Z}_m^2 \backslash \{0, 0\}$ into pairs $(\mathbf{x}, -\mathbf{x})$ and from each pair, take one as a column of a 2 by n parity check matrix H. Generalizing the idea of Hamming codes, define a linear code C by:

$$\mathbf{c} \in C \Leftrightarrow \mathbf{c}H^\top = \mathbf{0}.$$

C is clearly 1-error-correcting and C is perfect because $|C| \cdot V_1 = m^n$.

Chapter 8

8.5.1. First note that in a selfdual quaternary code of length 6, words with an odd coordinate have four such coordinates. Since the code must have 4^3 words, the generator matrix in the form (8.1.3) has $2k_1 + k_2 = 6$. Clearly, $k_1 = 3$ is impossible.

For $k_1 = 0$ we find the trivial example $\mathbb{F}_2^6$. Its Lee weight enumerator $(x^2+y^2)^6$ is one of two independent solutions to the equation of Theorem 3.6.8 in the case of selfduality. The other is $x^2y^2(x^4 - y^4)^2$. (These are easily found using the fact that all weights are even.) All linear combinations with a term x^{12} also have a term y^{12}, so the code contains (2, 2, 2, 2, 2, 2).

We now try $k_1 = k_2 = 2$. The rows of $(A \quad B)$ must each have three odd entries. One easily finds an example, e. g.

$$A = \begin{pmatrix} 1 & 1 \\ 1 & 1 \end{pmatrix}, \quad B = \begin{pmatrix} 1 & 0 \\ 2 & 1 \end{pmatrix}, \quad C = \begin{pmatrix} 1 & 1 \\ 1 & 1 \end{pmatrix}.$$

The weight enumerator is the solution to the equation of Theorem 3.6.8 with no term $x^{10}y^2$, namely $x^{12} + 15x^8y^4 + 32x^6y^6 + 15x^4y^8 + y^{12}$. The two rows of A could cause nonlinearity because $\sigma(1) + \sigma(1) \neq \sigma(2)$. The codeword (002200) compensates for that. Note that the binary images of the first two basis vectors are not orthogonal.

For $k_1 = 1$, $k_2 = 4$, we easily find a solution. Take $A = (1100)$, $B = (1)$, $C = (1100)^\top$. The binary image is linear and selfdual.

REMARK. We observed above that $k = 3$ is impossible. From (8.1.3) and (8.1.4) it is immediately obvious that the quaternary code with generator $(I_3 \quad 2J_3 - I_3)$ and its dual have the same weight enumerator (as above with 15 words of weight 4). The code is of course not selfdual. It is interesting that its binary image *is* a selfdual code. It is equivalent to the code with generator $(I_6 \quad J_6 - I_6)$.

8.5.2. Let the vectors $\mathbf{v}_i$ $(0 \le i \le m-2)$ be basis vectors of $\mathscr{R}(1, m-1)$. The codewords of $\mathscr{R}(2, m)$ are linear combinations of wordsof type $(\sum \alpha_i \mathbf{v}_i, \mathbf{0})$, $(\mathbf{0}, \sum \beta_i \mathbf{v}_i)$, $(\sum \gamma_{ij} \mathbf{v}_i \mathbf{v}_j, \sum \gamma_{ij} \mathbf{v}_i \mathbf{v}_j)$ and (ϵ_1, ϵ_2), where $\epsilon_i = \mathbf{0}$ or $\mathbf{1}$.

By Theorem 8.2.7 (and the fact that $\mathscr{R}(2, m)$ is linear), we only have to check whether the coordinatewise product of words of type $(\sum \delta_i \mathbf{v}_i + \epsilon, \sum \delta_i \mathbf{v}_i + \epsilon)$ is in the code and clearly it is.

8.5.3. Try $\xi^a + 2\xi^b$. One easily finds $a = n - i$, $b = n - 2i + j$.

8.5.4. By Lemma 4.6.4, the Nordstrom-Robinson code has weight enumerator

$$1 + 112z^6 + 30z^8 + 112z^{10} + z^{12}.$$

Apply Theorem 3.5.3. Here, calculating with the homogeneous form is easier.

8.5.5. The number of pairs $(\mathbf{x}, \mathbf{c})$ with $\mathbf{c} \in \mathscr{P}$, $w(\mathbf{x}) = 2$, $d(\mathbf{x}, \mathbf{c}) = 3$ is $10A_5$. By (7.3.1) it is also equal to $\binom{n}{2}(r-1) = \binom{n}{2}\frac{n-3}{3}$. Hence

$$A_5 = \frac{n(n-1)(n-3)}{6}.$$

There are $\binom{n}{3}$ words of weight 3. Of these, $10A_5$ have distance 2 to the code and therefore distance 3 to $\frac{n-6}{3}$ words of weight 6. The remaining $\binom{n}{3} - 10A_5$ words of weight 3 have distance 3 to $r - 1 = \frac{n-3}{3}$ words of weight 6. Since a word of weight 6 has distance 3 to 20 words of weight 3, double counting yields

$$A_6 = \frac{n(n-1)(n-3)(n-5)}{18}.$$

For $n = 15$ we find $A_5 = 40$, $A_6 = 70$. We know that the Nordstrom-Robonson code is distance invariant. So the weight enumerator has $A_i = A_{16-i}$. We know $A_6 = A_{10} = 112$ and therefore $A_8 = 30$.

8.5.6. This follows from Example 8.4.1

8.5.7. From Theorem 7.4.6 it follows that the linear span of the extended Preparata code is contained in the extended Hamming code and therefore has minimum distance 4.

Consider $GR(4^m)$. We know that $1 + \xi$ can be written as $\xi^i + 2\xi^j$. Hence C_m has a codeword $\mathbf{a}$ with coordinates 1,1,1,3,2 in the positions corresponding to $0, 1, \xi, \xi^i, \xi^j$. By a cyclic shift, we find a similar codeword $\mathbf{b}$ with a 1 in position 0 and the other odd coordinates in positions differing from those where $\mathbf{a}$ has odd coordinates. The linear span of C'_m contains the word $\phi(\mathbf{a}) + \phi(\mathbf{b}) + \phi(\mathbf{a} + \mathbf{b})$ which has weight 2.

Note that this argument does not work for $m = 3$.

Chapter 9

9.8.1. In Theorem 9.3.1 it was shown that if we replace $g(z)$ by $\hat{g}(z) := z + 1$ we get the same code. So $\Gamma(L, g)$ has dimension at least 4 and minimum distance $d \geq 3$. As was shown in the first part of Section 9.3, d might be larger. We construct the parity check matrix $H = (h_0 h_1 \ldots h_7)$ where h_i runs through the values $(\alpha^j + 1)^{-1}$ with $(j, 15) = 1$. We find that H consists of all column vectors with a 1 in the last position, i.e. $\Gamma(L, g)$ is the [8, 4, 4] extended Hamming code.

9.8.2. Let **a** be a word of even weight in C. By (6.5.2) the corresponding Mattson-Solomon polynomial $A(X)$ is divisible by X. By Theorem 9.6.1 the polynomial $X^{n-1} \circ A(X)$, i.e. $X^{-1}A(X)$, is divisible by $g(X)$. Since C is cyclic we find from (6.5.2) that $X^{-1}A(X)$ is also divisible by $g(\alpha^i X)$ for $0 < i \leq n - 1$. If $g(X)$ had a zero different from 0 in any extension field of $\mathbb{F}_2$ we would have $n - 1$ distinct zeros of $X^{-1}A(X)$, a contradiction since $X^{-1}A(X)$ has degree $< n - 1$. So $g(z) = z^t$ for some t and C is a BCH code (cf. (9.2.6)).

9.8.3. This is exactly what was shown in the first part of Section 9.3. For any codeword $(b_0, b_1, \ldots, b_{n-1})$ we have $\sum_{i=0}^{n-1} b_i \gamma_i^r = 0$ for $0 \leq r \leq d_1 - 2$, where $\gamma_i = \alpha^i$ (α primitive element). So the minimum distance is $\geq (d_2 - 1) + (d_1 - 2) + 2 = d_1 + d_2 - 1$.

9.8.4. Let $G^{-1}(X)$ denote the inverse of $G(X)$ in the ring $(T, +, \circ)$. The definition of GBCH code can be read as

$$P(X) \cdot (\Phi a)(X) = Q(X)G(X) + R(X)(X^n - 1),$$

where $Q(X)$ has degree $< n - t$. This is equivalent to

$$(G^{-1}(X) \circ P(X))(\Phi a)(X) = Q(X) + R^*(X)(X^n - 1),$$

for a suitable $R^*(X)$. The same condition, including the requirement that degree $Q(X) < n - t$, is obtained if we take the pair $(\hat{P}(X), X^t)$, where

$$\hat{P}(X) = X^t \circ G^{-1}(X) \circ P(X).$$

Second solution: To ensure that we have the same code, we see to it that we obtain the same parity check matrix as in §9.6. We have $h_i = p_i g_i^{-1}$, where $p(x) = \sum p_i x^i = (\Phi^{-1}P)(x)$ and $g(x) = \sum g_i x^i = (\Phi^{-1}G)(x)$. We must find $\hat{p}_i$ such that $\hat{p}_i \hat{g}_i^{-1} = p_i g_i^{-1}$. Since $\sum \hat{g}_i x^i = (\Phi^{-1}X^t)(x)$ is known, we also know $\hat{p}(x)$. Then $P(X) = \Phi \hat{p}$.

9.8.5. In (6.6.1) take $l = 5$ and $\delta = 2$. We find that C is a BCH code with minimum distance $d \geq 2$. Since $(x + 1)(x^2 + x + 1) = x^3 + 1 \in C$ we have $d = 2$. If in (9.2.4) we take $g(x)$ of degree > 1 then by Theorem 9.2.7 the Goppa code $\Gamma(L, g)$ has distance at least 3. If $g(z)$ has degree 1 then Theorem 9.3.1 yields the same result. So C is not a Goppa code.

Chapter 10

10.10.1. To study X in (1:0:0), we take (y, z) as affine coordinates. The equation becomes $z = y^2$. So, we see that y is a local parameter (and z is not). So in (1;0:0) we take y/x as local parameter. Since $x/z = (x/y)^2$, we see that there is a pole of order 2 in (1:0:0).

10.10.2. If f and the three partial derivatives are zero in a point $(x : y : z)$, then $xyz \neq 0$ and we find three equations $2x^3 = y^2z$, etc. These three give us $8(xyz)^3 = (xyz)^3$, so $p = 7$. We may take $x = 1$. The equations $y^2z = 2$ and $y = 2z^3$ then give $y = 2$ and $z = 4$. So there is one singular point.

10.10.3. X has five points: $P = (0 : 0 : 1)$, $Q := (1 : 0 : 0)$, and $R_i := (\alpha^i : \alpha^{2i} : 1)$ $(0 \le i \le 2)$. Clearly, every point R_i is a zero of multiplicity 1. In Q we have y/x as a local parameter and

$$g = \left(\frac{y}{x}\right)^3 \frac{x^3 + y^3}{x^3},$$

so Q is a zero with multiplicity 3. In P, a local parameter is y/z and

$$g = \left(\frac{z}{y}\right)^6 \frac{z^3 + y^3}{z^3},$$

so P is a pole of order 6. Hence $(g) = -6P + 3Q + R_1 + R_2 + R_3$.

10.10.4. We only have to look at the three points where two coordinates are 0, $P = (0 : 0 : 1)$, $Q = (1 : 0 : 0)$, and $R = (0 : 1 : 0)$. The easiest is Q, where z/x is a local parameter. So Q is a pole of order 1. In P, we have the local parameter y/z and

$$\frac{x}{z} = \left(\frac{y}{z}\right)^4 \frac{z^4}{x^3y + z^4},$$

where the second factor is a unit. So P is a zero with multiplicity 4. In R, a local parameter is x/y. From

$$\frac{x}{z} = \left(\frac{y}{x}\right)^3 \frac{y^4 + z^3x}{y^4}$$

we see that R is a pole of order 3. We find

$$(f) = 4P - Q - 3R.$$

10.10.5. Just substitute the coordinates of the points P_1 to P_6 in the three basis functions. By multiplication of certain rows and columns by suitable constants and a permutation, the two generator matrices are shown to belong to equivalent codes.

10.10.6. Since $g = 3$, we find from Theorem 10.5.1 that $l(3Q) \geq 1$. From Corollary 10.5.3 it follows that $l(5Q) = 3$. From Example 10.7.5 we see that the functions 1 and z/x are in $\mathscr{L}(3Q)$ and also that they are a basis.

Chapter 11

11.4.1. The words of C_α have the form $(a(x), \alpha(x)a(x))$ where $a(x)$ and $\alpha(x)$ are polynomials mod $x^6 + x^3 + 1$. To get $d > 3$ we must exclude those $\alpha(x)$ for which a combination $a(x) = x^i$, $\alpha(x)a(x) = x^j + x^k$ is possible and also the inverses of these $\alpha(x)$. Since $(1 + x)^8 = 1 + x^8 = x^{-1}(x + x^9) = x^{-1}(x + 1)$ it is easily seen that each nonzero element of $\mathbb{F}_{2^6}$ has a unique representation $x^i(1 + x)^j$ where

$$i \in \{0, \pm 1, \pm 2, \pm 3, \pm 4\}, \qquad j \in \{0, \pm 1, \pm 2, \pm 4\}.$$

So the requirement $d > 2$ excludes nine values of $\alpha(x)$ and $d > 3$ excludes the remaining 54 values of $\alpha(x)$. This shows that for small n the construction is not so good! By (3.8.14) there is a [12, 7] extended lexicographically least code with $d = 4$. In (4.7.3) we saw that a nonlinear code with $n = 12$, $d = 4$ exists with even more words.

11.4.2. Let $\alpha(x)$ be a polynomial of weight 3. Then in $(a(x), \alpha(x)a(x))$ the weights of the two halves have the same parity. So $d < 4$ is only possible if there is a choice $a(x) = x^i + x^j$ such that $\alpha(x)a(x) \equiv 0 \bmod (x^6 - 1)$. This is so if $\alpha(x)$ is periodic, i.e. $1 + x^2 + x^4$ or $x + x^3 + x^5$. For all other choices we have $d = 4$.

11.4.3. Let the rate R satisfy $1/(l + 1) < R \leq 1/l$ $(l \in \mathbb{N})$. Let s be the least integer such that $m/[(l + 1)m - s] \geq R$. We construct a code C by picking an l-tuple $(\alpha_1, \alpha_2, \ldots, \alpha_l) \in (\mathbb{F}_{2^m})^l$ and then forming $(\mathbf{a}, \alpha_1\mathbf{a}, \ldots, \alpha_l\mathbf{a})$ for all $\mathbf{a} \in \mathbb{F}_2^m$ and finally deleting the last s symbols. The word length is $n = (l + 1)m - s$.

A nonzero word $\mathbf{c} \in C$ corresponds to 2^s possible values of the l-tuple $(\alpha_1, \ldots, \alpha_l)$. To ensure a minimum distance $\geq \lambda n$ we must exclude $\leq 2^s \sum_{i<\lambda n} \binom{n}{i}$ values of $(\alpha_1, \ldots, \alpha_l)$. We are satisfied if this leaves us a choice for $(\alpha_1, \ldots, \alpha_l)$, i.e. if

$$2^s \sum_{i<\lambda n} \binom{n}{i} < 2^{ml}.$$

From Theorem 1.4.5 we find

$$s + nH(\lambda) < ml,$$

i.e.

$$H(\lambda) < \frac{ml - s}{n} = 1 - \frac{m}{n} = 1 - R + o(1), \qquad (m \to \infty).$$

Chapter 12

12.5.1. Consider the sequence $r, r^2, r^3, \ldots$. There must be two elements in the sequence which are congruent mod A, say $r^n - r^m \equiv 0 \pmod{A} (n > m)$.

12.5.2. Let $m = r^n - 1 = AB$, where A is a prime $> r^2$. Suppose that r generates a subgroup H of $\mathbb{F}_A^*$ with $|H| = n$ which has $\{\pm c | c = 1, 2, \ldots, r-1\}$ as a complete set of coset representatives. Consider the cyclic AN code C of length n and base r. Clearly every integer in the interval $[1, m]$ has modular distance 0 or 1 to exactly one codeword. So C is a perfect code. (Since $w_m(A) \geq 3$ we must have $A > r^2$.) A trivial example for $r = 3$ is the cyclic code $\{13, 26\}$. Here we have taken $m = 3^3 - 1$ and $A = 13$.

The subgroup generated by 3 in $\mathbb{F}_{13}^*$ has index 4 and the coset representatives are ± 1 and ± 2.

12.5.3. We have $455 = \sum_{i=0}^{5} b_i 3^i$ where $(b_0, b_1, \ldots, b_5) = (2, 1, 2, 1, 2, 1)$. The algorithm described in (10.2.3) replaces the initial 2, 1 by -1, 2. In this way we find the following sequence of representations:

$$(2, 1, 2, 1, 2, 1) \to (-1, 2, 2, 1, 2, 1) \to (-1, -1, 0, 2, 2, 1)$$
$$\to (-1, -1, 0, -1, 0, 2) \to (0, -1, 0, -1, 0, -1).$$

So the representation in CNAF is

$$455 \equiv -273 = -3 - 3^3 - 3^5.$$

12.5.4. We check the conditions of Theorem 10.3.2. In $\mathbb{F}_{11}^*$ the element 3 generates the subgroup $\{1, 3, 9, 5, 4\}$; multiplication by -1 yields the other five elements. $r^n = 3^5 = 243 = 1 + 11.22$. So we have $A = 22$, in ternary representation $A = 1 + 1.3 + 2.3^2$. The CNAF of 22 is $1 - 2.3 + 0.3^2 + 1.3^3 + 0.3^4 \pmod{242}$. The code consists of ten words namely the cyclic shifts of $(1, -2, 0, 1, 0)$ resp. $(-1, 2, 0, -1, 0)$. All weights are 3.

Chapter 13

13.7.1. Using the notation of Section 11.1 we have

$$G(x) = (1 + (x^2)^2) + x(1 + x^2) = 1 + x + x^3 + x^4.$$

The information stream 1 1 1 1 1... would give $I_0(x) = (1 + x)^{-1}$, and hence

$$T(x) = (1 + x^2)^{-1} G(x) = 1 + x + x^2,$$

i.e. the receiver would get 1 1 1 0 0 0 0....

Three errors in the initial positions would produce the zero signal and lead to infinitely many decoding errors.

13.7.2. In Theorem 13.4.2 it is shown how this situation can arise. Let $h(x) = x^4 + x + 1$ and $g(x)h(x) = x^{15} - 1$. We know that $g(x)$ generates an irreducible cyclic code with minimum distance 8. Consider the information sequence 1 1 0 0 1 0 0 0 0..., i.e. $I_0(x) = h(x)$. Then we find

$$T(x) = h(x^2)g(x) = (x^{15} - 1)h(x),$$

which has weight 6. By Theorem 13.4.2 this is the free distance. In this example we have $g(x) = x^{11} + x^8 + x^7 + x^5 + x^3 + x^2 + x + 1$. Therefore $G_0(x) = 1 + x + x^4$, $G_1(x) = 1 + x + x^2 + x^3 + x^5$. The encoder is

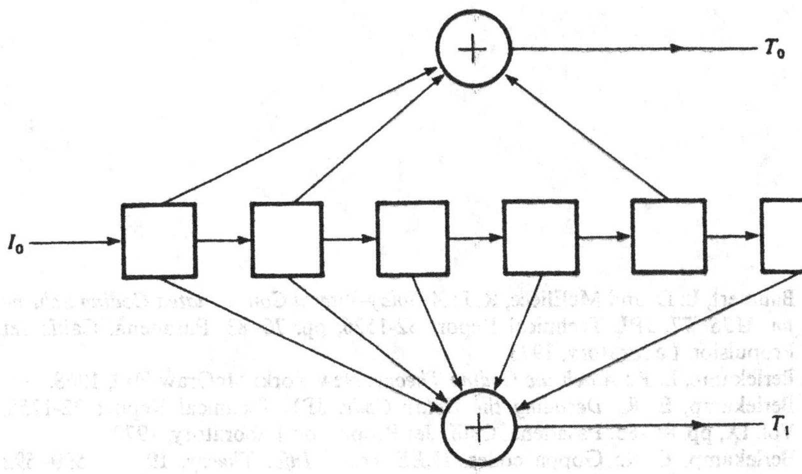

Figure 8

and

$$I_0 = 1\ 1\ 0\ 0\ 1\ 0\ 0\ 0 \ldots \quad \text{yields as output}$$

$$T = 11\ 00\ 10\ 00\ 00\ 00\ 00\ 01\ 10\ 01\ 00\ 00 \ldots$$

13.7.3. Consider a finite nonzero output sequence. This will have the form $(\mathbf{a}_0 + \mathbf{a}_1 x + \cdots + \mathbf{a}_l x^l)G$, where the $\mathbf{a}_i$ are row vectors in $\mathbb{F}_2^4$. We write G as $G_1 + xG_2$ as in (13.5.7). Clearly the initial nonzero seventuple in the output is a nonzero codeword in the code generated by m_3; so it has weight ≥ 3. If this is also the final nonzero seventuple, then it is (11...1) and the weight is 7. If the final nonzero seventuple is $\mathbf{a}_l G$, then it is a nonzero codeword in the code generated by $m_0 m_1$ and hence has weight at least 4. However, if $\mathbf{a}_l = (1000)$, then $\mathbf{a}_l G = \mathbf{0}$ and the final nonzero seventuple is a nonzero codeword in the code generated by m_1 and it has weight ≥ 3. So the free distance is 6. This is realized by the input $(1100) * (1000)x$.

References*

1. Baumert, L. D. and McEliece, R. J.: *A Golay-Viterbi Concatenated Coding Scheme for MJS '77*. JPL Technical Report 32-1526, pp. 76–83. Pasadena, Calif.: Jet Propulsion Laboratory, 1973.
2. Berlekamp, E. R.: *Algebraic Coding Theory*. New York: McGraw-Hill, 1968.
3. Berlekamp, E. R.: *Decoding the Golay Code*. JPL Technical Report 32-1256, Vol. IX, pp. 81–85. Pasadena, Calif.: Jet Propulsion Laboratory, 1972.
4. Berlekamp, E. R.: Goppa codes. *IEEE Trans. Info. Theory*, **19**, pp. 590–592 (1973).
5. Berlekamp, E. R. and Moreno, O.: Extended double-error-correcting binary Goppa codes are cyclic. *IEEE Trans. Info. Theory*, **19**, pp. 817–818 (1973).
6. Best, M. R., Brouwer, A. E., MacWilliams, F. J., Odlyzko, A. M. and Sloane, N. J. A.: Bounds for binary codes of length less than 25. *IEEE Trans. Info. Theory*, **23**, pp. 81–93 (1977).
7. Best, M. R.: *On the Existence of Perfect Codes*. Report ZN 82/78. Amsterdam: Mathematical Centre, 1978.
8. Best, M. R.: Binary codes with a minimum distance of four. *IEEE Trans. Info. Theory*, **26**, pp. 738–742 (1980).
9. Bussey, W. H.: Galois field tables for $p^n \leq 169$. *Bull. Amer. Math. Soc.*, **12**, pp. 22–38 (1905).
10. Bussey, W. H.: Tables of Galois fields of order less than 1,000. *Bull. Amer. Math. Soc.*, **16**, pp. 188–206 (1910).
11. Cameron, P. J. and van Lint, J. H.: *Designs, Graphs, Codes and their Links*. London Math. Soc. Student Texts, Vol. 22. Cambridge: Cambridge Univ. Press, (1991).
12. Chen, C. L., Chien, R. T. and Liu, C. K.: On the binary representation form of certain integers. *SIAM J. Appl. Math.*, **26**, pp. 285–293 (1974).
13. Chien, R. T. and Choy, D. M.: Algebraic generalization of BCH-Goppa-Helgert codes. *IEEE Trans. Info. Theory*, **21**, pp. 70–79 (1975).
14. Clark, W. E. and Liang, J. J.: On arithmetic weight for a general radix representation of integers. *IEEE Trans. Info. Theory*, **19**, pp. 823–826 (1973).

* References added in the Second Edition are numbered 72 to 81 and references added in the Third Edition are numbered 82 to 100.

15. Clark, W. E. and Liang, J. J.: On modular weight and cyclic nonadjacent forms for arithmetic codes. *IEEE Trans. Info. Theory*, **20**, pp. 767–770 (1974).
16. Curtis, C. W. and Reiner, I.: *Representation Theory of Finite Groups and Associative Algebras*. New York–London: Interscience, 1962.
17. Cvetković, D. M. and van Lint, J. H.: An elementary proof of Lloyd's theorem. *Proc. Kon. Ned. Akad. v. Wetensch.* (A), **80**, pp. 6–10 (1977).
18. Delsarte, P.: An algebraic approach to coding theory. *Philips Research Reports Supplements*, **10** (1973).
19. Delsarte, P. and Goethals, J.-M.: Unrestricted codes with the Golay parameters are unique. *Discrete Math.*, **12**, pp. 211–224 (1975).
20. Elias, P.: *Coding for Noisy Channels*. IRE Conv. Record, part 4, pp. 37–46.
21. Feller, W.: *An Introduction to Probability Theory and Its Applications*, Vol. I. New York–London: Wiley, 1950.
22. Forney, G. D.: *Concatenated Codes*. Cambridge, Mass.: MIT Press, 1966.
23. Forney, G. D.: Convolutional codes I: algebraic structure. *IEEE Trans. Info. Theory*, **16**, pp. 720–738 (1970); *Ibid.*, **17**, 360 (1971).
24. Gallagher, R. G.: *Information Theory and Reliable Communication*. New York: Wiley, 1968.
25. Goethals, J.-M. and van Tilborg, H. C. A.: Uniformly packed codes. *Philips Research Reports*, **30**, pp. 9–36 (1975).
26. Goethals, J.-M.: The extended Nadler code is unique. *IEEE Trans. Info. Theory*, **23**, pp. 132–135 (1977).
27. Goppa, V. D.: A new class of linear error-correcting codes. *Problems of Info. Transmission*, **6**, pp. 207–212 (1970).
28. Goto, M.: A note on perfect decimal AN codes. *Info. and Control*, **29**, pp. 385-387 (1975).
29. Goto, M. and Fukumara, T.: Perfect nonbinary AN codes with distance three *Info. and Control*, **27**, pp. 336–348 (1975).
30. Graham, R. L. and Sloane, N. J. A.: Lower bounds for constant weight codes. *IEEE Trans. Info. Theory*, **26**, pp. 37–40 (1980).
31. Gritsenko, V. M.: Nonbinary arithmetic correcting codes, *Problems of Info. Transmission*, **5**, pp 15–22 (1969).
32. Hall, M.: *Combinatorial Theory*. New York–London–Sydney–Toronto: Wiley (second printing), 1980.
33. Hartmann, C. R. P. and Tzeng, K. K.: Generalizations of the BCH bound. *Info. and Control*, **20**, pp. 489–498 (1972).
34. Helgert, H. J. and Stinaff, R. D.: Minimum distance bounds for binary linear codes. *IEEE Trans. Info. Theory*, **19**, pp. 344–356 (1973).
35. Helgert, H. J.: Alternant codes. *Info. and Control*, **26**, pp. 369–380 (1974).
36. Jackson, D.: *Fourier Series and Orthogonal Polynomials*. Carus Math. Monographs, Vol. 6. Math. Assoc. of America, 1941.
37. Justesen, J.: A class of constructive asymptotically good algebraic codes. *IEEE Trans. Info. Theory*, **18**, pp. 652–656 (1972).
38. Justesen, J.: An algebraic construction of rate $1/v$ convolutional codes. *IEEE Trans. Info. Theory*, **21**, 577–580 (1975).
39. Kasami, T.: An upper bound on k/n for affine invariant codes with fixed d/n. *IEEE Trans. Info. Theory*, **15**, pp. 171–176 (1969).
40. Levenshtein, V. I.: Minimum redundancy of binary error-correcting codes. *Info. and Control*, **28**, pp. 268–291 (1975).
41. van Lint, J. H.: Nonexistence theorems for perfect error-correcting-codes. In: *Computers in Algebra and Theory*, Vol. IV (SIAM–AMS Proceedings) 1971.
42. van Lint, J. H.: *Coding Theory*. Springer Lecture Notes, Vol. 201, Berlin–Heidelberg–New York: Springer, 1971.

43. van Lint, J. H.: A new description of the Nadler code. *IEEE Trans. Info Theory*, **18**, pp. 825–826 (1972).
44. van Lint, J. H.: A survey of perfect codes. *Rocky Mountain J. Math.*, **5**, pp. 199–224 (1975).
45. van Lint, J. H. and MacWilliams, F. J.: Generalized quadratic residue codes. *IEEE Trans. Info. Theory*, **24**, pp. 730–737 (1978).
46. MacWilliams, F. J. and Sloane, N. J. A.: *The Theory of Error-correcting Codes*. Amsterdam–New York–Oxford: North Holland, 1977.
47. Massey, J. L.: *Threshold Decoding*. Cambridge, Mass.: MIT Press, 1963.
48. Massey, J. L. and Garcia, O. N.: Error-correcting codes in computer arithmetic. In: *Advances in Information Systems Science*, Vol. 4, Ch. 5. (Edited by J. T. Ton). New York: Plenum Press, 1972.
49. Massey, J. L., Costello, D. J. and Justesen, J.: Polynomial weights and code construction. *IEEE Trans. Info. Theory*, **19**, pp. 101–110 (1973).
50. McEliece, R. J., Rodemich, E. R., Rumsey, H. C. and Welch, L. R.: New upper bounds on the rate of a code via the Delsarte–MacWilliams inequalities. *IEEE Trans. Info. Theory*, **23**, pp. 157–166 (1977).
51. McEliece, R. J.: *The Theory of Information and Coding*. Encyclopedia of Math. and its Applications, Vol. 3. Reading, Mass.: Addison-Wesley, 1977.
52. McEliece, R. J.: The bounds of Delsarte and Lovasz and their applications to coding theory. In: *Algebraic Coding Theory and Applications*. (Edited by G. Longo, CISM Courses and Lectures, Vol. 258. Wien–New York: Springer, 1979.
53. Peterson, W. W. and Weldon, E. J.: *Error-correcting Codes*. (2nd ed.). Cambridge, Mass.: MIT Press, 1972.
54. Piret, Ph.: Structure and constructions of cyclic convolutional codes. *IEEE Trans. Info. Theory*, **22**, pp. 147–155 (1976).
55. Piret, Ph.: Algebraic properties of convolutional codes with automorphisms. Ph.D. Dissertation. Univ. Catholique de Louvain, 1977.
56. Posner, E. C.: Combinatorial structures in planetary reconnaissance. In: *Error Correcting Codes*. (Edited by H. B. Mann). pp. 15–46. New York–London–Sydney–Toronto: Wiley, 1968.
57. Preparata, F. P.: A class of optimum nonlinear double-error-correcting codes. *Info. and Control*, **13**, pp. 378–400 (1968).
58. Rao, T. R. N.: *Error Coding for Arithmetic Processors*. New York–London: Academic Press, 1974.
59. Roos, C.: On the structure of convolutional and cyclic convolutional codes. *IEEE Trans. Info. Theory*, **25**, pp. 676–683 (1979).
60. Schalkwijk, J. P. M., Vinck, A. J. and Post, K. A.: Syndrome decoding of binary rate k/n convolutional codes. *IEEE Trans. Info. Theory*, **24**, pp. 553–562 (1978).
61. Selmer, E. S.: Linear recurrence relations over finite fields. Univ. of Bergen, Norway: Dept. of Math., 1966.
62. Shannon, C. E.: A mathematical theory of communication. *Bell Syst. Tech. J.*, **27**, pp. 379–423, 623–656 (1948).
63. Sidelnikov, V. M.: Upper bounds for the number of points of a binary code with a specified code distance. *Info. and Control*, **28**, pp. 292–303 (1975).
64. Sloane, N. J. A. and Whitehead, D. S.: A new family of single-error-correcting codes. *IEEE Trans. Info. Theory*, **16**, pp. 717–719 (1970).
65. Sloane, N. J. A., Reddy, S. M. and Chen, C. L.: New binary codes. *IEEE Trans. Info. Theory*, **18**, pp. 503–510 (1972).
66. Solomon, G. and van Tilborg, H. C. A.: A connection between block and convolutional codes. *SIAM J. Appl. Math.*, **37**, pp. 358 – 369 (1979).
67. Szegö, G.: *Orthogonal Polynomials*. Colloquium Publications, Vol. 23. New York: Amer. Math. Soc. (revised edition), 1959.

68. Tietäväinen, A.: On the nonexistence of perfect codes over finite fields. *SIAM J. Appl. Math.*, **24**, pp. 88–96 (1973).
69. van Tilborg, H. C. A.: Uniformly packed codes. Thesis, Eindhoven Univ. of Technology, 1976.
70. Tricomi, F. G.: *Vorlesungen uber Orthogonalreihen*. Grundlehren d. math. Wiss. Band 76. Berlin–Heidelberg–New York: Springer, 1970.
71. Tzeng, K. K. and Zimmerman, K. P.: On extending Goppa codes to cyclic codes. *IEEE Trans. Info. Theory*, **21**, pp. 712–716 (1975).
72. Baker, R. D., van Lint, J. H. and Wilson, R. M.: On the Preparata and Goethals codes. *IEEE Trans. Info. Theory*, **29**, pp. 342–345 (1983).
73. van der Geer, G. and van Lint, J. H.: *Introduction to Coding Theory and Algebraic Geometry*. Basel: Birkhäuser, 1988.
74. Hong, Y.: On the nonexistence of unknown perfect 6- and 8-codes in Hamming schemes $H(n, q)$ with q arbitrary. *Osaka J. Math.*, **21**, pp. 687–700 (1984).
75. Kerdock, A. M.: A class of low-rate nonlinear codes. *Info and Control*, **20**, pp. 182–187 (1972).
76. van Lint, J. H. and Wilson, R. M.: On the Minimum Distance of Cyclic Codes. *IEEE Trans. Info. Theory*, **32**, pp. 23–40 (1986).
77. van Oorschot, P. C. and Vanstone, S. A.: *An Introduction to Error Correcting Codes with Applications*. Dordrecht: Kluwer, 1989.
78. Peek, J. B. H.: Communications Aspects of the Compact Disc Digital Audio System. *IEEE Communications Magazine*, Vol. 23, No. 2 pp. 7–15 (1985).
79. Piret, Ph.: *Convolutional Codes, An Algebraic Approach*. Cambridge, Mass.: The MIT Press, 1988.
80. Roos, C.: A new lower bound for the minimum distance of a cyclic code. *IEEE Trans. Info. Theory*, **29**, pp. 330–332 (1983).
81. Tsfasman, M. A., Vlăduţ, S. G. and Zink, Th.: On Goppa codes which are better than the Varshamov–Gilbert bound. *Math. Nachr.*, **109**, pp. 21–28 (1982).
82. Barg, A. M., Katsman, S. L., and Tsfasman, M. A.: Algebraic Geometric Codes from Curves of Small Genus. *Probl. of Information Transmission*, **23**, pp. 34–38 (1987).
83. Conway, J. H. and Sloane, N. J. A.: Quaternary constructions for the binary single-error-correcting codes of Julin, Best, and others. *Designs, Codes and Cryptography*, **41**, pp. 31–42 (1994).
84. Duursma, I. M.: *Decoding codes from curves and cyclic codes*. Ph. D. dissertation, Eindhoven University of Technology (1993).
85. Feng, G.-L. and Rao, T. R. N.: A simple approach for construction of algebraic-geometric codes from affine plane curves. *IEEE Trans. Info. Theory*, **40**, pp. 1003–1012 (1994).
86. Feng, G.-L., Wei, V., Rao, T. R. N., and Tzeng, K. K.: Simplified understanding and efficient decoding of a class of algebraic-geometric codes. *IEEE Trans. Info. Theory* **40**, pp. 981–1002 (1994).
87. Garcia, A. and Stichtenoth, H.: A tower of Artin-Schreier extensions of function fields attaining the Drinfeld-Vlăduţ bound. *Invent. Math.* **121**, pp. 211–222 (1995).
88. Hammons, A. R., Vijay Kumar, P., Calderbank, A. R., Sloane, N. J. A., and Solé, P.: The $\mathbb{Z}_4$-Linearity of Kerdock, Preparata, Goethals, and Related Codes. *IEEE Trans. Info. Theory*, **40**, pp. 301–319 (1994).
89. Høholdt, T. and Pellikaan, R.: On the decoding of algebraic-geometric codes. *IEEE Trans. Info. Theory* **41**, pp. 1589–1614 (1995).
90. Høholdt, T., van Lint, J. H., and Pellikaan, R.: Algebraic Geometry Codes. In: *Handbook of Coding Theory*, (edited by V. S. Pless, W. C. Huffman, and R. A. Brualdi). Elsevier Science Publishers, Amsterdam 1998.
91. Justesen, J., Larsen, K. J., Elbrønd Jensen, H., Havemose, A., and Høholdt, T.: Construction and decoding of a class of algebraic geometry codes. *IEEE Trans. Info. Theory* **35**, pp. 811–821 (1989).

92. van Lint, J. H.: Algebraic geometric codes. In: *Coding Theory and Design Theory I*, The IMA Volumes in Math. and Appl. **20**, (edited by D. Ray-Chaudhuri). Springer Verlag 1990.
93. van Lint, J. H. and Wilson, R. M.: *A Course in Combinatorics*. Cambridge University Press 1992.
94. Long, R. L.: *Algebraic Number Theory*. Marcel Dekker Inc., New York 1977
95. Pellikaan, R.: On a decoding algorithm for codes on maximal curves, *IEEE Trans. Info. Theory*, **35**, pp. 1228–1232 (1989).
96. Serre, J.-P.: Sur le nombre des points rationnels d'une courbe algébrique sur un corps fini. *C. R. Acad. Sc. Paris*, **296**, pp. 397–402 (1983).
97. Skorobogatov, A. N. and Vlăduţ, S. G.: On the decoding of algebraic-geometric codes. *IEEE Trans. Info. Theory*. **36**, pp. 1051–1060 (1990).
98. Stichtenoth, H.: *Algebraic function fields and codes*. Universitext, Springer Verlag, Berlin 1993.
99. Tsfasman, M. A., Vlăduţ, S. G. and Zink, T.: Modular curves, Shimura curves and Goppa codes, better than Varshamov-Gilbert bound. *Math. Nachrichten*, **109**, pp. 21–28 (1982).
100. Uspensky, J. V.: *Theory of Equations*. McGraw-Hill, New York 1948.

Index

$\alpha(\delta)$, 65
adder (mod 2), 182
admissible pair, 176
affine
- curve, 149
- geometry, 18
- permutation group, 5
- plane, 18
- subspace, 18
- transformation, 18
$\mathrm{AG}(m,q)$, 18
$\mathrm{AGL}(l,q^m)$, 97
$\mathrm{AGL}(m,2)$, 18, 59
algebra, 5
alphabet, 33
$A(n,d)$, 64
$A(n,d,w)$, 72
arithmetic
- distance, 173
- weight, 173, 175
automorphism group, 59, 97, 192

Baker, 121
Barg, 164
Barrows, 179
basis, 4
Bell Laboratories, 31
Berlekamp, 31, 99, 147
- decoder, 99
Best, 53, 79, 138
Bézout's theorem 153
binary
- entropy, 20
- image 129
- symmetric channel, 24
binomial distribution, 20
bit, 23
block
- design, 17
- length, 33
Bose, 91
bound
- BCH, 91
- Carlitz-Uchiyama, 110
- Elias, 70
- Gilbert-Varshamov, 66, 143, 165, 168, 187
- Grey, 80
- Griesmer, 69
- Hamming, 69
- Johnson, 72
- linear programming, 74
- McEliece, 77
- Plotkin, 67
- Singleton, 67
- sphere packing, 69
burst, 99
byte, 23

Cameron, 62
CCC, 192
character, 14
- principal, 14
characteristic
- numbers, 117
- polynomial, 115, 117
Chebychev's inequality, 20
check polynomial, 84
Chen, 180
Chien, 145, 180
Choy, 145
Christoffel-Darboux formula, 16
Clark, 180
CNAF, 178
code, 34
- algebraic geometry, 148
- alternant, 146
- AN, 173
- arithmetic, 173
- asymptotically good, 167
- BCH, 91, 140
- narrow sense, 91

– primitive, 91
– Best, 53, 138
– block, 33
– catastrophic, 184
– completely regular, 115
– concatenated, 168
– constacyclic, 81
– convolutional, 33, 181, 185
– cyclic, 81
– – AN, 175
– – convolutional, 192
– cyclic over $\mathbb{Z}_4$ 136
– direct product, 45
– double circulant, 172
– dual, 36
– Elias, 54
– equidistant, 68, 179
– equivalent, 35, 128
– error-correcting, 22
– error-detecting, 23
– extended, 38, 51
– generalized BCH, 145
– generalized Reed-Muller, 108
– generalized Reed-Solomon, 100
– geometric generalized Reed-Solomon, 160
– geometric Goppa, 161
– Golay
– – binary, 47, 106
– – ternary, 51, 62, 63
– Goppa, 140
– group, 35
– Hadamard, 47, 62, 120, 167
– Hamming, 37, 58, 85, 107
– inner, 168
– irreducible cyclic, 83
– Justesen, 168
– Kerdock, 60, 130
– lexicogtraphically least, 46
– linear, 35
– Mandelbaum-Barrows, 179
– maximal, 64
– – cyclic, 83
– maximum distance separable, 64, 197
– MDS, 64, 67
– minimal cyclic, 83, 180
– modular arithmetic, 174
– Nadler, 52
– narrow sense BCH, 91
– nearly perfect, 118
– negacyclic, 81
– Nordstrom-Robinson, 52, 123, 127
– optimal, 64
– outer, 168
– perfect, 34, 38, 48, 112, 175, 180
– Preparata, 122, 130, 137
– primitive BCH, 91
– projective, 38
– punctured, 52
– QR, 103
– quadratic residue, 103
– quasi-perfect, 37
– quaternary, 128
– Reed-Muller, 54, 58
– Reed-Solomon, 99, 168
– regular, 115
– repetition, 24
– residual, 53
– RM, 58
– self-dual, 36
– separable, 36
– shortened, 52
– Srivastava, 147
– symmetry, 202
– systematic, 35
– ternary, 34
– trivial, 34
– two-weight, 119
– uniformly packed, 118
– uniquely decodable, 46
codeword, 23
coding gain, 29
collaborating codes, 53
conference matrix, 18
constraint length, 183
Conway, 50
coordinate ring, 150
coset, 3
– leader, 37
– representative, 3
covering radius, 34
curve,
– Hermitian, 163
– nonsingular, 151
– smooth, 151
Cvetković, 112
cyclic nonadjacent form, 178
cyclotomic coset, 86

decision (hard, soft), 24
decoder, 23
– Berlekamp, 99, 145
decoding
– BCH codes, 98
– complete, 23
– Goppa codes, 144
– incomplete, 23
– majority logic, 39

– maximum likelihood, 26
– multistep majority, 60
– RM codes, 59
– Viterbi, 50
defining set, 89
Delsarte, 74, 127
derivative, 11
design
– block, 17
– $t-$, 17, 48, 202
differential, 157
direct product, 45
distance, 33
– arithmetic, 173
– distribution, 75
– enumerator, 115
– external, 118
– free, 184
– Hamming, 33
– invariant, 41
– minimum, 34
divisor, 155
– canonical, 157
– degree of 155
– effective, 155
– principal, 155

Elias, 54
encoder, 23
entropy, 20, 65
erasure, 24
error, 23
– locator polynomial, 98, 144
Euler indicator, 2
expected value, 19
external distance, 118

factor group, 3
Feller, 20
Feng, 165
field, 4
finite field, 4, 6
flat, 18
Forney, 168, 185
Fukumara, 180
function field, 150

Galois ring, 132
Garcia, 166, 180
Gaussian distribution, 21
Geer, van der, 148
generator
– of a cyclic group, 3
– of AN code, 175
– matrix, 35
– – of a cyclic code, 83
– polynomial
– – of a convolutional code, 183
– – of a cyclic code, 83
genus, 158
Gilbert, 31, 66
Goethals, 50, 118
Golay, 61
Goppa, 147
– polynomial, 140
Goto, 180
Graeffe's method, 133
Gray map, 129
Griesmer, 68
Gritsenko, 180
group, 3
– abelian, 3
– algebra, 5, 115
– commutative, 3
– cyclic, 3
– Mathieu, 50
– transitive, 6

Hadamard matrix, 18
Hall, 17, 19
Hamming, 31, 44
Hartmann, 94
Hasse derivative, 11
Hasse-Weil bound, 162
Helgert, 52, 147
Hensel's lemma, 133
hexacode, 164
Hocquenghem, 91
Hong, 123
$H_q(\chi)$, 65
hyperplane, 18

ideal, 3
– maximal, 6
– prime, 6
– principal, 6
idempotent, 86
– of a QR code, 104
incidence matrix, 17
independent
– variables, 19
– vectors, 4
information
– rate, 26, 34
– symbol, 35
inner
– distribution, 75
– product, 4

integral domain, 4
irreducible polynomial, 7

Jackson, 14
Jet Propulsion Laboratory, 22
Justesen, 165, 168, 189

Kasami, 143
Kerdock, 60
Klein quartic, 152
Krawtchouk
– expansion, 16
– polynomial, 14, 74, 113
Kronecker product, 18

Lee
– distance, 42
– metric, 43
– weight, 43
– weight enumerator, 43
Lenstra, 180
Levenshtein, 79
Liang, 180
linear
– programming bound, 74
– recurring sequence, 90
Lint, van, 62, 92, 111, 112, 123, 201
Liu, 180
Lloyd, 31, 112
– theorem, 112, 119
local ring 6, 150

MacWilliams, 31, 41, 110
Mandelbaum, 179
Mariner, 22
Massey, 44, 45, 55, 180
Mattson-Solomon polynomial, 89
McEliece, 77, 79, 95
mean, 19
memory, 182, 184
minimal polynomial, 10
modular
– distance, 174
– weight, 174
Moebius
– function, 2
– inversion formula, 2
monic polynomial, 8
Moreno, 147
Muller, 53
multiplicative group of a field, 9

NAF, 176
nonadjacent form, 176
normal distribution, 21

order, 3
orthogonal parity checks, 40
outer distribution, 115

Paley, 19
– matrix, 19,47
parameter
– local, 152
– uniformizing, 152
parity check
– equation, 36
– matrix, 36
– symbol, 35
permutation, 5
– matrix, 5
Peterson, 180
$PG(n,q)$, 18
Piret, 191, 194
Pless, 202
Plücker formula, 158
point
– at infinity, 151
– nonsingular, 151
– rational, 152
– simple, 151
pole, 152
polynomials, 11
Post, 60
Preparata, 111
primitive 133
– element, 9
– idempotent, 87
– polynomial, 10
– root of unity, 9
principal
– character, 14
– ideal, 6
– ideal ring, 6
projective
– geometry, 18
– plane, 18
$PSL(2,n)$, 104

quadratic residue, 13
quotient field, 4

Rao, 180
Ray-Chaudhuri, 91
redundancy, 22
Reed, 53
regular function, 150
representative, 3
residue 158
– class, 6
– ring, 6

residue theorem, 158
Riemann-Roch theorem, 158
ring, 3
Rodemich, 77
Roos, 94, 191
$\mathscr{RM}(r,m)$, 58
Rumsey, 77

Schalkwijk, 194
Serre, 164
Shannon, 25, 31
– theorem, 22
shift register, 182
Sidelnikov, 79
Signal to Noise Ratio, 29
Skorobogatov, 165
Slepian, 31, 44
Sloane, 31
Solomon, 194
sphere, 27, 34
– packing condition, 34
Srivastava, 147
standard
– deviation, 19
– form, 35
state diagram, 182
Steiner system, 17
Stichtenoth, 166
Stinaff, 52
Stirling's formula, 20
subspace, 4
symbol error probability, 31
symmetric group, 5
symmetrized weight enumerator, 43
symplectic form, 60
syndrome, 36, 144
Szegö, 14

Tietäväinen, 123
Tilborg, van, 118, 123, 194
trace, 13, 89
Tricomi, 14
Tsfasman, 148
Turyn, 49
Tzeng, 94, 147

Vandermonde determinant, 90
variance, 19
variety 149
– affine, 149
– projective, 150
Varshamov, 66
vector space, 4
Vinck, 194
Viterbi algorithm, 185
Vlăduţ, 148, 165
$V_q(n,r)$, 61

weight, 33
– distribution, 40
– enumerator, 40
Weil, 110
Welch, 77
Weldon, 180
Wilson, 92, 121
word, 23
– length, 33

Zariski topology, 149
zero divisor, 3
zero of a cyclic code, 84
Zimmermann, 147
Zink, 148

Graduate Texts in Mathematics

continued from page II

65 WELLS. Differential Analysis on Complex Manifolds. 2nd ed.
66 WATERHOUSE. Introduction to Affine Group Schemes.
67 SERRE. Local Fields.
68 WEIDMANN. Linear Operators in Hilbert Spaces.
69 LANG. Cyclotomic Fields II.
70 MASSEY. Singular Homology Theory.
71 FARKAS/KRA. Riemann Surfaces. 2nd ed.
72 STILLWELL. Classical Topology and Combinatorial Group Theory.
73 HUNGERFORD. Algebra.
74 DAVENPORT. Multiplicative Number Theory. 2nd ed.
75 HOCHSCHILD. Basic Theory of Algebraic Groups and Lie Algebras.
76 IITAKA. Algebraic Geometry.
77 HECKE. Lectures on the Theory of Algebraic Numbers.
78 BURRIS/SANKAPPANAVAR. A Course in Universal Algebra.
79 WALTERS. An Introduction to Ergodic Theory.
80 ROBINSON. A Course in the Theory of Groups.
81 FORSTER. Lectures on Riemann Surfaces.
82 BOTT/TU. Differential Forms in Algebraic Topology.
83 WASHINGTON. Introduction to Cyclotomic Fields.
84 IRELAND/ROSEN. A Classical Introduction to Modern Number Theory. 2nd ed.
85 EDWARDS. Fourier Series. Vol. II. 2nd ed.
86 VAN LINT. Introduction to Coding Theory. 2nd ed.
87 BROWN. Cohomology of Groups.
88 PIERCE. Associative Algebras.
89 LANG. Introduction to Algebraic and Abelian Functions. 2nd ed.
90 BRØNSTED. An Introduction to Convex Polytopes.
91 BEARDON. On the Geometry of Discrete Groups.
92 DIESTEL. Sequences and Series in Banach Spaces.
93 DUBROVIN/FOMENKO/NOVIKOV. Modern Geometry – Methods and Applications. Vol. I. 2nd ed.
94 WARNER. Foundations of Differentiable Manifolds and Lie Groups.
95 SHIRYAYEV. Probability. 2nd ed.
96 CONWAY. A Course in Functional Analysis.
97 KOBLITZ. Introduction in Elliptic Curves and Modular Forms. 2nd ed.
98 BRÖCKER/TOM DIECK. Representations of Compact Lie Groups.
99 GROVE/BENSON. Finite Reflection Groups. 2nd ed.
100 BERG/CHRISTENSEN/RESSEL. Harmonic Analysis on Semigroups: Theory of Positive Definite and Related Functions.
101 EDWARDS. Galois Theory.
102 VARADARAJAN. Lie Groups, Lie Algebras and Their Representations.
103 LANG. Complex Analysis. 2nd ed.
104 DUBROVIN/FOMENKO/NOVIKOV. Modern Geometry – Methods and Applications. Part II.
105 LANG. $SL_2(R)$.
106 SILVERMAN. The Arithmetic of Elliptic Curves.
107 OLVER. Applications of Lie Groups to Differential Equations.
108 RANGE. Holomorphic Functions and Integral Representations in Several Complex Variables.
109 LEHTO. Univalent Functions and Teichmüller Spaces.
110 LANG. Algebraic Number Theory.
111 HUSEMÖLLER. Elliptic Functions.
112 LANG. Elliptic Functions.
113 KARATZAS/SHREVE. Brownian Motion and Stochastic Calculus. 2nd ed.
114 KOBLITZ. A Course in Number Theory and Cryptography. 2nd ed.
115 BERGER/GOSTIAUX. Differential Geometry: Manifolds, Curves, and Surfaces.
116 KELLEY/SRINIVASAN. Measure and Integral. Vol. I.
117 SERRE. Algebraic Groups and Class Fields.
118 PEDERSEN. Analysis Now.

Graduate Texts in Mathematics

119 ROTMAN. An Introduction to Algebraic Topology.
120 ZIEMER. Weakly Differentiable Functions: Sobolev Spaces and Functions of Bounded Variation
121 LANG. Cyclotomic Fields I and II. Combined 2nd ed.
122 REMMERT. Theory of Complex Functions. *Readings in Mathematics.*
123 EBBINGHAUS ET AL. Numbers. *Readings in Mathematics.*
124 DUBROVIN/FOMENKO/NOVIKOV. Modern Geometry – Methods and Applications. Part III.
125 BERENSTEIN/GAY. Complex Variables. An Introduction
126 BOREL. Linear Algebraic Groups.
127 MASSEY. A Basic Course in Algebraic Topology.
128 RAUCH. Partial Differential Equations.
129 FULTON/HARRIS. Representation Theory. A First Course. *Readings in Mathematics.*
130 DODSON/POSTON. Tensor Geometry.
131 LAM. A First Course in Noncommutative Rings.
132 BEARDON. Iteration of Rational Functions.
133 HARRIS. Algebraic Geometry. A First Course.
134 ROMAN. Coding and Information Theory.
135 ROMAN. Advanced Linear Algebra.
136 ADKINS/WEINTRAUB. Algebra: An Approach via Module Theory.
137 AXLER/BOURDON/RAMEY. Harmonic Function Theory.
138 COHEN. A Course in Computational Algebraic Number Theory.
139 BREDON. Topology and Geometry.
140 AUBIN. Optima and Equilibria. An Introduction to Nonlinear Analysis.
141 BECKER/WEISPFENNIG/KREDEL. Gröbner Bases. A Computational Approach to Commutative Algebra.
142 LANG. Real and Functional Analysis. 3rd ed.
143 DOOB. Measure Theory.
144 DENNIS/FARB. Noncommutative Algebra.
145 VICK. Homology Theory. An Introduction to Algebraic Topology. 2nd ed.
146 BRIDGES. Computability: A Mathematical Sketchbook.
147 ROSENBERG. Algebraic *K*-Theory and Its Applications.
148 ROTMAN. An Introduction to the Theory of Groups. 4th ed.
149 RATCLIFFE. Foundations of Hyperbolic Manifolds.
150 EISENBUD. Commutative Algebra with a View Toward Algebraic Geometry.
151 SILVERMAN. Advanced Topics in the Arithmetic of Elliptic Curves.
152 ZIEGLER. Lectures on Polytopes.
153 FULTON. Algebraic Topology: A First Course.
154 BROWN/PEARCY. An Introduction to Analysis.
155 KASSEL. Quantum Groups.
156 KECHRIS. Classical Descriptive Set Theory.
157 MALLIAVIN. Integration and Probability.
158 ROMAN. Field Theory.
159 CONWAY. Functions of One Complex Variable II.
160 LANG. Differential and Riemannian Manifolds.
161 BORWEIN/ERDÉLYI. Polynomials and Polynomial Inequalities.
162 ALPERIN/BELL. Groups and Representations.
163 DIXON/MORTIMER. Permutation Groups.
164 NATHANSON. Additive Number Theory: The Classical Bases.
165 NATHANSON. Additive Number Theory: Inverse Problems and the Geometry of Sumsets.
166 SHARPE. Differential Geometry: Cartan's Generalization of Klein's Erlangen Program
167 MORANDI. Field and Galois Theory.
168 EWALD. Combinatorial Convexity and Algebraic Geometry.
169 BHATIA. Matrix Analysis.
170 BREDON. Sheaf Theory. 2nd ed.
171 PETERSEN. Riemannian Geometry.
172 REMMERT. Classical Topics in Complex Function Theory.
173 DIESTEL. Graph Theory.
174 BRIDGES. Foundations of Real and Abstract Analysis.
175 LICKORISH. An Introduction to Knot Theory.

Graduate Texts in Mathematics

176 LEE. Riemannian Manifolds.
177 NEWMAN. Analytic Number Theory.
178 CLARKE/LEDYAEV/STERN/WOLENSKI. Nonsmooth Analysis and Control Theory.
179 DOUGLAS. Banach Algebra Techniques in Operator Theory. 2nd ed.
180 SRIVASTAVA. A Course on Borel Sets.
181 KRESS. Numerical Analysis.
182 WALTER. Ordinary Differential Equations.
183 MEGGINSON. An Introduction to Banach Space Theory.
184 BOLLOBAS. Modern Graph Theory.
185 COX/LITTLE/O'SHEA. Using Algebraic Geometry.
186 RAMAKRISHNAN/VALENZA. Fourier Analysis on Number Fields.
187 HARRIS/MORRISON. Moduli of Curves.
188 GOLDBLATT. Lectures on Hyperreals.
189 LAM. Lectures on Rings and Modules.

9783642636530